Mass Notification and Crisis Communications

Planning, Preparedness, and Systems

Mass Notification and Crisis Communications

Planning, Preparedness, and Systems

Denise C. Walker, D.B.A.

CRC Press
Taylor & Francis Group
Boca Raton London New York

CRC Press is an imprint of the
Taylor & Francis Group, an **informa** business

CRC Press
Taylor & Francis Group
6000 Broken Sound Parkway NW, Suite 300
Boca Raton, FL 33487-2742

© 2012 by Taylor & Francis Group, LLC
CRC Press is an imprint of Taylor & Francis Group, an Informa business

No claim to original U.S. Government works

Printed in the United States of America on acid-free paper
Version Date: 20111017

International Standard Book Number: 978-1-4398-7438-7 (Hardback)

This book contains information obtained from authentic and highly regarded sources. Reasonable efforts have been made to publish reliable data and information, but the author and publisher cannot assume responsibility for the validity of all materials or the consequences of their use. The authors and publishers have attempted to trace the copyright holders of all material reproduced in this publication and apologize to copyright holders if permission to publish in this form has not been obtained. If any copyright material has not been acknowledged please write and let us know so we may rectify in any future reprint.

Except as permitted under U.S. Copyright Law, no part of this book may be reprinted, reproduced, transmitted, or utilized in any form by any electronic, mechanical, or other means, now known or hereafter invented, including photocopying, microfilming, and recording, or in any information storage or retrieval system, without written permission from the publishers.

For permission to photocopy or use material electronically from this work, please access www.copyright.com (http://www.copyright.com/) or contact the Copyright Clearance Center, Inc. (CCC), 222 Rosewood Drive, Danvers, MA 01923, 978-750-8400. CCC is a not-for-profit organization that provides licenses and registration for a variety of users. For organizations that have been granted a photocopy license by the CCC, a separate system of payment has been arranged.

Trademark Notice: Product or corporate names may be trademarks or registered trademarks, and are used only for identification and explanation without intent to infringe.

Library of Congress Cataloging-in-Publication Data

Walker, Denise C.
  Mass notification and crisis communications : planning, preparedness, and systems / Denise C. Walker.
      p. cm.
   Includes bibliographical references and index.
   ISBN 978-1-4398-7438-7 (hbk. : alk. paper)
   1. Emergency management. 2. Emergency communication systems. 3. Risk communication. 4. Crisis management. I. Title.

HV551.2.W34 2012
384.5--dc23   2011041794

Visit the Taylor & Francis Web site at
http://www.taylorandfrancis.com

and the CRC Press Web site at
http://www.crcpress.com

This book is dedicated to you.

*This book is also dedicated to my husband Alfred,
my children Charles, Anthony, Crystal, and Marvin, and
my father Samuel Chatam
whose steadfast love and support made this book possible.*

Contents

	Preface	xxi
Chapter 1	History of Communications	1
	History of Communications	1
	Communications—3500 BC to 1 BC	2
	Heliographs	3
	Paper, Newspapers, and Magazines	3
	Telegraph Service and Morse Code	4
	Telegraph Service	4
	Morse Code	5
	Mail and Parcels	6
	Facsimile (Fax)	7
	Telephone	8
	Mobile Phone	9
	Copy Machine	9
	Loudspeakers	10
	Magnetic Recording	10
	Radio	11
	Amateur Radio	14
	AM Broadcasting	17
	FM Broadcasting	18
	Telephony—Cellular and Satellite Phones	20
	Cellular Phones	20
	Satellite Phones	21
	Navigation	22

RADAR	22
Data (Digital Radio)	23
Unlicensed Radio Services	25
Radio Control (RC)	26
Wireless	26
Other Radio	27
Television	27
Satellites	29
Information Science and Computers	30
Internet	32
Communications for People with Disabilities and Others with Functional and Access Needs	32
Hearing Impaired	33
Hearing Aid	33
Text Telephone	33
Speech Impaired	34
Visually Impaired	34
Braille	34
Elderly	34
Summary	35
Endnotes	35

Chapter 2 Disaster Communications—Two-Way Conversations   37

Emergency Broadcasting	39
Universal Emergency Calling Number	40
History of the Universal Emergency Calling Number	40
Universal Emergency Calling Number—United Kingdom	41
Universal Emergency Calling Number—North America	44
9-1-1 Emergency Operator Procedures	51
Enhanced 9-1-1	51
9-1-1 A Vulnerable Target	52
9-1-1 and First Amendment Rights	52
Silent 9-1-1 Calls	53

	9-1-1 Internet Address Purchased for Profit	54
	Blocking Inbound Calls to 9-1-1	54
	Other Vulnerabilities to Emergency Calling Systems	56
	Next Generation 9-1-1 (NG 911)	57
	Emergency Requests for Help around the World	57
	Australia	57
	Europe	57
	Summary	58
	Endnotes	58
Chapter 3	**Dynamics of Communications and Concepts of Operations**	**61**
	Effective Communications	62
	Crisis Communications	66
	Crisis Communications Planning	68
	Emergency Communications Characteristics	71
	How to Recognize When a Message Is Not Being Communicated	73
	Crisis Communications in the Midst of Violent Events	74
	Emergencies Related to Violence by Journalists	74
	Crisis Communications for War Preparation	76
	Propaganda	77
	Summary	78
	Endnotes	79
Chapter 4	**Challenges to Effective Communications**	**81**
	The Cost of Quality	83
	Communications for People with Disability and Others with Functional and Access Needs	87
	Primary Groups	89
	Speech Disabilities	89
	Mentally Disabled, Developmentally Disabled, and the Cognitively Impaired	90
	Hearing Impaired	95
	Visually Impaired	98
	Mobility Challenged	101

	Other Groups	101
	Non-English Speaking	101
	Limited English Proficiency (LEP)	102
	Culturally or Geographically Isolated	104
	Homeless	105
	Medically or Chemically Dependent	106
	Children	107
	Elderly	107
	Communications Methods for Populations with Disabilities and Others with Functional and Access Needs	108
	Summary	109
	Endnotes	112
Chapter 5	**Changing Realities in Emergency Communications**	**115**
	2009 Worldwide Catastrophes Statistics	116
	Disaster Types Predicted over the Next 25 Years	119
	Earthquakes	119
	Volcanic Disruptions	121
	Tsunamis	123
	Atmospheric Conditions	125
	Superstorms	126
	Solar Flares	127
	Comets, Asteroids, and Meteor Strikes	129
	Shortages of Key Resources	130
	Water Shortages	130
	Food Shortages	133
	Power Shortages—Blackouts and Brownouts	133
	Health Disasters	135
	Aging Infrastructure	135
	Pipeline	135
	Waste Treatment Facilities	137
	Wastewater Sewer Lines	137
	Roads, Bridges, and Transit Systems	138

	Cyber Attacks and Cyber Warfare	139
	War/Conflict	141
	Disaster Planning	143
	The New Mobility	147
	Communications in the Future	148
	Satellites	151
	The New Mobility Pushes the Federal Communications Commission (FCC) Changes to 9-1-1 Services	152
	Summary	153
	Endnotes	154
Chapter 6	**Emergency Communications Framework**	**157**
	Putting the Framework into Operations	160
	Sending Messages	162
	Receiving Warnings from Trusted Sources	164
	The National Warning System (NAWAS)	164
	National Weather Service (NWS)	164
	State Government and Local Notification	165
	Notification of Local Officials	165
	Organizations and Individuals	165
	Dissemination of Warnings to the Public	166
	When to Activate	166
	Activations for People with Disabilities and Others with Functional and Access Needs	169
	Warnings to Outside Agencies	170
	Phases of Emergency Management	171
	Preparedness	172
	Response	173
	Recovery	174
	Mitigation	174
	Prevention	175
	Organization and Responsibilities	176
	Assignment of Responsibilities	177
	Department Heads	177
	Office of Emergency Management (OEM)	177

	The Emergency Management Coordinator (EMC) or Director	177
	External Entities	178
	Joint Information Center (JIC)	178
Direction and Control		179
Readiness Levels		179
	Green (Level III)—Normal Operations	180
	Yellow (Level II)—Increased Readiness or Partial Activation	181
	Red (Level I)—Maximum Readiness or Full-Scale Activation	181
Emergency Communications and Record Keeping		182
Training and Education		182
Security, System Maintenance, and Capacity Building		184
Summary		185
Endnotes		186

Chapter 7	A Crisis Occurs … Now What?	187
	The Ws of Effective Communications	188
	Why	189
	The Scenario	189
	Who	190
	What	191
	Situational Crisis Communication Theory and Attribution Theory	192
	What to Avoid Saying	194
	When	194
	Where	195
	How	196
	The Message and the Messenger	198
	Moral and Ethical Reflection	201
	Preparation Is Key	203
	Lessons Learned and Continuous Improvement	204
	Summary	207
	Endnotes	207

Chapter 8	**International, Federal, State, and Local Laws, Regulations, Systems, Plans, and Structures**	**209**
	National Response Framework	210
	National Incident Management System	212
	Incident Command System	215
	Communications	221
	Debriefing	222
	National Emergency Communications Plan	228
	Governance	230
	Standard Operating Procedures	230
	Technology	230
	Data Elements	230
	Voice Elements	231
	Training and Exercise	231
	Usage	232
	National Telecommunications and Information Administration	233
	Federal Spectrum Use	234
	Emergency Communications for Which an Immediate Danger Exists to Human Life or Property	234
	National Security and War Emergency Communications	234
	Emergency Use of Nonfederal Frequencies	238
	Coordination and Use of Emergency Networks	238
	Interoperability between Federal Entities and Nonfederal Public Safety Licensees	238
	Federal Aviation Administration (FAA)	239
	Federal Communications Commission	241
	Public Safety and Homeland Security Bureau (PSHSB)	242
	The Common Air Interface	243
	9-1-1 and E9-1-1 in Tribal Lands	243
	Other FCC Emergency Communications Initiatives	243

	Accessibility Act	246
	2-1-1	246
	Stafford Act	247
	Other Laws and Regulations	248
	Health Insurance Portability & Accountability Act (HIPAA)	248
	Gramm–Leach–Bliley Act (GLBA)	249
	Fair and Accurate Credit Transactions Act (FACTA) and the Red Flag Rule	250
	Family Educational Rights and Privacy Act (FERPA)	250
	Clery Act	251
	Summary	252
	Endnotes	252
Chapter 9	**Ripple Effect of Social Media and Social Networking**	**255**
	Introduction to Social Media	255
	More about Social Media	259
	Commonly Used Applications	263
	Twitter	263
	Facebook	265
	Blogs	266
	Widgets	267
	Mobile Phone Applications and Mobile Web Widgets	267
	Really Simple Syndication (RSS Feeds)	268
	Wiki Sourcing	269
	Geo-Location Systems	270
	Geo-Targeting	271
	Geotagging	272
	Geocoding	273
	Quick Response (QR) Codes	273
	Shared Content	274
	Social Storage	275
	Social Bookmarking	276
	Internet Radio, Blogs, Talk Radio	276
	Maps	276

	SMS Text Messaging and MMS	277
	Shared Video/Streaming Video	278
	Forums	279
	Alternate Reality Games	279
	Using Social Media before It Is Needed in an Emergency	280
	Social Media Monitoring	280
	Software Integration	280
	Mashup	281
	Fundamental Change	282
	Social Media as a Technological Hazard	284
	Summary	287
	Endnotes	288
Chapter 10	**Solutions—Some Solutions Are Better than None**	**289**
	Systems You Probably Already Have	290
	Cellular Phones and Smartphones	290
	Bandwidth and Speed	292
	Fire and Gas Detector and Alarm Systems	294
	Voicemail and Voice Systems	295
	E-Mail and Instant Messaging	296
	Instant Messaging	298
	Flash Messages	298
	Web Page	299
	Public Media	300
	In a Classroom or Conference Room	301
	Alert Systems	302
	Emergency Automated Telephone Notification System	302
	Public Address Systems	303
	In-Building Voice Announcements	303
	Outdoor Sirens and Speaker Arrays (Including Voice Warning)	304
	Community Outdoor Warning Sirens (COWS) and Community Activated Lifesaving Voice Emergency Systems (CALVES)	305

	Speakers and Video Surveillance Emergency Phone Towers	306
	Billboards, Video Displays, Digital Signage, and Community Access Television (CATV)	307
	Two-Way Radios	310
	Weather Radios	313
	Weather Radio Distribution Ensuring Adequate Hazard Warning	313
	Weather Radios—Enabling People to Hear and Heed Severe Weather Warnings	314
	Evacuation Warning System—Weather Warning System—CodeRED	315
	Regional Resources	316
	Highway Alerting	316
	Reverse 9-1-1	317
	Earthquake and Tsunami Monitoring	318
	Earthquake Monitoring Project TriNet	318
	Measuring Earthquake Potential with GPS	319
	Other Solutions to Consider	320
	Combined Alerting System (Software to Allow Delivery to Multiple Platforms)	320
	3-D Meeting Software	321
	Other Crisis Communications Solutions	322
	Common Operating Picture (COP)	324
	Summary	327
	Endnotes	327
Chapter 11	**Learning Your Systems Requirements**	**329**
	The Challenge	329
	Fusion Center and the EOC	331
	Solution Selection for Your Organization	332
	Analysis	332
	Stakeholders and Target Audiences	332
	Emergency Communication Needs	335
	Design Brief	336
	Goal Setting	336
	SMART Goals	336

	Problem Statement	337
	Assumptions	337
	Constraints	337
	Context	338
	Budget	339
	Time	342
	Solution Analysis	342
	Prototypes	342
	Use Cases	343
	Functional Requirements	343
	System Requirements Specification	345
	Requirements Specification	346
	Effective Governance of the Project	353
	Solution Analysis	354
	Risks and Benefits	354
	Summary	356
	Endnote	357
Chapter 12	**Picking a Solution That Fits**	**359**
	Financial Considerations	360
	Selecting a Product	362
	Evaluation Process	363
	Acceptance	364
	Use Cases	365
	Test Plan	367
	Test Types	368
	Test Deliverables	368
	Testing the Environment and Resources	369
	System Installation and Implementation	369
	Implementation Success or Failure	372
	User Documentation	374
	Training for Users and Administrator	375
	Summary	376
	Endnotes	377

Chapter 13	**Effective Communication Plans**	379
	What the Plan Is and Why You Should Have One	379
	Plan, Then Detail Your Plan—Its Purpose and Objectives	381
	Put Your Team in Place	383
	Prepare to Manage the Message and the Media	387
	Crisis Communications Team Expectations	388
	Prioritize Your Target Audience and Tailor the Message to Them	392
	Training and Exercises	393
	Plan Reviews and Updates	395
	Plan Appendices	395
	News Conference Guidelines	396
	Media Relations Reminders	397
	Summary	398
	Endnotes	399
Chapter 14	**Getting the Message Right the First Time**	401
	After the Crisis	401
	Crisis Communications Planning—Things You Need to Do Now	402
	Communication Mistakes Communicators Still Make	403
	Getting the Message Right	407
	Effective Communications Is a Human Issue	407
	After-Action Review	409
	Drafting Your Message—What to Avoid	411
	Other Messaging Guidelines	413
	Summary	419
	Endnote	419
Chapter 15	**Prepare to Prosper**	421
	Keys to Success	421
	Tips in Preparing Your Emergency Operations Center	425
	Marketing Emergency Services and Your Plan	427
	Components of Effective Messaging	431
	Other Critical Plans to Support Emergency Operations	433

	The Human Factor—Mental Health and First Aid	434
	Opportunities in Public/Private Partnerships	437
	The Future Starts Now	440
	Summary	441
	Endnotes	442
Chapter 16	Conclusions	445

Appendix A: List of Acronyms	451
Appendix B: Sample Mass Notification System Activation and Criteria Guidance Sheet	459
Appendix C: Sample Messages	461
Appendix D: List of Questions to Ask a Vendor before You Buy	467
Appendix E: Sample Social Media Strategy	473
Appendix F: Emergency Call Numbers	477
Glossary	487
Additional Resources	507
Index	511

Preface

As I finished this manuscript, I was reminded about what had inspired me to share what I know about crisis communications and mass notification. I came across a massive amount of information, from many diverse sources, and as expensive "must have" solutions. Too many organizations I spoke with said they simply could neither afford the "must haves" nor hire a consultant to mine though the web of data to find the best fitting solution. This was an interesting thought. I spent many months hiding away like a hermit and overcoming challenges in writing this book to address this issue. My aim was to offer a trusted resource to guide you through the mountains of data on mass notification and crisis communications planning, preparedness, and systems. My writing is based on sound research, real-world case studies, and my own experiences.

This book is more than the ABCs of creating an on-point crisis communiqué in the midst of a storm. It is about reaching the masses in a targeted and timely manner using what you have and buying what you can afford. Information included is a review of challenges organizations face with communications and the types of disasters predicted in the future. The information here assists in defining your target audiences—your stakeholders—those with every means of communication available to them and those with none. It describes different communication techniques used throughout history during peace and wartime, and provides methods on consistently getting your message right the first time.

This book helps you to create a crisis communications plan that works and analyzes the technology used for crisis communications and mass notification. The material provides a comprehensive appraisal of the technology you already have and other solutions available—from e-mail and social media to sonic buoys and early earthquake warning systems. This book goes a step further; it looks at the legal landscape, the processes for product selection, identifying requirements, designing your integrated solution, testing your program, and maintaining systems. This book offers a broadminded view of key decision-making considerations, the financial and mental health aspects of crisis communications, marketing and message mapping, and pathways for building new relationships. The approach used throughout the book aligns with the U.S. Department of Homeland Security National Emergency Communications Plan, the U.S. strategic plan to improve emergency response communications, and public expectations.

Learning is a passion of mine; sharing what I have learned is my way of giving back. It is with this spirit, I share the experiences I have learned in an organized and straightforward way about crisis communications and mass notification. I thank my husband, children, and parents for their support. I thank you for the privilege of sharing this information with you.

Denise Chatam Walker, D.B.A.
Richmond, Texas
dmcwalker@yahoo.com

CHAPTER 1

History of Communications

The single biggest problem in communication
is the illusion that it has taken place.

George Bernard Shaw

HISTORY OF COMMUNICATIONS

The history of communications began with the start of life, and it has been a part of our lives ever since. *Communication* means information that is shared by two or more people. Communications includes a sender and a receiver. The *sender* is the one that triggers the message. The *receiver* is the party who receives the message.

For communications to occur between a sender and receiver, you first need media to send the message (the output) from the sender. *Media* is the resource used for send a message, such as paper, television, telephone, or smoke signals. Once the sender has selected a media, the sender will transmit the message. *Transmission* is the method used for sending a message, such as over the radio or cellular phones, by airwaves, or using a physical transporter such as a homing pigeon or postal service. The media and transmission used by the sender will determine how well the message is received, called *reception*. Is the radio transmission clear enough to be intelligible or is the handwriting legible? As the message is received, it is interpreted by the receiver and becomes the input. Interpretation by the receiver is the most critical component in the formation of a common understanding between the sender and the receiver. *Interpretation* determines how well the receiver understands what has been sent (see Figure 1.1 Communications process).

Does the sender write the message in English not knowing whether the receiver can read, write, speak, or understand English? Must the receiver hear the message in order to interpret its meaning even though the sender is unsure whether the receiver is hearing-impaired or otherwise

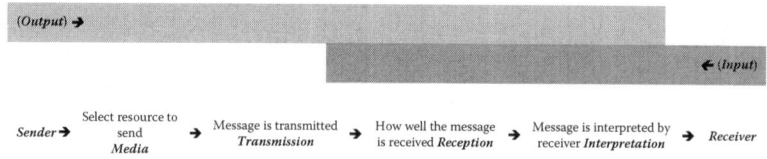

FIGURE 1.1 Communications process.

unable to interpret the message? To have a common understanding between the sender and receiver the media, transmission, reception, and interpretation considerations play a pivotal role in communications.

COMMUNICATIONS—3500 BC TO 1 BC

There are records dating back to 3500 BC showing a time when paintings were made by indigenous tribes as a means to communicate. Around that period, the Sumerians developed *pictographs* of events that were written on clay tablets. The Egyptians also created *hieroglyphic writing* (see Figure 1.2 Early Egyptian hieroglyphic alphabet). In 1500 BC, the Phoenicians created an alphabet. In 1400 BC, bones were used for writing in China, the oldest record of writing. In 1300 BC, drum beat codes sounded alarms during the Shang Dynasty in China. The Chinese government introduced the first *postal service* in 900 BC. The postal service was one of the first processes used to deliver communications over a distance to a specific individual.

In 776 BC, homing pigeons were used to send messages including an announcement of the winner of the Olympic Games to the Athenians. Before homing pigeons, human messengers running on foot or horseback were the only way to send messages from town to town or to relay orders and

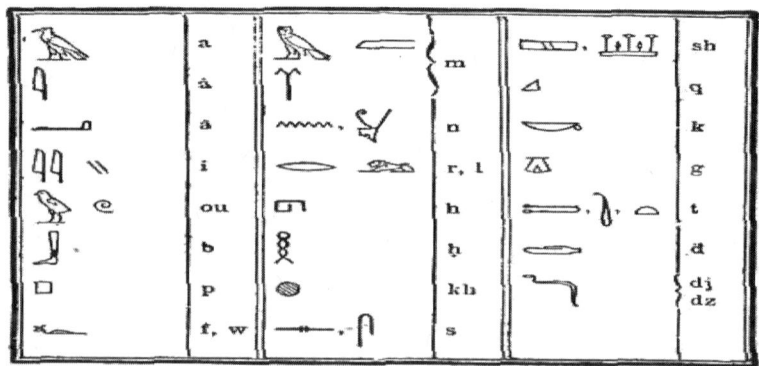

FIGURE 1.2 Early Egyptian hieroglyphic alphabet.

intelligence during wartime. This was a dangerous task for human messengers. Messengers were killed, bribed, or their messages intercepted. About 400 years later, more effective communication methods were introduced.

Between 1200 BC and 100 BC, fire messages were used from relay station to station instead of human messengers in Egypt and China. In ancient Greece, the Greek were reported to have used fire signals to send a message from Troy to the city of Argos (chief town in eastern Peloponnese), about 325 miles (600 km) in the late eleventh century BC. Troy was located on the top of the Hisarlik, a mountain in western Anatolia, the area now known as Turkey. It was said that alone on the Argos palace roof, the watchman awaited the fire signal that would tell the household that the Greeks had captured Troy. The message reached the city of Argos after a few hours.[1] Smoke signals were also used by Native American tribes.

HELIOGRAPHS

Sending messages using mirrors or shiny metals and the rays from the sun has been done by flashing reflected rays to another location up to 50 miles away. This form of communicating is known as *heliography,* with the first recorded use in 405 BC. The ancient Greeks used their polished shields to signal in battle. The Roman emperor Tiberius was thought to have sent coded orders using heliographs daily to the mainland, about 8 miles (12.8 km) away in 37 AD. The Egyptians were also known to have use heliographs that could be seen as far as 80 miles away.[2]

Thousands of years later, heliographs remain in use. As late as 1979, Afghan forces were reported to use heliographs to warn of approaching enemy troops during the Soviet invasion of Afghanistan. Today, heliographs are included in some survival kits used by hikers and pilots for *emergency* signaling and for the search and rescue of aircraft. An *emergency* is a situation or event that presents an immediate risk to life, health, property, or the environment. Heliographs are also used to measure long distances by triangulations by the U.S. Coast Guard and for geological surveys in the early 1900s.[3]

Other primitive forms of communications used for telegraphy over distances include smoke signals, torch signaling, and signal flags. The word *telegraph,* with its origins from the union of Greek words *tele* and *graph,* that essentially mean "long-distance writing."

PAPER, NEWSPAPERS, AND MAGAZINES

The Chinese were the first to use paper in 104 AD. In 1450, when paper was easily produced and widely available, newspapers were created in

Europe. In Renaissance Europe, newsletters were handwritten and privately circulated among merchants. They disclosed information about wars, economic conditions, social customs and "human interest" features. Printed news pamphlets, also known as *broadsides,* were the forerunners of the newspaper. German broadsides of the late 1400s had highly sensationalized content.

Newspaper circulations have grown over the years. The United States had nearly 2,150 daily newspapers in circulation in 1900 and peaked at 2,200 by 1910. Daily newspapers were a way to share information on events in the coverage area and world news. By 1967, most newspaper and magazine production was digitized (Media History Project). In 1910, there were an estimated 1,800 magazines in publications. A *magazine* is a publication published periodically, less frequently than newspapers.

TELEGRAPH SERVICE AND MORSE CODE

Telegraph Service

A Frenchman named Claude Chappe invented the first long-distance semaphone telegraph line in 1793. Telegraph services was used during the Revolutionary and Napoleonic wars when communication systems were simplistic, relying mainly on mounted dispatch riders. As telegraph service was adopted, communication towers were erected in the line of sight of each other. Once the towers were installed, the French sent a signal using the telegraph system from Paris to Lille, approximately 118 miles (191 km) in five minutes. Each tower had 196 combinations, also known as *signs*, and each was worked by a series of pulleys and levers. An operator could send three signs in a minute, provided visibility was good.

There were a number of significant communication "firsts" during the eighteenth century. Many of these can be attributed to the military needs of the time. Major wars include the French Revolution starting in 1789, the Napoleonic Wars that began in 1803, the Mexican–American War of 1846, and the U.S. Civil War in 1861. The first optical telegraph system was invented in the mid-1800s that covered approximately 3,100 miles (5,000 km) and encompassed more than 550 stations. These early systems included the naval semaphore system, the railroad semaphore system, and the "wig-wag" system.

William Cooke and Charles Wheatstone developed an early form of telegraphy system called the English Needle Telegraph in 1837. This system used pointing needles rotating over an alphabetical chart to indicate the letters that had been sent. The major drawbacks of this system were that it had a complex configuration and it was slow. These were common issues among electrical telegraphing systems of this era.

Optical and visual telegraphy systems enabled information to be transmitted more quickly than the fastest form of transportation. This is at a time when advances in communications had outpaced earlier messenger systems. Telegraphy systems enabled the use of error control (resending lost characters), message priority, and the flow control (send faster or slower) for the first time. These were significant milestones in communications. These concepts remain in use today for crisis and emergency communications. Telegraphy systems continued to evolve with encoded shutter system developments in Sweden and England in the late 1800s.

Morse Code

Samuel Finley Breese Morse invented the *Morse code* in 1835, a landmark in using technology to communicate electronically. Morse code is a method for transmitting textual information using a series of indentation marks (dots and dashes) on paper tape that can be directly understood by a skilled listener or observer without special equipment (see Figure 1.3 Morse code).[4] The system sent pulses of electric current along wires that were controlled by the receiving end of the telegraph system to deflect an electromagnet.

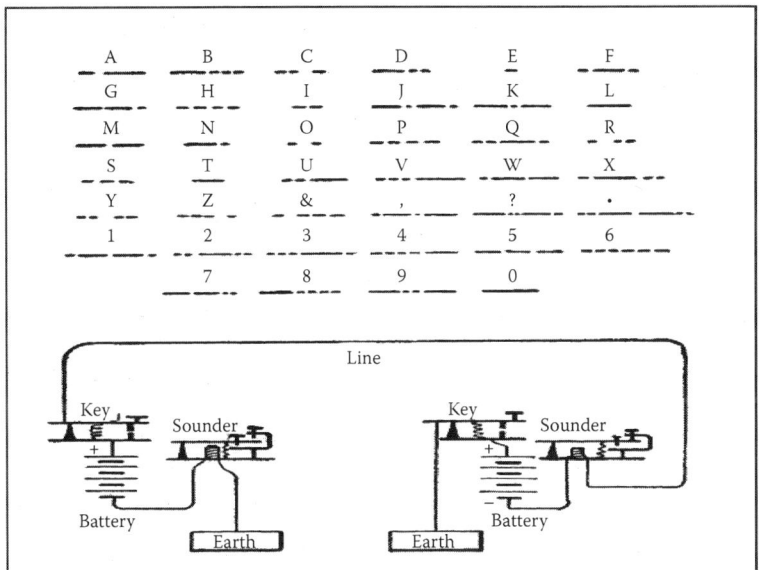

FIGURE 1.3   Morse code.

Samuel Morse entered into an agreement with Alfred Vail to expand the original code to include the alphabet, numbers, and special characters for broader appeal. A group of "dots" also known as *dit* (·) or short marks and "dashes," also known as *dah* (-) or longer marks are assigned to a character. Pauses, referred to as *gaps,* are used between letters, words, and sentences. A short gap is used between letters, a medium gap is used between words, and a long gap is used between sentences.

Eight years later, in 1843, Samuel Morse invented the first long-distance electric telegraph line. The first telegraph message was sent electrically from the U.S. Supreme Court chamber in Washington, DC, to the railway depot in Baltimore, Maryland. Congress funded construction of this experimental telegraph line. The U.S. Morse code was sometimes known as the "Railroad Morse." A trained operator could now send or receive 40–50 words per minute. By 1914, automated transmission was in use. A trained operator could handle more than twice the original rate.

The modern International Morse code invented by Friedrich Gerke was introduced in 1848. Morse code gained popularity in Europe while the United States continued to use the former version. The International Morse code eliminated the use of spaced dots. This system had an advantage over other forms of communications during this time because it had an easy working principle, and it could function efficiently with low-quality wires common in rural areas. As the number of users and undersea cables increased, transmission errors occurred.

Morse code with its many improvements became a mainstay in communications during disasters and emergencies for more than 160 years. A *disaster* is a natural situation or event that overwhelms the local capacity to respond, recover, prevent, or mitigate damage and may require a request for external help. Morse code was the backbone of early emergency communications technology.

MAIL AND PARCELS

The Chinese introduced the first *postal service* in 900 BC. The Pony Express was used for U.S. mail delivery during the Wild West days; by 1912, the first mail was carried by airplane.

Toronto, Canada, first used numbered postal zones in 1925. By the 1940s other Canadian cities were using the system. For example, the postal zone used for Toronto was 5. Mail would be addressed giving the postal zone name and postal zone number, and province, that is, Toronto 5, Ontario. By 1943, the City of Toronto was divided into 14 zones using the numbers 1 through 6, 8 through 10, and 12 through 15. The numbers 7 and 11 were not used and a 2B zone was added.

Today the postal code contains six alphanumeric characters in the form of A#A #A#. The letter "A" represents an alphabetic character and the # represents a numeric character. The postal code consists of two three-character segments. The first section is for the forward sortation area A#A, and the second section is for the local delivery unit #A#. Similar systems were implemented elsewhere around the world over the next 20 years.[5]

In 1943, the U.S. Postal Service (USPS) started using postal zones for large U.S. metropolitan areas. In 1963, 20 years later, the USPS introduced *ZIP codes* to leverage technology in managing the explosive growth of mail. ZIP is the acronym for *Zoning Improvement Plan*. The use of ZIP codes was optional until 1967 when USPS required all second-class and third-class mail to be presorted by ZIP code. In 1983, the ZIP+4 system, also called the "plus-four codes" or "add-on codes," was implemented by USPS. The additional four digits represent a geographic area within a ZIP code delivery area, such as an apartment complex, neighborhood, or other small community. A postal bar code is now used for automated sorting of mail called the *Postnet*. There are long Postnet bar that contain the ZIP+4 code and the short-Postnet bar that contains only the five-digit ZIP code.

In addition to ZIP codes, globally ISO2 and ISO3 codes are now used to define a country such as the US (ISO2) and USA (ISO3) country codes. ISO represents the *International Organization for Standardization*. ISO 3166 standard provides a listing of two and three character codes that represent countries, dependent territories, and unique areas of geographical interest.[6]

FACSIMILE (FAX)

Alexander Bain of England patented the first facsimile (fax) machine in 1843. Bain's fax machine consisted of two pens attached to pendulums that were connected by a telegraph wire. When an electrical charge was sent through the telegraph wire, the pendulums would pass over chemically treated paper and would make stains as directed. Five years later, Frederick Bakewell invented a conducting roller. The transmission began as revolving drums covered with treated tinfoil turned and on the receiving end, the receiver would receive a recorded image. The first transatlantic fax service was provided by the company RCA in 1922.

Fax services were used primarily by newspapers to transmit photographs, the weather service to fax weather charts, and later by the military to transmit maps, orders and weather charts. In 1960, USPS experimented with facsimile mail. Xerox, in 1966, emerged and sold the first successful *telecopier* in the United States, a fax machine called the Telecopier I (Magnafax).[7] Telecopier, also called fax machine, is the equipment used for transmitting a copy of a document to the receiving fax machine. Fax machines remain in use, with facsimile servicers delivering facsimiles via the Internet and IP fax services. Today, the Internet's global electronic communications system connects computer networks and organizational computer facilities, providing public and private access to information to its users.

By 1983, the popularity of faxing grew exponentially. The Comité Consultatif International Téléphonique et Télégraphique (or CCITT Group 3) increased the standard protocol to 9,600 bits per second (bps). The CCITT was an organization that set international communications standards. Fax machines are commonly use at work and home. Faxes are widely accepted as legal documents, particularly since it can capture from where, when, and who sent the fax. A fax can also include handwritten information such as a signature.

In the 1800s, advancements in communications technology were occurring in many different areas. Coleman Sellers invented a machine that flashed a series of still photographs onto a screen, called a *kinematoscope*, in 1861. Early motion picture cameras and projectors that could photograph motion pictures were called *kinetograph*.[8] Kinetographs were later followed by the kinematoscope in 1890. Thomas Edison and his assistant William Dickson were credited with the development of the kinetograph.

The kinetograph, a motor-driven camera, could capture movement with a synchronized shutter and sprocket system. This system would move film through the camera using an electric motor. The film, made by Eastman and according to Edison's specifications, was 35 mm wide and had sprocket holes on the sides to advance the film. Edison's Kinetograph camera essentially has became the standard for theatrical motion picture cameras used today.

The United States' involvement in World War II led to a proliferation of patriotism and propaganda movies. Two U.S. films, *The Desperate Journey* (1942) and *Forever and a Day* (1943) were movies of patriotism and propaganda. The need for wartime propaganda also occurred in Britain with war dramas such as *The Forty-Ninth Parallel* (1941) and *Went the Day Well?* (1942).

TELEPHONE

In 1875, Alexander Graham Bell developed "harmonic telegraphy" with which he could hear a sound over a wire. In 1876, Bell patented the electric *telephone*, devices that can transmit speech electrically. Elisha Gray also invented a different electric telephone. Both men rushed to the patent office with Bell patenting his telephone first, just hours before Gray.

A famous legal battle ensued between the two men over the invention of the telephone with Bell emerging the winner.

By 1905, approximately 2.2 million telephones were active within the Bell System. The first coast-to-coast telephone line, from New York to San Francisco was completed by the Bell System in 1915. Combining "harmonic telegraphy" and traditional Morse code telegraphy led to a new generation of communication devices that could be used in times of peace and for disaster communications.

Shortly thereafter, overseas cable installations began to connect the United States to other countries. In 1918, a major business disruption occurred when the U.S. government took over the telephone service. After the attack on Pearl Harbor during WWII, the volume of calls increased dramatically and telephone technicians were called into service by the thousands. Western Electric shifted 85% of its output to war-related projects leaving home phone service as a low priority, and many users without regular service. By 1956, the first transatlantic cable between England and Newfoundland opened.

MOBILE PHONE

Lars Magnus Ericsson developed the first mobile telephone application in the early twentieth century. Today, Ericsson is a global provider of telecommunications equipment and related services to mobile and fixed network operators. This was a portable phone handset with a crank that could be hooked to bare phone wires. The connection was made using a pair of metal hooks that were placed over the wires using an extension wand. Once contact with the wires was made, the magneto in the handbox was cranked. This would make a signal that could be answered by someone on the line. Around 1907 this system was used to report a train robbery that led to the arrest of the bandits. In 1946 the first commercial mobile phones were installed in St. Louis.

Telephone service has served as the basic communication link. Phone services can be completed using cellular phones, satellite phones, and computers that have each revolutionized personal and commercial communications. Telephone service has served as the basic communication link worldwide since its inception and remains a true statement today and for the near future.

COPY MACHINE

Thomas Edison patented the office copying machine called a *mimeograph* in 1876. A mimeograph machine, also called "mimeo," is a printing press that forces ink through a stencil onto paper. Early versions had an electric pen that was used for making a stencil and a flatbed duplicating press. It was used to print short-run classroom materials, classroom and local bulletins, and office work. Many developing countries continue to use this device. In developed countries, this machine has been replaced by less expensive photocopying and offset printing that was developed in the 1960s.

LOUDSPEAKERS

In 1877, Ernst Siemens of Germany patented the first *loudspeaker*. Loudspeakers were later redesigned to use electricity. By 1916, the electric loudspeaker was introduced. In 1924 Chester Rice and Edward Kellogg, two GE researchers, patented the modern loudspeaker that contained a moving coil-driven mass-controlled diaphragm in a baffle and direct radiator. This design remains in use today.

The quality of loudspeaker systems has made significant audible improvements over the last 50 years. These improvements are due to continuous developments in enclosure design and materials using computer-aided design and finite element analysis. Loudspeakers are used to give updates at sporting events and to provide warnings of impending danger. Loudspeakers can offer both a voice and a tone.

MAGNETIC RECORDING

Valdemar Poulsen invented the first magnetic recordings using magnetized steel tape for the recording medium. The use of magnetic recording was the start in using disk and tape for mass data storage. The music recording industry and radio used tape as early as the 1930s, and tape continued for years to come. In 1963, Phillips developed the compact cassette with Dolby noise reduction technology. Cassettes markedly improved the quality of recorded sound. In 1979, Sony developed the Walkman, which used compact cassettes.

These developments led to the widespread use of magnetic audio tape. Revolutionary changes for radio and the recording industry occurred now that sound could be recorded, erased, and rerecorded using the same tape multiple times. These tapes also got smaller. Preproduced programming on tape cartridges could have endless loops. This helped the broadcasting industry and led to growth in the monitoring of the industry. Hitachi and New Energy and Industrial Technology Development Organization (NEDO),[18] have developed technologies for microwave magnetic recording that will expand the recording density of *hard disk drives* (HDD) to greater than 1.5 TB for even rugged mini external HDD.

RADIO

At the start of the twentieth century, more advancement occurred in the use of technology to communicate to larger audiences over longer distances. The *radio* came into use with its modulation of electromagnetic waves to transmit signals with frequencies below those of visible light.

An Italian inventor named Guglielmo Marconi in 1894 built his first radio equipment, which could ring a bell from 30 feet away—the first radio signal. In 1899, he sent a wireless signal across the English Channel.[9] In 1901 Marconi had his staff in Poldhu, Cornwall, England, transmit the Morse code letter "s" (three dots) at an appointed time. The letter "s" was chosen since it was easy to distinguish. Marconi pressed his ear to the telephone headset attached to his receiver and successfully heard "pip, pip, pip" 1,700 miles away from the transmitter—across the Atlantic from Cornwall to Newfoundland. Marconi's telegraphy sparked gap technology within a broad segment of the *radio spectrum*. The radio spectrum is the radio frequency (RF) portion of the electromagnetic spectrum.[10] This transoceanic signaling became a part of his business empire of ship-to-ship and ship-to-shore communications, including the communication system used on the *Titanic*.[11]

The sinking of the *Titanic* led to a U.S. Senate investigation into the practices of Marconi's business. It was found that there were several ships that were responding to the *Titanic* distress signals. The ship that was closer did not receive signals from the *Titanic* since the vessel's only radio operator was off duty. The investigation concluded that with the earlier arrival of the closer ship more lives could have been saved.

The U.S. Radio Act of 1912 was enacted, which required that at least two radio operators are on board all vessels with more than 50 passengers and at least one operator had to be on duty in the Marconi room at all times the vessel was at sea. This was landmark legislation in a largely unregulated industry, where unfettered development and use of communications led to the early control of communications and the formation of some important principles in the use of communications technology.

Marconi was only one individual who influenced the early years of radio. Others include Nikola Tesla, Alexander Popov, Sir Oliver Lodge, Reginald Fessenden, and Heinrich Hertz among others listed below. The key accomplishments by year, the name of the inventor, and the invention or notable works are mentioned in Table 1.1 (Radio Inventors and Their Notable Works) for the early years of radio.

TABLE 1.1 Radio Inventors and Their Notable Works

Year	Name	Inventions/Notable Works
1864	James Clerk Maxwell	• A Scottish physicist and mathematician. • Maxwell predicted the existence of radio waves. It was on this basis that radio waves were discovered and Einstein's theory of relativity gained traction.
1868	Mahlon Loomis	• A Washington, DC, dentist. • The first wireless telegrapher to demonstrate a wireless communication system that could operate between two sites. The sites were 14 to 18 miles apart.
1882	Amos Dolbear	• An American physicist, inventor, and Tufts University professor. • Received a U.S. patent for a wireless telegraph.
1888	Heinrich Hertz	• A German physicist. • The first individual to validate the presence of electromagnetic waves by constructing a system to create and detect UHF radio waves in 1888. • Hertz's name is used to represent radio frequencies and was officially added to the metric system in 1933.
1892	Nikola Tesla	• An inventor, mechanical engineer, and electrical engineer. • Tesla is the designer of the early radio in 1892. • Tesla patented a radio-controlled robot boat that was directed by radio waves in 1898.
1894	Alexander Popov	• A Russian physicist. • Popov constructed the first radio receiver containing a "coherer" in 1894. A coherer is a primitive form of radio signal detector used in the first radio receivers. • Popov's radio was modified to a lightning detector that was demonstrated before the Russian Physical and Chemical Society in 1895. • He sent the transmission of radio waves across different campus buildings at St. Petersburg in 1896.
1894	Sir Oliver Lodge	• A British physicist and writer. • Lodge perfected the design of the coherer. His was a radio-wave detector and the basis of the early radiotelegraph receiver. • He is considered the first to transmit a radio signal.

TABLE 1.1 (continued)   Radio Inventors and Their Notable Works

Year	Name	Inventions/Notable Works
1895	Guglielmo Marconi	• An Italian inventor. • Marconi conducted the first experimental transmission of wireless signals. • He filed a U.S. patent for wireless communication in 1896. • Marconi demonstrated the first transatlantic wireless transmission between Poldhu, England, and St. John's, Newfoundland, by using Morse code in 1901.
1898	Nathan Stubblefield	• An American inventor and Kentucky melon farmer. • Thought to have actually invented the radio before Tesla and Marconi. • Stubblefield's device worked by audio frequency induction or audio frequency earth conduction rather than radio frequency radiation for radio transmission telecommunication. • There are also reports that Stubblefield is the inventor of wireless telephony or wireless transmission of the human voice.
1900	Reginald Fessenden	• A Canadian inventor. • Fessenden sent the first audio transmission by radio in 1900. • He sent the first two-way transatlantic radio transmission in 1906. • He designed a high-frequency alternator and transmitted human voice over the radio in 1906. • He went on the airways with the first radio broadcast of entertainment and music in 1906. • He stated that he had a better spark-gap transmitter and coherer-receiver combination than those developed by Lodge and Marconi.
1903	Valdemar Poulsen	• A Danish engineer. • Poulsen began arc transmission to create high-frequency alternators for sending radio waves. The *New York Times* and the *Times* of London knew about the Russo-Japanese War due to radio transmissions in 1903.
1906	Lee de Forest	• An inventor with more than 180 U.S. patents. • De Forest made the detection, transmission, and amplification of sound possible.

The term *radio* was adopted by the U.S. Navy in 1912 and was commonly used by the public when the first U.S. commercial broadcasts occurred in the 1920s. Between 1911 and 1930, the growth in radio exploded in the United States and across the Atlantic Ocean. The Radio Corporation of America was formed by combining General Electric (GE), Western Electric, AT&T, and Westinghouse. In France, battery-powered receivers with headphones and valves were seen. Radio broadcasting began in Asia and the Caribbean—in Shanghai and Cuba—and the first regular broadcasts occurred in Australia, Belgium, Finland, Germany, Norway, and Switzerland. By the 1930s, radio was prevalent around the world.

Telephone and radio communications were enhanced with the use of the electronic amplifying tube also known as the *triode*, which enabled all electronic signals to be amplified. Lee de Forest introduced the triode in 1906. In less than eight years after the introduction of the triode, the first cross-continental telephone call was made in 1914. *Radio telephony*, also known as voice communications using radio waves, was developed to support safety and military communications.

Early radio supported maritime vessels and was used for sending telegraphic messages using Morse code between ships and land. The Japanese Navy scouting the Russian fleet during the Battle of Tsushima in 1905 was one of the early adopters of radio. Radio was used to pass on orders and communications between armies and navies on both sides in World War I. Germany used radio communications for diplomatic messages once it discovered that the British had tapped its submarine communications cables. The United States passed on President Woodrow Wilson's Fourteen Points to Germany via radio during the war.

Amateur Radio

The countries that were using radio and had the technology were also adopting the concept of amateur radio. Much of the implementation of amateur radio was completed using a combination of scientists, hobbyists, and others familiar with Marconi's radio communications systems. As use of the radio spectrum grew in the first quarter of the twentieth century, a portion of the spectrum was allocated for amateur use. The need for systematic relay of messages became evident as the portion of the radio spectrum allotted for amateur use could carry messages only over relatively short distances. Shortly after a portion of the radio spectrum was allocated for amateur use the United States entered in to World War I. The U.S. Navy seized controls of radio and shut down the amateur radio transmitters in 1917. Within a year, the amateur radio transmitters were restored for use.

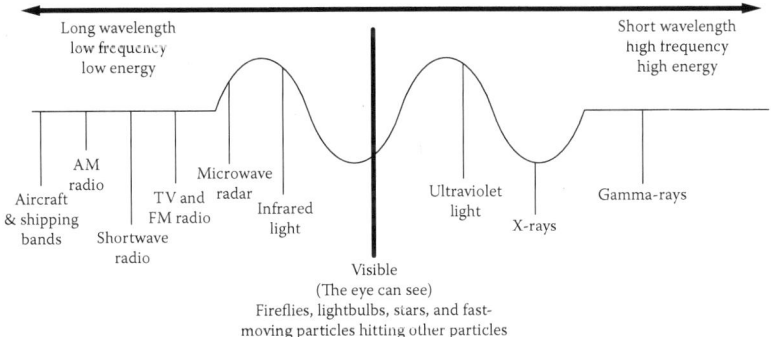

FIGURE 1.4 Electromagnetic spectrum.

In 1919, *shortwave radio*, referring to the high frequency (HF) portion of the radio spectrum was discovered. Shortwave radio is used by amateur radio, for broadcasting voice and music, and used for long-distance communication in remote areas, aircraft in flight, ships at sea, and other areas where wired communication or other radio services may be out of reach. (See Figure 1.4 Electromagnetic spectrum.)

Hiram Percy Maxim, in 1919, originated the American Radio Relay League (ARRL). The ARRL was a voluntary network of associated radio amateurs whose purpose was to facilitate the long-range relay of *radiograms*. Radiograms are written messages, telegram style messages, routed through a network of amateur radio operators who are on the air to relay messages over the radio. Similar associations were established outside the United States to foster long-distance passage of radiograms.

As ARRL and the use of radiograms grew, the reliability and accuracy in relaying these messages increased in importance. Radiograms were adopted for military, commercial, and public service relay of information concurrently. Standardized message formats were established. Military and commercial services were staffed by amateur radio operations.

As processes evolved, the National Traffic System was created and in conjunction with the ARRL, radio messages were passed nationwide for commercial, health and welfare, and disaster information in this manner. The National Traffic System format for messages or radiograms, including voice messages, remains active today. A movement to redesign the format so that it can easily interface with the technologies and techniques used today to transmit and receive disaster communications are under way under the IPAWS system.

The ARRL developed an Emergency Corps in 1940 before the U.S. involvement in World War II. The Emergency Corps trained and drilled even on frequencies that were closed for casual amateur use. Hundreds of these amateur operators would staff listening and direction finding stations. ARRL became the War Emergency Radio Service in 1942 where its primary purpose was air raid protection and notification. In 1948, the Military Affiliate Radio System was established.

The Military Affiliate Radio System was an integration of amateur operators, also known as *hams*, and military operators using a common set of frequencies worldwide. Hams are radio hobbyists licensed to transmit radio signals as a leisure-time interest, and if desired for emergency and public service communications. Hams are licensed to use frequencies in a wide range of *narrowbands* throughout the radio spectrum. To become a member of the Military Affiliate Radio System, an operator is required to receive a base level of training and continuing active participation in practice nets and drills are required. This requirement continues today.

Narrowband refer to the bandwidth of a radio message not exceeding its channel's maximum bandwidth. Narrowband communications provides stable long-range communication where the carrier purity of transmission spectrum is good, and able to manage an operation of many radio devices within same frequency bandwidth simultaneously. Narrowband is ideal for areas where many radio-control devices are used such as at a construction site.

Around 1952, the Radio Amateur Civil Emergency Services (RACES) was formed in concurrence with the federal Civil Defense effort. This was just as the Cold War began. Organizations like RACES were established around the world over the next 25 years. Federal and local authorities realized the need for disaster and emergency communications that went beyond military and nuclear disasters but comprised all facets of civil life. In 1972, the U.S. Civil Defense name was changed to the Federal Emergency Management Agency (FEMA).

FEMA and the FCC, in accordance with FCC Part 97, Section 407, govern all U.S. RACES organizations today. All transmitted communications in RACES must be authorized by the civil defense organization for the area served. The only types of messages that may be sent are concerning:

1. Impending or actual conditions jeopardizing public safety during civil emergencies
2. Immediate safety of life, protection of property, maintenance of law and order, alleviating human suffering and need, and combating an attack
3. Public information essential to activities of civil defense or other government or relief agencies
4. Communications for RACES training, testing and drills.[12]

By the 1970s, transistors and integrated circuits were in existence. Ham radio operators were using this technology to develop new ways to use the most advanced communications technologies on a range of frequencies from the lowest to microwaves. As technology evolved so did emergency preparedness activities.

AM Broadcasting

AM broadcasting uses amplitude modulation for radio broadcasting. An AM receiver is used to detect amplitude variations in radio waves at a specific frequency. Using the specified frequency the receiver amplifies changes in the signal voltage to push sound through a loudspeaker or earphones. An example of AM broadcasting is if an AM radio station is broadcasting at 440 Hz sound, the signal gets louder and quieter 440 times every second. The signal adjusts in Hertz's, as the frequency is increased or decreased. This results in the listener hearing the encoded information rather than the actual radio signal.

AM radio broadcasting was the primary method for broadcasting for the first 75 years of the 1900s and used today. Reginal Fessenden, a Canadian, launched the first experimental AM radio broadcast in 1906 on Christmas Eve. In 1909, Charles "Doc" Herrold began weekly broadcasts in San Jose, California, on radio station KCBS. KCBS claims to be the world's oldest broadcasting station, having celebrated its 100th anniversary in 2009.

Radio programming mushroomed between the 1920s and 1950s, following World War I, an era known as the "Golden Age of Radio." Comedies, dramas, soap operas, and music were developed and broadcasted. In Britain and the United States, the institutionalization of radio began in 1927. In America, establishment of the Federal Radio Commission ended chaotic noise and harshness in the sound of words or phrases on the airwaves.

The 1930s was a time of widespread want and deprivation due to the Great Depression, and it is a time that is marked as an important stage in the evolution of a mass consumer society in the United States. Many in this country dealt with economic hardship. Simultaneously, the U.S. appetite for a variety of consumer goods grew, thanks to media, specifically radio. Radio moved the culture towards a national consumer culture. The "American way of life" and the "American Dream" took root during this era. The early radio soap opera were daytime serials that appealed to American homemakers who were most likely to purchase many of the household products.[13]

At the start of the 1940s, radio was used as a primary medium to prepare people for the possibility that the war in Europe could soon command American participation, fueling anxieties about national cohesiveness. This was exacerbated during a decade of economic depression. Although the national networks dominated the radio airwaves at the beginning of the 1940s, by the end of the decade, some of that power shifted as changes occurred in the industry. National demographics allowed for the emergence of stronger independent urban radio stations and more politically daring broadcasts.

AM radio stations use ranges from 250 to 50,000 watts. Experimental licenses were issued before World War II for 500,000 watts. These stations were established for wide-area communications during disasters. These stations were decommissioned in the early 1940s in the United States and Canada. In Mexico, a few stations operated at 250,000 watts, and other countries authorize high power operations.

Radio was distinguished as the desired medium for mass communication since radio programs came directly into the home. This gave radio a convenience and an intimacy that no other communication agency had at this time. The 1950s began with President Truman delivering a message on the U.S. position of strengthening the defense of Europe and the nature of Germany's contribution to the war.

AM radio remains vulnerable to atmospheric disturbances such as lightning strikes, static electricity, and other sources that can cause brief spikes or troughs to the signal. These disturbances cause a crackly or muffled sound. Today AM radio is used primarily for talk and sports radio due to the quality of the sound.

FM Broadcasting

Edwin Howard Armstrong in 1902 pioneered *FM broadcasting*, a broadcast technology that uses frequency modulation (FM) to provide high-fidelity sound over broadcast radio. Like AM broadcasting, FM radio uses electromagnetic radiation to broadcast sound information. A receiver is used to pick up the radio signal, looking for small changes in the carrier signal on a specific frequency carrying the sound information called the carrier wave. This process enables the listener to hear the information rather than the radio signal like AM radio broadcasting.

Cornelius Ehret, Philadelphia, Pennsylvania, submitted a patent application defining the use of frequency modulation in both radiotelegraphy and radiotelephony that includes circuitry for both FM transmission and reception.[14] Armstrong, in 1934, began testing FM from the Empire State Building using 41 MHz with cooperation from RCA in New York City. He transmitted using both AM and FM. After some resistance, Armstrong was granted an experimental license for a FM station in 1936.

The first commercial FM broadcasting stations were in the United States. These stations were primarily used to broadcast classical music to an affluent listenership in urban areas and for educational programming. FM broadcasting expanded into other music categories and news by the 1960s as the FCC awarded more FM permits and approved FM stereo multiplex standards.

The advantage FM stations had over AM stations was better quality sound over longer distances. The FM signal is not subject to static, lightning, and spark interference like AM signals. The reason for improved sound quality is that FM is transmitted in the Very High Frequency (*VHF*), 30 MHz to 300 MHz, radio spectrum. VHF radio waves are similar to light since the waves travel in straight lines limiting the reception range to 50 or 100 miles. Under good conductions, such as flat terrain, the waves can travel 150 miles or more. FM stations are spaced 200 kilohertz (KHz) between each other, while AM stations are spaced 9 to 10 KHz away from each other. Since FM is designed to handle audio from 20Hz to 15 KHz, more space is available to place in the audio spectrum twice over, meaning more information on one channel without infringing on the other channel.

The United States, Belgium, Denmark, and Germany were among the first countries to adopt FM widespread. AM stations, in the medium wave band, had become overcrowded after World War II. Top 40 music and country and western music was migrating from AM to FM stations. Many stations simulcast in both AM and FM. With the proliferation of FM stations problems began to occur. A primary issue is that FM receivers are subject to the capture effect causing the radio to receive only the strongest signal when multiple signals appeared on the same frequency. Another concern was that public service broadcasters from Ireland, Australia, and the United States were slow to adopt FM. This created a need for standards on how to use the available FM bandwidth.

The purpose of the standard was to deliver stereo sound using both the left and the right channel of a stereo broadcast on the same FM channel. The process of combining multiple signals onto one composite signal so that the receiver reconstitutes the original signals is called *multiplexing*. The growth in FM stereo broadcasting spurred any many new FM stations starting in 1961.

FM broadcasting is not without its drawbacks. To cover a geographically large area, particularly where terrain is difficult, having a large enough number of FM transmitting stations is expensive. This makes FM more suited for local broadcasting than for nationwide networks in many countries.

Small-scale transmitters can be used to transmit a signal from an audio device such as a MP3 player to a standard FM radio receiver. A transmitter can also be used in near-professional-grade broadcasting systems to transmit audio throughout a property such as a neighborhood or campus radio stations. Campus radio stations are often run over carrier current, also considered a secondary signal transmitted in a "piggyback" fashion along with the main program. This form of broadcasting is called *microbroadcasting*. The reach of FM microbroadcasters is generally less than their AM competitors.

FM subcarrier services are secondary signals transmitted in a "piggyback" fashion along with the main program. Special receivers are required to utilize these services. Analog channels may contain alternative programming, such as reading services for the blind, background music or stereo sound signals. In some extremely crowded metropolitan areas, the subchannel program might be an alternate foreign language radio program for various ethnic groups. Subcarriers can also transmit digital data, such as station identification, the current song's name, web addresses, or stock quotes. In some countries, FM radios automatically re-tune themselves to the same channel in a different district by using subbands.

Radio transmission also occurs by air and water. Aviation voice radios use VHF AM so that multiple stations on the same channel can be received. Aircraft fly at altitudes high enough for their transmitters to receive a signal hundreds of miles away even though they are using VHF. Marine voice radios use a single sideband voice (SSB) in the shortwave High Frequency (HF) radio spectrum (3 MHz to 30MHz) for longer distances than VHF. The reason for the longer distances is that SSB does not transmit the unused carrier and sideband needed for other frequencies. Marine voice radios can also use narrowband FM in the VHF spectrum for shorter ranges.

Early police radios used AM receivers for one-way dispatches. Today, government, police, fire, and commercial voice services use narrowband FM on special frequencies. Civil and military HF voice services also use shortwave radio for communications with aircraft, isolated settlements, and ships at sea.

TELEPHONY—CELLULAR AND SATELLITE PHONES

Cellular Phones

Mobile phones, also known as cellular phones or cell phones, are an electronic device that transmits its signal to a local cell site (transmitter/receiver). The cell site connects to the *public switched telephone network* (PSTN) through an optic fiber, microwave radio, or other network elements. A PSTN, also known as the *plain old telephone service* (POTS), is a global network of public circuit-switched telephone networks. When the mobile phone nears the edge of the cell site's radio coverage area, the central computer switches the phone to a new cell. Cell phones enable seamless telephone calls whether an individual is stationary are moving around wide areas, such as across the country or a town. Cordless phones, sometimes confused with cell phones, provide telephone services within a limited range using a single base station that is attached to a fixed landline such as that used within a home or office.

Early cell phones used FM; most cell phones today use various digital modulation schemes. Public radiotelephone service on the high seas was introduced in 1929. By 1965, mobile radiotelephone service was widely available in the United States. Dr. Martin Cooper invented the first portable handset in 1973 when he made a call using a portable cellular phone. Dr. Cooper was a former Motorola systems manager. Dr. Cooper setup a base station in New York using the Motorola DynaTac, a cellular telephone prototype to show the public. In 1977, cell phones went public and testing of public cell phones began in Chicago the same year. The test group consisted of 2,000 customers. The testing was expanded to Washington, DC, and Baltimore and then to Japan in 1979.

Demand grew around the world. The cellular phone business started 50 years ago. Today, the cellular business has become a $30 billion per year industry and has more than 60 million users.[15] Globally, mobile phones are considered a necessity. Millions in India, China, Nigeria, and other emerging markets will continue to acquire mobile phones according to the International Telecommunication Union (ITU). In mature markets such as Europe and North America mobile phone use will also continue to increasing replacing fixed-line telephone.

Recent advances in cell phone technology enable an individual, on demand, to download digital material from a radio broadcast (such as a song) to a mobile phone, pay for a service at the point of sale, or request emergency services through an Enhanced 9-1-1 system. *Terrestrial Trunked Radio* (TETRA) is a digital-trunked mobile radio standard developed by the European Telecommunications Standards Institute (ETSI), an advancement in mobile telephony solutions. The purpose of the TETRA standard was to meet the needs of traditional Professional Mobile Radio (PMR) user organizations such as government, military, law enforcement, and EMS.

Satellite Phones

Satellite phones use satellites rather than cell towers to communicate. The two types of satellite phones are *Inmarsat* and *Iridium*. Both satellite types provide global coverage. Inmarsat uses geosynchronous satellites with aimed high-gain antennas on the vehicles. Inmarsat offers a comprehensive range of global mobile satellite services for use on land, at sea and in the air. Inmarsat services include voice, broadband data communications, and fax via its satellite system. Iridium constellation uses 66 low earth-orbiting (LEO) cross-linked satellites as the cells. Iridium is a global communications network with coverage for the entire planet. Iridium offers mobile voice and data communications.

Navigation

Satellites with precision clocks are used for satellite navigation system. A receiver listens to four satellites, each transmitting its position and the time of its transmission. The receiver then calculates using this information for its position based on a line that is tangent to a spherical shell around each satellite. This is as determined by the time-of-flight of the radio signals from the satellite.

The oldest form of radio navigation is radio direction-finding. Marine radio-location beacons and moveable loop antennas were used to locate commercial AM stations near a metropolitan area. Radio-location beacons were also shared with amateur radio operations for frequency ranges just above AM radio.

Aircraft used *VOR (Very High Frequency Omni-directional Range)* systems that have an antenna array transmitting two signals simultaneously. A directional signal rotates like a lighthouse at a fixed rate. An omni-directional signal pulses when the directional signal is facing north. An aircraft can determine its bearing by measuring the difference in phase of these two signals to establish a line of position. An aircraft can get a "fix" by obtaining readings from two VORs and then locate its position at the intersection of the two radials. Aircraft can also get a fix from one ground station when the VOR station is collocated with *DME (Distance Measuring Equipment)*. This type of station is called a *VOR/DME*. A similar system of navaids is operated by the military called *TACAN*. TACANs are often built into VOR stations, known as VORTACs. VOR/DME and VORTAC stations are similar in navigation potential to civil systems since TACANs include distance-measuring equipment.

RADAR

Radio was used in the prewar years for detecting and locating aircraft and ships via *radar (Radio Detection and Ranging)*. Radar detects objects at a distance by bouncing radio waves off the objects. The delay caused by the echo from the object measures the distance. The direction of the beam determines the direction of the reflection. The type of surface can be sensed through the polarization and frequency of the return. Navigational radars, common on commercial ships and long-distance commercial aircraft, can scan a wide area two to four times per minute using short waves that reflect from earth and stone.

Less sophisticated radars are general-purpose radars that use navigational radar frequencies. These devices modulate and polarize the pulse so the receiver can determine from the reflector its type of surface. The more advanced general-purpose radars distinguish the rain of heavy storms, land, and vehicles, and can superimpose sonar data and map data from Global Positioning System (GPS).

Other types of radar include search, targeting, and weather radars. Search radars can scan a wide area with pulses of short radio waves on average two to four times a minute.

Search radars sometimes use the *Doppler effect* to separate moving vehicles from clutter. The Doppler effect is the change in the frequency of energy in the form of waves, such as light or sound, that results from the motion from the source or the receiver of the waives. The Doppler effect was named after Christian Doppler, an Austrian scientist, who demonstrated the effects of sound. When the source of the wave and the receiver are approaching each other, the wave frequency will increase and the wavelength shortened. In short, sounds become higher pitched and light bluer. Conversely, when the source of the wave and the receiver are moving away from each other, the wave frequency will decrease and the wavelength lengthens. This means sounds become lower pitched and light redder.

Targeting radars are much like search radar scanning a smaller area more often, usually several times a second or more. Weather radars are similar to search radars, but use radio waves with circular polarization and a wavelength to reflect from water droplets. Some weather radar use the Doppler effect to measure wind speeds.

Data (Digital Radio)

Radio systems today primarily use digital broadcasting. The oldest form of digital broadcasting was spark gap telegraphy pioneered by Marconi. An operator of the Marconi system would press a key to send messages in Morse code by energizing a rotating commutating spark gap. The receiver would produce a tone from the rotating commutator, such as a hiss, that was indistinguishable from static. A spark gap transmitter broadcasts using a span covering several hundred megahertz (MHz).

These transmitters are now illegal since they waste radio frequencies and power.

Continuous wave (CW) telegraphy was the next advancement in telegraphy. CW uses a pure radio frequency that is produced by a vacuum tube electronic oscillator. The oscillator had an on/off switched that was operated by a key. A whistle-like audio tone was generated by a receiver having a local oscillator that would "heterodyne" with the pure radio frequency. CW use less than 100 Hz of bandwidth and are used by amateur radio operators today.

The military and weather services use radio teletypes that operate on short wave (HF) since it can create written information without a skilled operator. These radio teletypes sends a bit using one of two tones. A group of five to seven bits represents a character printed by a teletype. Aircraft radio-teletype service uses 1,200 Baud over VHF to send its identification, altitude and position. It is also used to obtain gate and connecting flight information. Radio teletype was a primary means of sending commercial messages to less developed countries between 1925 and 1975.

Other areas where radio-teletype service is used are microwave dishes on satellites, telephone exchanges and TV stations leveraging quadrature amplitude modulation (QAM). QAM sends data by changing the phase and amplitude of the radio signal. Engineers like QAM since it bundles the most bits into a radio signal when using an exclusive (non-shared) fixed narrowband frequency range. These bundles of bits are sent in "frames" that repeat. A unique bit pattern is used to locate the start of a frame.

The various radio-teletype communication systems described using fixed narrowband frequency range is vulnerable to jamming. To counter this threat a number of jamming-resistant spread spectrum techniques have been developed. *Global Positioning System* (GPS) satellite transmissions as used by the military are one example. Other spread spectrum for commercial use was deployed in the 1980s. Today, many cell phones, Bluetooth, and Wi-Fi (802.11b version) use various forms of spread spectrum.

GPS is a U.S. space-based radio-navigation system that provides positioning, navigation, and timing services to civilian users on a continuous worldwide basis. This is a free service provided by the U.S. government to anyone. A GPS receiver can receive location and time information from the system for an unlimited number of people, anytime, anywhere, regardless of time of day or weather conditions. The GPS consists of three segments:

- Satellites orbiting Earth
- Control and monitoring stations on Earth
- GPS receivers owned by users

Each GPS receiver gets the latitude, longitude, altitude (location), and time.

Bluetooth, named after Harold Bluetooth, a king in Denmark over 1,000 years ago, is the specification for using low-power radio communications to wirelessly link devices such as personal phones, PDAs, headsets, computers, and other networked devices. Bluetooth devices link up to distances of 30 feet or less and communicate at less than 1 Mbps.

The Institute of Electrical and Electronics Engineers (IEEE) established the standard protocol 802.11 for wireless network. There are three types—802.11a, 802.11b, and 802.11g. The first version is 802.11b, which is slower and less expensive than the other types and operates on frequency 2.4 GHz and transfer speeds up to 11 Mbps. 802.11a is the next generation wireless network operating on frequency 5 GHz with transfer speeds up to 54 Mbps. The third type is 802.11g, a combination of 802.11a and 802.11b wireless networks operating on frequency 2.4 GHz with a transfer speed up to 54 Mbps.

Both Bluetooth and Wi-Fi are wireless technologies that use radio frequency waves to create networks. Bluetooth temporarily connects an individual personal device over a short distance creating what is called a *personal area network* (PAN). The PAN enables devices in the immediate area to join the network. Wi-Fi connects multiple computers, desktops, laptops, and PDAs, over longer distances. The Wi-Fi Alliance, an international association, certifies interoperability of wireless Local Area Network (LAN) products based on IEEE 802.11. According to the Alliance on every continent, one in ten people around the world use Wi-Fi at home and work in countless ways.

Another resource is *coded orthogonal frequency-division multiplexing (COFDM)*. COFDM is a computer-aided system that makes and decodes signals, resists fading and ghosting, has error-correction coding, can resist interference, and is an adaptive system. COFDM is used for Wi-Fi, some cell phones, *Digital Radio Mondiale*, Eureka 147, local area networks (LANs), digital TV (DTV), and radio standards.

Digital Radio Mondiale™ (DRM) is the universal, open-standard, digital broadcasting system for all broadcasting frequencies up to 174 MHz, including AM and FM bands. The Eureka 147 System is the worldwide standard for both terrestrial and satellite delivery and will eventually replace traditional FM services. A LAN is a computer network connecting computers and devices within a limited geographical area such as a home, school, or office building. DTV is the transmission of video and audio by discrete (digital) signals.

Unlicensed Radio Services

Throughout North America, government authorized personal radio services such as Family Radio Service, Citizens' Band Radio, Multi-Use

Radio Service, and other radio services offer short-range, easy-to-use communications for small groups without licensing overhead. Similar services are used around the world.

Unauthorized, unlicensed radio broadcasting does occur and is called *pirate* or *free radio*. *Free radio* does not advertise or make money, while *pirate radio* broadcasting does advertise and make money to exist. Pirates are usually hobby broadcasters operating for entertainment by their owners and broadcasts are seldom political in nature.

Radio Control (RC)

Radio remote controls use radio waves to transmit control data to a remote object as in some early forms of guided missile, some early TV remotes and a range of model boats, cars and airplanes. Large industrial remote-controlled equipment such as cranes and switching locomotives now usually use digital radio techniques to ensure safety and reliability.

In Madison Square Garden, at the Electrical Exhibition of 1898, Nikola Tesla successfully demonstrated a radio-controlled boat. He was awarded U.S. patent No. 613,809 for a "Method of and Apparatus for Controlling Mechanism of Moving Vessels or Vehicles."

Wireless

In recent years the term "wireless" has gained renewed popularity through the rapid growth of short-range computer networking, for example, Wireless Local Area Network (WLAN), Wi-Fi, and Bluetooth, as well as mobile telephony, for example, GSM and UMTS. Today, the term *radio* often refers to the actual transceiver device or chip, whereas *wireless* refers to the system or method used for radio communication. Hence, one who talks about radio transceivers and Radio Frequency Identification (RFID) may actually be talking about wireless devices and wireless sensor networks.

One of the first developments in the early twentieth century (1900–1959) was that aircraft used commercial AM radio stations for navigation. In the early 1930s, amateur radio operators invented single sideband (SSB) and frequency modulation (FM). By the end of the decade, they were established commercial modes. In 1954, Regency introduced a pocket transistor radio, the TR-1, powered by a "standard 22.5 V Battery." In 1960, Sony introduced its first transistorized radio, small enough to fit in a vest pocket and able to be powered by a small battery. It was durable, because there were no tubes to burn out.

Today, a digital cell phone system, called *Terrestrial Trunked Radio* (TETRA), is used for military, police and emergency medical services. Commercial services such as *XM* (a pay-for-service satellite radio operator in the United States and Canada), WorldSpace (International satellite radio operator), and Sirius (a satellite radio (SDARS) service operating in the United States and Canada) offering encrypted digital satellite radio. In 2008, XM and Sirius satellite companies merged. The new organization called for the provision of satellite digital audio radio service (or "SDARS") in the United States, which would benefit consumers by offering more programming choices at various price points and greater choice and control over the programming subscription selected.

Other Radio

Energy autarkic radio technology consists of a small radio transmitter powered by environmental energy (push of a button, temperature differences, light, vibrations, etc.). A number of schemes have been proposed for wireless energy transfer. Various plans included transmitting power using microwaves. *Microwave power transmission* is the process of using microwaves to transmit power through space without wires. These schemes include, for example, solar power stations in orbit beaming energy down to terrestrial users.

Early radio systems relied on the energy collected by an antenna to produce signals for the operator. Radio became more useful after the invention of electronic devices such as the vacuum tube. The transistor later extended the use of radio, making it possible to amplify weak signals. Today, radio systems are used for applications from walkie-talkie children's toys to the control of space vehicles, as well as for broadcasting, and many other applications.

Radio assumes many forms today, including wireless networks and mobile communications of all types and radio broadcasting. Before the advent of television, commercial radio broadcasts included news and music, dramas, comedies, variety shows, and many other forms of entertainment. Radio was unique among methods of dramatic presentation in that it used only sound. For more, see radio programming in Section B, AM Broadcasting.

Television

Radio became an old medium with the arrival of television. Television is a telecommunications medium for transmitting and receiving moving images. The term *television* was a new word coined by the French and Constantin Perskyi, a Russian scientist, in 1900. After much effort, John Logie Baird transmitted the first experimental television signal in 1925 using radio to send pictures visible as television. Television offered live broadcasting of what was being recorded. In 1938, television broadcasts were taped, edited, and transmitted later. Scheduled television programming was introduced in 1939. Commercial television transmissions across North America and Europe occurred in the 1940s.

Television sends the picture as AM and the sound as FM, with the sound carrier a fixed frequency (4.5 MHz in the National Television System Committee [NTSC] system) away from the video carrier as used by many countries in the Americas. NTSC system establishes video bandwidth at 4.2 MHz and sound carrier at 4.5 MHz. Analog televisions uses a vestigial sideband on the video carrier to reduce the bandwidth required.

Digital television uses *8VSB* modulation in North America (under the Advanced Television Systems Committee (ATSC) digital television standard), and coded orthogonal frequency-division multiplexing (COFDM) modulation elsewhere in the world (using the Digital Video Broadcasting–Terrestrial (DVB-T) standard). 8VSB, denoted as 8-level Vestigial Side Band (VSB), is the transmission method for HDTV. *Terrestrial* means to send over the air. A Reed-Solomon error correction code adds redundant correction codes and allows reliable reception when moderate data loss occurs.

Many *codecs*, a computer program able to encode or decode a digital signal or data stream, can be sent in the MPEG-2 transport stream container format. *MPEG-2* represents digital television signals that are broadcast terrestrial, cable, or by direct broadcast satellite. High-definition television (HDTV) is possible simply by using a higher-video resolution picture. HD has one or two million pixels per frame. Standard-definition TV (SDTV) has 720 pixels for 480 lines, roughly five less than HDTV. With compression and improved modulation offered by HD over SD, a single channel can have a high-definition program and several standard-definition programs.

With the majority of the world's population focused on agriculture and manufacturing, illiteracy remained a problem in the early 1900s. In 1910, only 24% of U.S. adults had 5 years of education, 13.5% had completed high school, and 2.7% had a college degree. In the early 1960s, an estimated 44% of the world's population was illiterate and in the United States, 90% of households had a television set. The high level of illiteracy in the United States and the growing use of televisions in the home for entertainment led then FCC chairman Newton Minow in 1962 to call television a "vast wasteland."

Despite Chairman Minow's views of television, the FCC continued its role in the regulation and the advancement of related technology. In 1962, the FCC required television sets to have UHF tuners. By 1963 television news was so much the choice of the people that most learned of President John F. Kennedy's assassinaton by this medum. TV news was coming of age. By 1965, most television broadcasts were in color. Television sets were in the homes and businesses of 200 million around the world at a time when the United States had a population of 78 million. With the use of satellites for broadcasting, the first words from the moon "That's one small step for man; one giant leap for mankind"[16] as said by Neil Armstrong, Apollo 11, in 1969, were heard through this technology.

The FCC promotes use of *digital television (DTV)*, a technology using digital signals to transmit television programs rather than analog signals. DTV that allows for the transmission of better quality sound and higher resolution pictures is referred to as *high-definition television* (HDTV). HDTV was first developed in Japan by NHK in 1964. Fifty years later, in 2009, FCC required all television stations to broadcast only in digital in the United States. Television has been used as a communication medium to deliver emergency communications through live and recorded coverage of an event.

Satellites

Live television broadcasting for global audiences began with the introduction of the first international communication satellite that could transmits images called Telstar in 1962. As Telstar and other satellites were launched into orbit an international satellite organization was formed called Intelsat in 1964. By 1965, satellites began domestic TV distribution in the Soviet Union and the commercial communications satellite "Early Bird" also known as "Intelsat I" went into orbit. By 1968 Intelsat I had completed its global communications satellite loop. Shortly thereafter, the first television programming was available nationwide using a satellite as placed into operations by Ted Turner.

Over the next 20 years, transistors replaced tubes almost completely except for very high-power uses. By 1963, color television was being regularly transmitted commercially, and the first (radio) communication satellite, TELSTAR, was launched. In the late 1960s, the U.S. long-distance telephone network began to convert to a digital network, employing digital radios for many of its links. In the 1970s, *LORAN* (acronym for *Long Range Navigation*) became the premier radio navigation system. LORAN is a terrestrial radio navigation system that uses low-frequency radio transmitters for navigation.

Soon, the U.S. Navy experimented with satellite navigation and launched the GPS constellation in 1987. It became a replacement system for the LORAN since GPS systems offered greater accuracy. In the early 1990s, amateur radio experimenters began to use personal computers with audio cards to process radio signals. In 1994, the U.S. Army and the *Defense Advanced Research Projects Agency* (DARPA) launched an aggressive, successful project to construct a software-defined radio that can be programmed to be virtually any radio by changing its software program. Digital transmissions began to be applied to broadcasting in the late 1990s. DARPA is the research and development section for the U.S. Department of Defense that is focused on technology for the military and preventing technological surprises from harming national security.

Information Science and Computers

At the end of World War II, the global population grew rapidly, and the many ways to communicate paralleled this increase. In the United States, populations were sprawling into suburban areas. There was a shift from an agricultural and manufacturing society to a technology and information society was well under way. In 1941, Konrad Zuse of Germany developed the Z3. This was a fully program-operational calculating machine introduced in Berlin. In the 1940s, the information science age began, when the first government-owned computers were placed into public service in 1944. By 1946, the ENIAC was completed (see Figure 1.5 ENIAC). ENIAC was among the first computers. It covered more than 136 square meters, used 18,000 vacuum tubes, and could compute 5,000 numbers ten digits in length per second. It was created for the army to compute World War II ballistic firing tables. The ENIAC was completed before the Manchester Mark 1.

FIGURE 1.5   ENIAC.

Information science is concerned with the ways in which the human experience is structured, represented, managed, stored, retrieved, and transferred later. Harvard University-IBM Mark I computer was among the first computers used for public service. By late 1949, the Harvard Mark III was 25 times faster than the Harvard Mark II, and 250 times faster than its predecessor.

The 1950s began with the first commercialized computers for business such as the UNIVAC. The first graphical computer game was also developed call "OXO" for tic-tac-toe by Alexander Sandy Douglas on an EDSAC. IBM mass-produced the IBM 650 and introduced an early formula translator called FORTRAN. Computers became faster, smaller, cheaper, more reliable, and generated less heat with the introduction of Bell Labs' transistor computers. The first video game was created by William Higinbotham called *Tennis for Two*. The number of computers that were in use by academia, the government and businesses grew astronomically. By 1968, Intel had introduced an Intel 1 KB RAM microchip into the marketplace.

People, now wanted to use computers like they had in the office at home. By 1976, Apple Computers had invented the Apple I, the first home computer. Unlike its predecessors, the Apple I was preassembled, included a TV interface with built-in support for a keyboard and later could store information to a cassette recorder. Eight years after the first home computer was released, in 1984, International Business Machines (IBM) introduced its personal computer called the IBM PC AT. The IBM PC AT had the power of a word processor, brand-name recognition and acceptance, and was far more powerful than the Apple I.

IBM in early 2000 sold its PC brand to Lenovo including the popular ThinkPad. ThinkPad was a durable laptop introduced in the 1990s. Apple Computers now offers a number of new computing technologies, iMacs (desktop) and iPads (tablets), computing devices you can fit neatly in a binder or purse.

The number of computers that were in use by academia, the government, and businesses grew astronomically. The need for increased power and processing speed to handle more data for research, commerce, and daily business processes led to the creation of a local area network (LAN). A *LAN* is a computer network that connects computers and devices that are in close proximity such as a home, school, computer laboratory, or office building. LANs were initially for atomic weapons research in 1964. These new capabilities lead to a revolution in information technology—the advent of the first Internet known as *ARPANET* (Advanced Research Projects Agency Network) in 1969. ARPANET was the first wide-area packet switching network. *Packets switching* are small blocks of data sent over a dedicated line. ARPANET was a computer networking research project that was to provide a secure and survivable communications system in case of war.

INTERNET

ARPANET began as connections among University of California–Los Angeles, Stanford Research Institute, the University of California–Santa Barbara, and the University of Utah. Twenty years later, the U.S. government released control of the Internet and the World Wide Web (WWW) was born. ARPANET expanded and other fields began to use it and the private sector including businesses like Dow Chemicals. A program to send electronic mail, known as *e-mail*, was developed in 1971. By 1973, Britain and Norway provided the first international connections to the system. Mailing lists, newsgroups, and electronic bulletin board systems (BBS) were established, and TCP/IP was adopted as the communications protocol in 1982. In 1986, the National Science Foundation established the NSFNET, a distributor of networks capable of handling more traffic. Within a year, more than 10,000 hosts were connected to the Internet.

In 1988 real-time conversations via the network was possible using the newly developed Internet Relay Chat protocols. Dial-up access for commercial access started in 1990 as ARPANET was deactivated. A year later, in 1991 the World Wide Web was publicly available using FTP. By 1997, there were more than 10 million hosts on the Internet and more than 1 million registered domain names. Traditional Internet access was granted by using dial-up across public telecommunications (telephone) network. Internet access was expanded to include radio signals, cable-television lines, satellites, and fiber-optic connections, though most traffic still uses a part of the public telecommunications (telephone) network.[17]

COMMUNICATIONS FOR PEOPLE WITH DISABILITIES AND OTHERS WITH FUNCTIONAL AND ACCESS NEEDS

Communications for people with disabilities and others with functional and access needs began to take shape at the start of the twentieth century. This segment of the population have historically be referred to as "special needs populations," "vulnerable populations," or "at risk" populations to defined groups having unique needs that may not be addressed using only traditional approaches. These populations include those who are physically disabled, mentally disabled, hearing impaired (deaf and hard of hearing), visually impaired (blind, unable to see at night or different colors or shapes), cognitively impaired, and mobility challenged. Other groups that may qualify include those who are non-English speaking (or not fluent in English), culturally or geographically isolated, homeless, medically or chemically dependent, children and the elderly. Technology developed to support the communication needs of this population included hearing aids for the hearing impaired and Braille for the visually impaired. The history of these and other adaptive technologies are discussed below.

Hearing Impaired

Two key technologies were commercialized in the twentieth century for the hearing-impaired. These technologies opened new and affordable methods for people who are hard of hearing or unable to hear to communicate with others—the hearing aid and Text Telephone services.

Hearing Aid

The *hearing aid* started as a big, bulky apparatus that amplified sound. The early hearing aids were huge, horn-shaped trumpets with a large, open piece at one end to collect sound. The trumpet gradually narrowed into a thin tube that funneled the sound into the ear. Using Thomas Edison's carbon transmitter invented in 1886 and Alexander Graham Bell's electronically amplified sound in his telephone using a carbon microphone and battery influenced a new generation of hearing aids.

A new concept emerged—an electronic apparatus changing sounds into electrical signals that could travel through wires and be converted back into sounds. This concept was adopted by hearing aid manufacturers and used in the first hearing aids of the twentieth century. By 1901, Alexander Graham Bell, improving communications for the hearing-impaired, developed a new hearing aid. Over time, the hearing aid has become smaller. By 1952, hearing aids went from an electronic apparatus with large batteries and vacuum tubes designed to fit within eyeglass frames to devices fitting behind the ear using transistors.

Text Telephone

A deaf physicist named Robert Weibrecht developed the *text telephone* called *TTY* in 1964. TTY is a device that enables the hearing impaired or speech-impaired to type messages back and forth with the caller rather than talking and listening. To use TTY both the sender and the receiver must have the equipment. TTY is sometimes called the "Telecommunication Device for the Deaf" or TDD. Of the three terms, TTY is most widely accepted.

Advancements in technology led to the *Telecommunications Relay Service* (TRS). TRS is a service providing a trained operator who types what is said so that receivers can read the message on their TTY display and type a response. The TRS operator will read aloud the message to the party on the line. TRS services are available 24 hours a day, 365 days a year, and are toll free.

Speech Impaired

In 1960, a voice communication device for individuals who could not speak called the *electronic larynx* was introduced. The larynx is the voice box of mammals, including humans, which manipulates the pitch and volume of our voice. An electronic larynx, also known as a "mechanical larynx" or "throat back" is a speech aid used to produce intelligible speech for those whose voice box is not functioning properly.

Visually Impaired

Braille

Louis Braille, born in 1809 near Paris, was injured while playing in his father's shop. The affected eye became infected and left him blind. Louis learned of a raised dots system developed by Charles Barbier de la Serre, a French army captain. To write and read messages at night without using a light that could give away their positions, soldiers used Barbier's system.

Barbier's system was built using phonetics with groups of raised twelve dots arranged in two columns of six dots each. Louis modified Barbier's system to develop his own simplified system known today as *Braille*. Braille uses the standard alphabet and reduced the number of dots by half that can be read by touch. The first Braille book was published in 1829. In 1837, math and music symbols were added. Braille is the standard form of reading and writing used by those who are visually impaired. It is a common form of communication in the blind culture, in virtually every language around the globe, and is unique to the blind.

Braille can be produced in-house with the right software, training, and embosser. Braille translation software can translate information on a computer screen into Braille that the user reads on an adapted keyboard. Various cassette recorders/players are used to record and listen to data or information.

Elderly

The *elderly* are often said to be those who are 65 years of age or older. At the start of the twentieth century, life expectancy of this age group was short and many lived in extended family settings with one or more of the children. The literacy rate was low; many had reading skills of the 6th grade or less. Some became "snowbirds"—those who temporarily move away from colder climates in the winter months to warmer climates. Others move to other regions of the world for a reduced cost of living and access to services and cultures not available in their home residence. This relocation sometimes takes older people far away from family and familiar settings when an emergency occurs. The 1970s brought push-button telephones with large displays and commercialization of life-alert devices. These devices were installed near the telephones in the homes. If one were to need emergency assistance, they could go to this device and press one button that would dial a calling center that would alert local first responders, usually summoning medical attention.

SUMMARY

Communications mediums have evolved as humankind has evolved. Early humankind was focused on the basics of survival and where communications was limited to small groups over short distances. Early communications began with the establishment of a system that two or more could use for comprehending what was said; later information could be shared in written forms starting with simple pictures and an alphabet. As humankind started developing and using tools, many of its tools began to take on meanings expanding from their original purpose, such as the horns of a buffalo or the metal shield used by soldiers, and were adapted as tools to warn people across distances. Electronic communications expanded this reach with the first transatlantic telegram using Morse code over wire and later with the telephone. The telephone transformed the world. With the twentieth century came significant advancements in broadcasting on the big screen live events to on-demand pre-recorded broadcasts. Communications were tailored so that multiple population subgroups with varying needs and capacities could comprehend a single message using the tool that was most effective for them such as the radio broadcast of a message that is also available on the web, on television, or on a cell phone. The history of communications mirrors the advancement of humankind.

ENDNOTES

1. Gransden, K. W. 1985. The fall of Troy. Second series *Greece & Rome* 32: 60–72. Cambridge, MA: Cambridge University Press.
2. Sylvester, Charles H. (Editor in Chief). 1909. Vol. 3 of *The New Practical Reference Library*. Chicago: The Dixon-Hanson Company.

3. Sylvester, Charles H. (Editor in Chief). 1909. Vol. 3 of *The New Practical Reference Library*. Chicago: The Dixon-Hanson Company.
4. Macomber, Hattie E. 1897. *Stories of great inventors—Fulton, Whitney, Morse, Cooper, Edison*. Boston: Educational Publishing Company.
5. Canada Post. 2008. http://web.archive.org/web/20080601173108/www.canadapost.ca/segment-e.asp (accessed January 11, 2011).
6. International Organization for Standardization. 2007. *ISO 3166-2:2007, Codes for the Representation of Names of Countries and Their Subdivisions—Part 2: Country Subdivision Code*, 2nd ed. Geneva: ISO Central Secretariat.
7. Xerox. 2010. *Xerox Online Factbook*. http://www.xerox.com. (accessed January 5, 2011).
8. Kinetograph. (2010). *Encyclopedia Britannica Online*. http://www.britannica.com/EBchecked/topic/318205/Kinetograph (accessed December 21, 2010).
9. Kurzweil, Raymond. 1990. *The age of intelligent machines*. Boston: MIT Press.
10. FCC, Office of Engineering and Technology. 2010. FCC Radio Spectrum Home Page, http://www.fcc.gov/oet/spectrum/ (accessed January 2, 2011).
11. 1996. *Marconi*. http://www.heritage.nf.ca/society/marconi.html. St. John's, NL, Canada: Memorial University.
12. U.S. Government Printing Office. (2002, October 1). Code of Federal Regulations, Title 47 Telecommunication Commission, Part 97—Amateur Radio Service, [47CFR97.407], pp. 592–593.
13. Lavin, Marilyn. 1995. Creating Consumers in the 1930s: Irna Phillips and the Radio Soap Opera. *The Journal of Consumer Research*. 22: 75–89.
14. Frost, Gary. 1969. *Early FM Radio: Incremental technology in twentieth-century America*.
15. *Cellphones.org*. 2008. Cell Phone History. http://cellphones.org/cell-phone-history.html (accessed January 2, 2008).
16. NASA, "Apollo 11" *Lunar Surface Journal*, (Texas: 1995).
17. *Encyclopedia Britannica Online*. http://www.answers.com/topic/internet#ixzz1BPica3OV (accessed December 21, 2010).
18. StorageNewsletter.com. 2011. LaCie Rugged Mini External HDD, Next Month at 1.5 TB. http://www.storagenewsletter.com/news/disk/lacie-rugged-mini (accessed August 15, 2011).

CHAPTER 2

Disaster Communications—Two-Way Conversations

> ... Never allow a crisis to go to waste. ... A crisis occurs at a breaking point that requires some urgent, usually drastic change. Not supplying the needed change is letting the crisis go to waste. To recognize the wake-up call to decide to act differently is the more effectual path ... members of our community learned good lessons from the crisis ...
>
> —Rahm Emmanuel, *former U.S. Congressman, presidential Chief of Staff, and Mayor of Chicago*[1]

Communications has evolved as humankind has evolved. Warning others started with early communications systems and tools that are still used today—smoke signals and heliographs, a type of *mass notification* system. Mass notification is the capability to provide information to all affected parties in a given area. When the concept of mass notification is applied to an emergency situation it includes notification before, during, or after an emergency situation. Mass notification systems used for emergency situations provide the capability to provide real-time information to building occupants or others in the immediate area during event.

Drums made from wood and animal skins, horns made from tree branches and animal tusks, and shakers were instruments adopted for disaster communications. The bugle was at times of war to signal the end of the day for troops or to signal a change in direction of an attack. It can be said that primitive man laid the foundation of all communication systems used today.

Throughout nature, each species has a fundamental need to communicate, to identify themselves from others, to give notice, and to warn—a form of crisis communications. *Crisis communication* is sending messages and using channels to share information to internal and external sources in identifying a threat or crisis. Ideally, crisis communications are handled in an efficient manner to minimize the damaging effects of the crisis.

The history of crisis communications using mass notification began with the use of hieroglyphics and man on foot. We advanced to the Pony Express as we looked to get the message out faster. We wanted to go further, do more, and reach more, so machines and technology were the next step. Railroads, Morse code, and teletype meant transatlantic communications occurred at speeds that started to tap into our imaginations. TVs, telephones, and airplanes meant more people, most anywhere, at anytime had access to crisis communication. Satellites, rockets, and missiles, the space shuttle, the Internet, and mobile computing meant that now a single message had a global reach. Advances in technology enable anyone, anytime, anywhere, and at the same time can receive the same message.

Over the last 20 years, telephones and computers have converged, producing more services and tools available for crisis communications. Video conferencing, fire alarm panels integrated with security cameras and speakers, alarm systems, and lighting managed from mobile devices anywhere a phone signal or the Internet is accessible are newer ways to share information and activate systems. Internet speeds of 100 megabits (as found in Korea) and social media have changed the way we communicate and share information. A *megabit* (mb), in terms of the Internet, is the rate at which information is transferred. Internet migration into the world of Web 2.0 has led to social networking, enabling anyone to be the "eyes" and "voice" of events as they occur.

When crisis communication is needed, it is most important to provide information regarding a disaster before, during, and after the disaster. The need to communicate quickly over distances is inextricably rooted in the motivation to avoid or at least mitigate the effects of a disaster. Today, the Internet and social networking tools are used to inform large numbers of people immediately when catastrophic events occur, such as the South Asia tsunami, the Haiti earthquake, and the trapped miners of Chile.

Communications today enable casual conversations, entertainment, research and training, and emergency information to targeted audiences including vulnerable populations such as the visually impaired or hearing-impaired. These messages can be prioritized for more urgent matters using technology. At the start of a disaster, first responders usually encounter chaos during the first phase, which is characterized by confusion, fear, panic, and inadequate leadership. For the more urgent matters today, communications can be prioritized, just as the injured can be triaged for needed services.

Microwave relays for delivering television programming over long distances, cellular phones, and fiber optic cables meant more services with less bulky infrastructure and clear and intelligible voice technologies.

Each of these solutions moved telecommunications to a wireless world, but still bound to a backbone of wire at some level. These solutions were also used to advance emergency services telecommunications including public safety radio systems and improvements in capabilities. Despite these radical advances, these systems remain vulnerable to disruption during a disaster, a time when they are most critical.

The military and amateur radio communities are two communities with the longest range and the most dependable emergency communications. *WinLink* and *Echolink*, two examples of global radio messaging systems, use Internet technology for amateur radio systems. *WinLink* uses Internet technology with amateur radio frequency (RF) technologies. *Echolink* is a computer-based amateur radio system that uses VoIP technology over the Internet. Internet radios and computers were interlinked, improving the dependability of global disaster communications. New methods of communications continue to be introduced that increase the dependability of systems, thus improving the effectiveness of emergency communications—before, during, and after the crisis.

This has been an abbreviated overview of communications throughout history. From the beginning of time, new communications methods were developed because we wanted to share common information. Many were created based on the need to warn. Numerous early forms of communication continue in use today such as heliographics. History has also demonstrated that even though significant advances have occurred, many old practices die hard. Communications used for disasters and emergencies requires dependability.

EMERGENCY BROADCASTING

U.S. President Harry Truman in 1951 established the *CONELRAD* (CONtrol of ELectronmagnetic RADiation) system, a method of emergency broadcasting, for speaking to the public in the event of enemy attack during the Cold War era. In the early 1950s, many believed that if the United States were attacked, commercial broadcasting stations would become valuable navigation aids for enemy aircraft. People believed stations would be silenced by an attacked. The purpose of CONELRAD was to prevent Soviet bombers from homing in on U.S. cities using radio or television stations as beacons. CONELRAD was also to provide essential civil defense information. In 1963 when the intercontinental ballistic missiles were developed, reducing the likelihood of a bomber attack, the Emergency Broadcast System (EBS) replaced CONELRAD.

Like CONELRAD, the EBS was designed for use by the president to quickly communicate with the American public in the event of enemy attack, war, threat of war, or grave national crisis. EBS was also used for

TABLE 2.1 Types of EAS Providers by End of 2007

AM	Amplitude Compandored Single Sideband (ACSSB($))	Land Mobile Radio Service (LM(R)) radio
FM	Digital radio broadcasters	Satellite radio (XM and Sirius)
VHF	Cable providers	Digital television
UHF	Worldspace	IBOC
DAB	Cable television (including low-power stations)	Satellite television providers (DIRECTV, DishNetwork, etc.)
Muzak	Other DBS providers	Music Choice
DMX Music	Video Dial Tone (OVS)	

state and local peacetime emergencies, civil emergency messages, and severe weather hazard warnings. Between 1976 and 1996, there were more than 20,000 activations. None of these activations were for war, threat of war, or grave national crisis. Most activation was related to local events. The EBS system was replaced in 1997 by the *Emergency Alert System (EAS)*.

EAS is a national warning system that is coordinated and regulated by the FCC. The system covers all states and several territories. Like its predecessors, EAS was designed to let the president speak to the nation within 10 minutes—a process that has never been activated. Systems and providers required to participate in the EAS by the end of 2007 are listed in Table 2.1 (Types of EAS Providers by End of 2007).

EAS has been superseded by the FEMA program *Integrated Public Alert and Warning System*, called *IPAWS*. The FCC, FEMA, and the National Weather Service (NWS) jointly coordinate IPAWS, a new digital message format. The primary goal of IPAWS is to expand the existing Emergency Alert System by enabling emergency management officials to reach as many people as possible using as many communication devices that are available and compatible. The earlier system relied predominately on radio and television for communications with people. IPAWS uses an *common alerting protocol* (CAP) enabling the use of mobile phones, personal computers, and network attacked speaker arrays, among other devices and social media. IPAWs will be discussed at length in future chapters.

UNIVERSAL EMERGENCY CALLING NUMBER

History of the Universal Emergency Calling Number

For the first 50 years of telephony all telephone calls were operator assisted. Operator-assisted calls were routed through a telephone switchboard operator who connected callers manually. The person placing a call would pick up the telephone receiver and then wait for the telephone operator to answer, usually with the phrase "Number please?" The caller would then ask to be connected to the number he or she wished to call, and then the operator would make the required connection using a *switchboard*. A switchboard was a device that connected several telephones together manually or to an external telephone exchange.

To place an emergency call, the requestor could say,

> "I need the police," "There is a fire next door," or "I need an ambulance/doctor."

With operator-assisted calls, it was unnecessary to request emergency services by number, even in a large city. The operator knew the calling party's number by looking at the number above the line jack of the calling party. In smaller centers, many telephone operators knew the locations of local emergency resources that were willing or able to assist in an emergency. Telephone operators could also activate the town's fire alarm, and act as an informational clearinghouse when an emergency occurred.

The need for operator-assisted calls declined when rotary dial telephones were widely adopted. The first rotary dial telephones were introduced in the 1920s. By the 1950s, these telephones moved telecommunications from operator-assisted to a self-service model. Operator assistance was an option. Accustomed to the personalized services offered by local operators, people were becoming anxious about the change and loss. This issue was partially resolved by telephone providers advising the public to dial "0" for local operator assistance if they did not know the telephone numbers of their local fire or police department.

Universal Emergency Calling Number—United Kingdom

In 1935, a fire on Wimpole Street in Britain resulted in five fatalities. Norman MacDonald, a neighbor, had tried to telephone the fire brigade and was so outraged at being held in a queue by the telephone exchange that he wrote a letter to the editor of the *Times*. Mr. MacDonald's outcry in the news prompted a government inquiry. After a careful investigation, it was found that the phone system in place for police, fire, and emergency medical services (EMS) led to delays in reporting.

An outcome from this investigation led to Britain implementing a three-digit emergency telephone system, 9-9-9. The new 9-9-9 provided access to police, fire, and EMS emergency services within the region. By the summer of 1937, 9-9-9 service was introduced in the London area, and later nationally.[2]

The reason 9-9-9 was chosen was based on the 1925 design of the "A" and "B" buttons on *public payphones*. Public payphones are telephones that required a user to pay before placing a call using the device. The user can dial "0" without inserting any money. This design was easily modified to allow free use of the "9" digit on the rotary dial without enabling free use of any other number combination. The digits "2" through "8" were already in use across the United Kingdom and would have required significant modification of payphones to be adapted as the universal emergency number. The "1" digit was established for international calls. 9-9-9 was adopted as the number to dial for requesting emergency services in the UK.

In 1937 the wife of John Stanley Beard of 33 Elsworthy Road, Hampstead, London, placed the first 9-9-9 call to report a burglar outside her home at 4:20 a.m. The burglar, Thomas Duffys, a 24-year-old, was apprehended.[3] This is the first recorded use of an assigned three-digit number used exclusively for emergency services. The implementation and use of a single number for requesting emergency services by telephone grew across Europe and Asia.

A call to 9-9-9 was placed when an individual

- was in immediate danger of injury or their life was at risk
- was suspicious that a crime was in progress
- required immediate emergency response services

When a 9-9-9 call was placed, the typical experience went as follows:

- Operator answers the call asking "Emergency. Which service?"
- Before the mid-1990s, operators asked, "Which service do you require?"
- If the caller is unsure the operator will default the call to the police.
- If an incident requires more than one service, for instance, an automobile accident with injuries and trapped persons, depending on the service the caller has chosen, the chosen service provider will alert the other service providers for the caller.
- The operator will also contact each emergency service individually, regardless of whether the caller has remained on the line.

Behind the scenes, the operator started the connection to the emergency service control room by stating the location of the operator, for instance, "Bangor connecting 01248 300 000." The operator's locations represented the caller's information. This process created confusion for the requester who did not understand that the operator was speaking with the agency that would be providing service. British Telecom (BT) introduced a new system in 1998 that could transmit electronically the location information of the calling telephone to the third party. This minimized operator error and eliminated the need for the operator to state the information with the caller on the line. The new system is called the Enhanced Information Service for Emergency Calls or *"EISEC."*

As of 2010, approximately 50% of the Emergency Authorities (EAs) have EISEC. *Emergency Authorities* are first responder agencies. Expansion continues. For those areas without EISEC, the operator continues to follow the process of passing the calling telephone location and number.

In recent years, the number 1-0-1 was introduced for placing non-urgent calls to the Police throughout Wales and a few pilot areas across England. In 2010, 1-1-1 was introduced as the number to dial for urgent but not life-threatening cases requiring health services in England.

Textphone and *RNID Typetalk relay service* are telecommunication tools used by the hearing impaired to access emergency services. The number was shorten from 0800 112999 to 18000 to make it easier to access these services. *Text Relay* is a national telephone relay service operated by BT. Text Relay connects a caller who is using a textphone with a person who is using a telephone or another textphone. A *textphone* is a telephone with a keyboard. Text relay enables the deaf, hard of hearing, and speech-impaired people to maintain contact with others and conduct business over the telephone, the same as those available on standard telephone systems. A trained Text Relay operator provides a communicative link between the textphone user and the hearing person. Textphone services are discreet, confidential, and convenient.

From 2008 to 2009, Nottinghamshire Police conducted a pilot of *Pegasus*. Pegasus is a database containing the detail information on persons with physical and learning disabilities or mental health problems. Individuals register with the Nottinghamshire Police Pegasus if their disabilities make it difficult for them to give spoken details when calling the police. Registered individuals are issued with a personal identification number (PIN) that can be used by calling 9-9-9 from a telephone or contacting the nonemergency 0300 300 9999 number. The caller is connected to the control room saying "Pegasus" and their PIN. The caller's information is retrieved from the database and then the caller begins explaining the purpose of their call. In person, the Pegasus PIN can be told or shown to a police officer. Expansion into other areas is in progress.

The growing use of mobile phones has generated an increase number of false calls for emergency services. The same-digit sequence of 9-9-9 is dialed accidently due to vibrations and other objects colliding with the mobile phone's keypad. This is of less concern with emergency numbers that use two different digits such as 1-1-2 or 9-1-1 used in other countries.

In 1982, the Confederation of European Posts and Telecommunications (CEPT) designing a pan-European mobile technology formed Groupe Spéciale Mobile (GSM). GSM defines the internationally accepted digital cellular telephony standard. One GSM standard mandates that the user of a GSM phone is able to dial the local emergency services number without unlocking the keypad. This feature can save time in emergencies and causes some accidental calls. By the end of 1992, BT introduced the pan-European 1-1-2 code in the UK, which connects to existing 9-9-9 circuits. In some countries, a valid subscriber identification module (SIM) card is required to make a 9-9-9 or 1-1-2 emergency call. In 2008, the world population was 6.7 billion with 1.2 billion people living in regions classified as more developed by the United Nations; 5.5 billion people in regions that are less developed.[4] In 2008, CEPT reported that GSM had surpassed three billion connections worldwide.[5]

Universal Emergency Calling Number—North America

In North America, people calling to request emergency services frequently faced telephone delays for emergency services organizations. There were many stories told of unnecessary harm to people and property due to these delays. This caused a movement where local people began their own emergency reporting system by forming their own telephone networks called *telephone co-ops*. These clusters of people eventually became telephone companies.

One of the initial catalysts for a U.S. nationwide emergency telephone number was in 1957 when the National Association of Fire Chiefs recommended use of a single number for reporting fires. The California Highway Patrol debuted its traffic emergency number "ZEnith 1-2000" or as dialed 931-2000. The letters "Z" and "E" represented the numbers "9" and "3" on a rotary dial telephone. Using a name to represent a telephone prefix, the first three digits, indicated the area being serviced.

The United States was one of many countries that adopted a universal emergency telephone number in the late 1950s. While the U.S. fire chiefs were making this suggestion, 9-9-9 was adopted for use in Sydney, Australia. New Zealand adopted 1-1-1 as the three-digits to dial for emergency services in 1958 nationwide. Winnipeg, Ontario, Canada implemented 9-9-9 emergency service across the region in 1959. At the same time the first U.S. system was implemented in 1959.

In the United States, the first step was to modify the telephone equipment and systems. Telephone companies installed equipment that could provide a method to hold the originating calling line long enough to check the line. This process worked even if the call was disconnected or the caller hung up. Telecommunication equipment could now mimic what telephone operators had been able to do, a service called *CLR holding*.

The benefits of CLR holding was that the modified telephone equipment could now ring back the calling party once the telephone receiver was back on the hook. In the early days when many callers had *party lines*, this feature was limited often sending the call to all on the party line or the ring-back feature did not function. Party lines are telephone lines shared by more than one household. The telephone equipment changes meant that the telephone companies needed a standard operating process throughout the United States.

Payphones also had to be modified. Payphones required a coin deposit before the caller would receive a dial tone and able to place a call. The emergency service system needed the payphones to be able to make emergency calls without the need for a coin deposit to place an emergency call and limited this process to emergency calls only.

Updating telephone equipment and payphones and standardizing operating procedures for handling emergency calls was a significant challenge for U.S. telephone companies. Each change required resources and time for development. Both large and small operating telephone companies needed to be involved in the development and rollout. It took years to develop a system that was easy for the caller to remember and use, and that all telephone companies, large and small, could implement successfully.

In 1967, President Lyndon Johnson's Commission on Law Enforcement and Administration of Justice issued a report recommending the following:

> Wherever practical, a single police telephone number should be established, at least within a metropolitan area and eventually over the entire United States, comparable to the telephone company's long-distance information number …

The use of different telephone numbers for each type of emergency was determined to be contrary to the purpose of a single, universal number. Other federal agencies and governmental officials supported and encouraged the recommendation. Since there was immense interest in this issue, the President's Commission on Civil Disorders turned to the Federal Communications Commission (FCC) for a solution. The recommendation of a single telephone number for police services came despite great opposition from the telephone industry. The telephone industry countered that the telephone operator number "0" (zero) was an adequate emergency number.

In November 1967, the FCC met with the American Telephone and Telegraph Company (AT&T) to find a means of establishing a universal emergency number that could be implemented quickly. AT&T and the United States Independent Telephone Association (USITA) had the challenge of agreeing on one number for the U.S. telephone system. The access code requirements were

- Limited to three digits
- The first digit could not be the numbers "0" or "1"
- The second and third digits had to be "1"

Once these requirements were defined, what remained would become the first digit. Ultimately, the number "9" was selected since it was the easiest to clear for access. In many systems the number was already clear; in others, only small equipment changes were needed. This effort resulted in the selection of 9-1-1. Work began to implement this emergency number.

A year later, in 1968, AT&T announced their designation of 9-1-1 as the universal emergency telephone number for the United States. Initially, AT&T's plan affected only the Bell companies, excluding the independent U.S. telephone companies. The three digits, 9-1-1, were chosen because they best fit the needs of all parties involved. The logic used to validate the decision to use 9-1-1 included:

- It met public requirements because it was brief, easily remembered, and could be dialed quickly.
- Since it is a unique number, never having been authorized as an office code, area code, or service code, it best met the long range numbering plans and switching configurations of the telephone industry.
- The location of these digits on the dial or keypad made it easier to place the call in the dark.
  - Using a rotary phone, the caller could place their finger in the dial, slide the finger from the one position to the zero position, back up one-step, and this would be the ninth position or digit 9. Then the call originator would again place the finger into the first position—this would be the digit one—and dial it two times. The outcome is 9-1-1.
  - Using a keypad, the call originator would locate the lower right-hand key position (the # or pound key) and move up to the next position, which is the digit 9. Next, the caller moves their finger to the upper-most left hand side, which is the digit 1.

Although AT&T was first to announce 9-1-1, Alabama Telephone Company was first to implement 9-1-1 within its territory in 1968. The town of Haleyville, Alabama, the northwest section of the state, is where the birth of 9-1-1 occurred. Bob Gallagher, president of the independent Alabama Telephone Company (ATC), read an article in the *Wall Street Journal* revealing that AT&T intended to announce its emergency number plan. Gallagher wanted to be first and quickly had his technicians complete the necessary central office work and installed a red 9-1-1 phone. On February 16, 1968, Alabama Representative Rankin Fite completed the first 9-1-1 call made in Haleyville City Hall in Haleyville, Alabama, to U.S. Rep. Tom Bevill at the city's police station. A few days later, February 22, 1968, Nome, Alaska, implemented its 9-1-1 service, and then other U.S. cities quickly followed.

By 1973, the White House's Office of Telecommunications issued National Policy Bulletin Number 73-1 denoting the benefits of 9-1-1, and promoted its nationwide adoption simultaneously in both urban and rural areas. Initially, 9-1-1 was promoted as 9-11. Some users of the service found this confusing because they would first select the number "9" on the telephone, then look for the number "11" which is not on the phone. To make it easier to remember, the promotion was changed from 9-11 to 9-1-1 to match the numbers dialed on the telephone.

To support this initiative, Bell Telephone Laboratories of New Jersey, Joseph Bernard Connell, and Alfred Zarouni invented an emergency reporting system that could selectively interconnect a plurality of telephone stations. Each assigned telephone station was assigned a directory number that designated the emergency service centers using the telephone communications-switching network. The U.S. Patent Office granted Patent Number 3,881,060 to this team of inventors in 1975. This was considered a modern selective routing and ANI 911 system as depicted in Figure 2.1 (New Jersey 9-1-1 Emergency Service Center).[6]

In March 1973, the White House's Office of Telecommunications issued a national policy statement that recognized the benefits of 9-1-1, encouraged the nationwide adoption of 9-1-1, and provided for the establishment of a Federal Information Center to assist units of government in planning and implementation. The intense interest in the concept of 9-1-1 can be attributed primarily to the recognition of characteristics of modern society, that is, increased incidences of crimes, accidents, and medical emergencies, inadequacy of existing emergency reporting methods, and the continued growth and mobility of the population.

In 1977, Robert M. Pirnie, III of Montgomery, Alabama, developed the Emergency Call Answering System. Pirnie was granted Patent No. 4,052,569 by the U.S. Patent Office. The Emergency Call Answering System had a specialized console that controlled emergency call handling

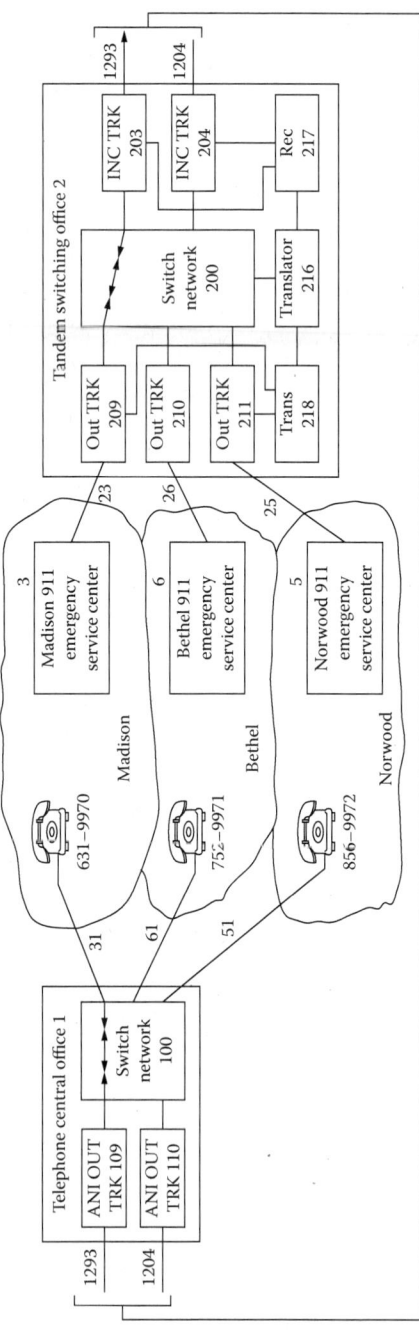

FIGURE 2.1 New Jersey 9-1-1 Emergency Service Center.

capabilities for 9-1-1 answer locations with answer, conference and transfer capabilities incorporated for each position. As calls are received by the system, they are automatically distributed to the appropriate emergency operator position. Calls are allotted to emergency operators on a sequential basis. The 9-1-1 Emergency Call Answering System provides emergency line outputs connected directly with the emergency response agencies.[7] (See Figure 2.1.)

In the early 1970s, AT&T began the development of sophisticated features for the 9-1-1 with a pilot program in Alameda County, California. The feature was "selective call routing." This pilot program supported the theory behind the Executive Office of Telecommunication's Policy. By the end of 1976, 9-1-1 was serving about 17% of the population of the United States. In 1979, approximately 26% of the U.S. population had 9-1-1 service, and nine states had enacted 9-1-1 legislation. The 9-1-1 service was growing at the rate of 70 new systems per year. By 1987, 50% of the U.S. population had access to 9-1-1 emergency service numbers.

Canada recognized the advantages of a single emergency number and adopted 9-1-1 rather than use a different universal emergency number. Canada's adoption of 9-1-1 unified the United States and Canada giving 9-1-1 international status.

Through the mid-1980s, some areas of the United States still used the seven-digit numbers for emergency calls. Dialing a seven-digit number was like calling any other number except the number was associated with local police, fire, or other responding agency. At that point, 9-1-1 began reaching beyond large urban areas into other parts of the country.

By the mid-1990s, 9-1-1 service was upgraded to the *Enhanced 9-1-1* (E911) service that could automatically retrieve the address assigned to the wireline telephone used to place the call in the computer system. The migration of emergency service providers to the E911 required every structure and road in the United States and Canada to have a street address.

President Clinton designated 9-1-1 as the nationwide emergency telephone number in 1999 (Senate Bill 800)—the designated Universal Emergency Number in place for more than 98% of the U.S. population. 9-1-1 gave the public quick and easy access to a *Public Safety Answering Point* (PSAP) as illustrated in Figures 2.2 (PSAP, Part 1) and 2.3 (PSAP, Part 2).

There are two types of PSAPs—a primary and a secondary PSAP. A primary PSAP directly routes 9-1-1 calls from a 9-1-1 Control Office, such as a selective router or 9-1-1 tandem to the responding agency. A secondary PSAP transfers a 9-1-1 call from a primary PSAP to the responding agency. The FCC maintains a Master PSAP Registry that was started in December 2003 to assist the FCC in the evaluation of PSAP readiness and the wireless Enhanced 9-1-1 (E9-1-1) deployment.

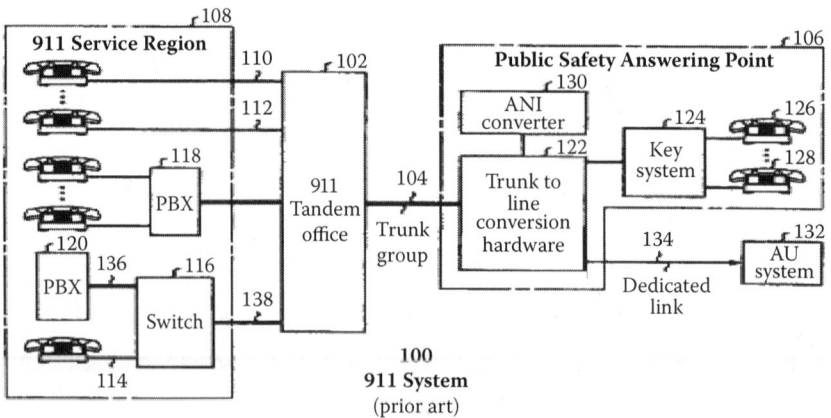

FIGURE 2.2 PSAP, Part 1.

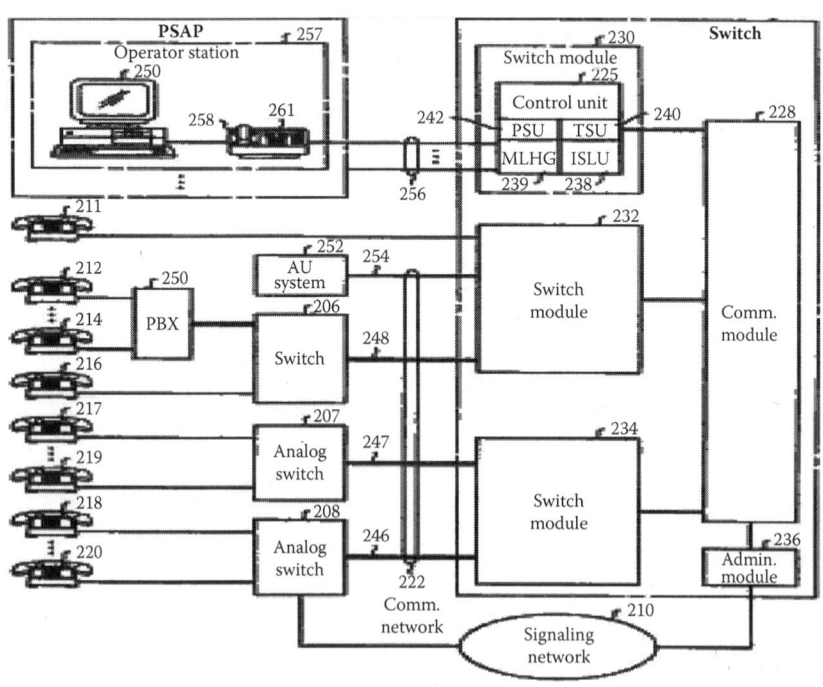

FIGURE 2.3 PSAP, Part 2.

9-1-1 Emergency Operator Procedures

A caller may place a call for emergency assistance when they witness an event that requires immediate attention by first responders or the caller needs emergency assistance. When placing a call, the requester can expect the following process:

1. Caller dials 9-1-1 from a telephone.
2. Operator answers the call by asking, "What's the nature of your emergency—police, fire, or medical?"
3. If known, the caller is transferred to the emergency operator of that service. (Go to Step 6.)
4. If the service is unknown, the operator will ask a few questions to determine where to route the call and obtain the caller's name and telephone number.
5. The operator will stay on the line until the connection has been confirmed.
6. The emergency services operator will confirm receipt of the call with the 9-1-1 operator.
7. The 9-1-1 operator disconnects from the call.
8. The emergency services operator gathers detail information from the caller.
9. The emergency services operator will dispatch the required services based on the information gathered.
10. If a high-priority call, the emergency services operator will stay on the telephone with the caller until a first responder arrives and indicates no further assistance is needed.
11. If a medium- or low-priority call, the emergency services operator will advise the caller that the request has been recorded and the appropriate agency dispatched for follow-up.

The number of U.S. 9-1-1 calls over the years has increased substantially to approximately 240 million calls in 2010. These calls were managed by a staff of nearly 100,000 as outlined in Table 2.2, National Average Statistics of 9-1-1 Public Safety Communications.

Enhanced 9-1-1

In October 1999, the Wireless Communications and Public Safety Act of 1999 (9-1-1 Act) went into effect. The purpose of the 9-1-1 Act was

- to improve public safety by promoting, and
- facilitate the prompt deployment of a nationwide, seamless communications infrastructure for emergency services.

TABLE 2.2 2010 National Average Statistics of 9-1-1 Public Safety Communications[12]

Annual 9-1-1 calls	240 million
Daily 9-1-1 calls	260,000
Primary and secondary public safety communication centers	6,170
Personnel staffing communication centers	99,900
Law enforcement officers supported	883,600
Firefighters supported	731,200
Emergency medical service personnel supported	210,700

Source: 2010, Bureau of Labor Statistics; National Emergency Number Association.

The FCC also introduced wireless Enhanced 9-1-1 (E9-1-1) rules

- to improve the effectiveness and reliability of wireless 9-1-1 services
- provide 9-1-1 dispatchers with additional information on wireless 9-1-1 calls

E9-1-1 rules apply to all wireless licensees, broadband Personal Communications Service (PCS) licensees, and certain Specialized Mobile Radio (SMR) licensees.

The FCC has divided its wireless E9-1-1 program into two parts—Phase I and Phase II. Phase I requires carriers to provide the PSAP with the telephone number of the originator of a wireless 9-1-1 call and the location of the cell site or base station transmitting the call. Phase II, requires wireless carriers to provide the PSAP with the latitude and longitude of the caller using FCC accuracy standards, generally within 50 to 300 meters, depending on the type of technology used. The deployment of E9-1-1 required the development of new technologies, upgrades to local 9-1-1 PSAPs, and coordination among public safety agencies, wireless carriers, technology vendors, equipment manufacturers, and local wire line carriers.[8] By the start of 2000, 95% of the United States was covered by E9-1-1.

9-1-1 A Vulnerable Target

9-1-1 and First Amendment Rights

By 1990, many countries had adopted a three-digit number for emergency services. Automation and connectivity to massive computer systems peaked the interest of those with malicious intent. One of the earlier cases regarding 9-1-1 services was a 1988 case involving hackers of the 9-1-1 system. This was a case between BellSouth and the U.S. Constitution's First Amendment. This case was widely published among free speech advocates, far more than among 9-1-1 administrators. Personal computers' (PCs') use and experimentation by hackers to access remote systems was growing. Although there was a fringe element with criminal intent, most of the hackers were focused on learning as much as they could about remote computer systems and exploiting them.

Hackers gave themselves unique names such as "Prophet" and "FryBoy," formed groups, and traded information on system access and data. One group called the "Legion of Doom" based in the United States would reroute telephone calls, steal phone codes, or just explore various system. Businesses saw hackers as a security problem, and governments viewed hackers as a threat to national security and criminals to identify, track, and arrest.

Robert Riggs of Decatur, Georgia, accessed a BellSouth computer system and downloaded "Bell South Standard Practice (BSP) 660-225-104SV—Control Office Administration of Enhanced 911 Services for Special Services and Major Account Centers." This document defined the procedures and methods for handling the provisioning of 9-1-1 services in BellSouth's territory. This document had not been classified, did not contain sensitive information, and was publicly available for purchase at $13. Hackers shared the document freely, even posting it on a computer bulletin board. Others accessed the document and posted on other bulletin boards along with the 9-1-1 glossary.

Riggs and his friend the magazine publisher were indicted in 1990 on

- interstate transportation of stolen property
- wire fraud
- violations of the federal Computer Fraud and Abuse Act of 1986

The government suddenly asked the court to dismiss all the charges during the trial. It is presumed that after further investigation it was found that the information in the document was available from other publicly available source. Riggs and another hacker did plead guilty to one count of conspiracy each for the original computer break-in. Riggs was sentenced to 21 months in prison on federal charges of breaking into BellSouth's computer network and passing the 9-1-1 document along over computer connections and across state lines. He and his friend were also ordered to serve probation and perform community service after their prison time. (See *United States v. Riggs*, 739 F.Supp. 414 (N.D.Ill.1990).)

Silent 9-1-1 Calls

With the introduction of mobile telephones, accidental or "silent" 9-1-1 calls have become an increasing problem. Hoax and improper use of emergency calling systems are also a rising issue. According to a news report conducted by Amanda Bellanger of KTAR in Phoenix, Arizona, on January 27, 2011, the 9-1-1 call center received approximately 5,700 calls within 24 hours. Nearly 1,000 of these calls were 9-1-1 hang-up calls, most from "pocket dialing." *Pocket dialing* is when a cell phone, usually in a purse or pocket, accidentally dials a number unbeknownst to the phone's owner. With most cellular phones programmed for one-button dialing for 9-1-1 services, it increases the likelihood that 9-1-1 emergency services are reached accidentally. Another reason for silent or accidental 9-1-1 calls is children playing with old cell phones that do not have a carrier. Since 1997, all cellular telephones are required to have carrier network access to make unlimited FREE 9-1-1 calls. Accidental calls could leave someone with a life-threatening emergency on hold.

9-1-1 Internet Address Purchased for Profit

Prior to 1995, an Internet address had to contain at least one alphabetic character. In 1995, there was also the opportunity to profit from 9-1-1 computing systems and access to it using 9-1-1 via the Internet. Web master Nick Lawrence began a campaign to have Internet administrators register several all-number web domains. Having Internet administrators registering all-number web domains meant that these numbers would only be available if the number was purchased from the domain name owner. Lawrence succeeded and was issued 9-1-1.com. This website is now operated as 9-1-1 Crime Reporter. 9-1-1 Crime Reporter does indicate that it is not affiliated with any public or law enforcement agency or with the 9-1-1 telephone service. Lawrence was also issued 411.com, 611.com, and other all-numeric domain names.

The reason behind Lawrence's interest in capturing these domain names was profit. Short domain names, particularly those representing popular phrases or business names, command significant dollars. For example, the domain name WQS could yield $20,000; IBM could yield $1,000,000. The longer the domain name and the less it is associated with something popular, the more likely that the cost to register the domain name will decrease significantly. A longer domain name can be registered for as little as $2.95 for a one-year subscription.

Blocking Inbound Calls to 9-1-1

Malicious attacks began with accessing sensitive data. By 1996, these attacks migrated toward pursuits to create artificial system outages. In January 1996, a Swedish telephone system hacker called from London making multiple, simultaneous 9-1-1 calls to various Southern Bell west-central Florida PSAPs. He did this from a computer connection. These calls tied up their trunks for legitimate callers. So many calls came that they were no longer capable of disconnecting the hacker to make the line available for real callers. News stories reported he generated approximately 60,000 telephone calls during his attack. Not all were directed to 9-1-1. The hacker was later apprehended and prosecuted by British authorities. The incident did not involve the unauthorized entry of a hacker into the 9-1-1 system itself; rather, the incident blocked inbound calls. The then FBI director Louis Freeh called the incident a "dress rehearsal for a national disaster."[9]

Hacker skills continued to grow and the use of e-mail to transmit their information mushroomed. Hackers increased their automated use of remote computers autonomously to conduct their attacks, a type of *botnet*. E-mail was the tool of choice to deliver viruses through infected attached files and web links.

In April 2000, the BAT.Chode.Worm was unleashed and spread by e-mail. A computer worm is a type of virus that is able to spread from computer to computer without human action. A worm uses file or information transport features on a computer to travel unaided. This particular worm would search a range of Internet Protocol (IP) addresses of known Internet Service Providers (ISPs) to find an accessible computer that had a shared C drive that was not password protected. Once accessed, the file would copy itself to the C drive of that computer. An infected computer with a connected modem would place a silent call to 9-1-1. The 9-1-1 worm could also automatically transmit itself to other computers on the same network or even to other computers over the Internet. It was not dependent upon e-mail or any user action to spread itself.

Since some emergency 9-1-1 services return phone calls, this worm could cause the emergency switchboards to become overloaded. A call could also lead to a dispatched vehicle to the location of the call. There were no reports of any PSAPs affected by false calls. The hacker remains anonymous. This attack was conducted in the greater Houston, Texas, area with 50 or fewer infections. On May 15, a federal grand jury indicted Franklin Adams, a programmer for a Houston area bank, on charges of knowingly causing the transmission of a program onto the Internet that caused damage to a protected computer system, and loss aggregated at least $5,000. Adams was also charged with unauthorized access to electronic or wire communications while those communications were in electronic storage. Adams pleaded guilty to charges of attempting to damage a protected computer system. In April 2001, Adam was sentenced to 5 years probation, fined more than $12,000 for restitution, and restricted to using a computer only for work and educational purposes.[10]

In July 2002, there were reports that a small number of WebTV devices were dialing 9-1-1. These devices had been re-programmed to dial 9-1-1 instead of dialing the WebTV provider for updates, an activity unknown to the owner. WebTV enables subscribers to access the Internet using their phone line and their television as a display device. David Jeansonne was accused of sending WebTV users e-mail messages containing an executable attachment. The attachment was disguised as a utility that allowed the user to adjust the WebTV screen colors. Instead, it reset the system's server dial-in code from a California seven-digit number to 9-1-1. The next time the user attempted to connect to the company server or when the system would check for updates nightly, the device would instead dial 9-1-1. This virus generated calls to 9-1-1 from California to New York some leading to the dispatch of emergency responders to the location of the call.

By December 2003, the FBI arrested Jeansonne, 43, at home in Metairie, Louisiana. Jeansonne was indicted by a federal grand jury on two counts of damaging protected computer systems "without authorization." In 2005, Jeansonne pleaded guilty to two counts of intentionally damaging a computer and causing a threat to public safety. Jeansonne was sentenced to 6 months in jail, 6 months home detention, and restitution of more than $27,000 to Microsoft, who operated WebTV.[11]

Other Vulnerabilities to Emergency Calling Systems

Hoax and improper use of emergency calling numbers are also an issue for the service. The 9-1-1 system is subject to routine or extraordinary outages that can reduce the public's incident reporting capability. Other vulnerabilities to an emergency calling system include:

- software upgrade by the telephone company
- failure of central office or tandem equipment
- lightning strike to communication center gear
- water damage to 9-1-1 trunks leading into communication center
- damage or cut to fiber-optic cable serving major central office
- insufficient backup power sources when extended power outages occur

For these reasons, there are frequent public information campaigns around the world on the correct use of emergency calling number and supporting systems. Many times the media will reference the term "9-1-1" (North America) or "112" (Europe) to indicate the entire public safety communications system, rather than the emergency telephone system. News accounts frequently say that "9-1-1 was affected," when in fact they mean that the radio or computer system was not operable. It sometimes takes considerable investigation to determine exactly what systems were affected.

Next Generation 9-1-1 (NG 911)

In the United States, the next generation 9-1-1 (NG911) services will enable the use of multimedia (voice, video, short text messages [SMS], and data) for emergency communications. A new architecture for emergency communications, under the guidance of IPAWS, is an architecture based on the Internet Protocol (IP) and open standards. In 2005, when Hurricane Katrina made landfall in Louisiana, Voice Over Internet Protocol (VOIP)–based communication services significantly improved the effectiveness of relief. Use of multimedia for emergency communications does present challenges for Public Service Answering Points (PSAPs) and new opportunities.

Emergency Requests for Help around the World

Australia

In 1961, Australia adopted Triple Zero (000) as its national number for emergency services introduced by the Office of the Postmaster General (PMG). Prior to this time police, fire and ambulance services possessed many phone numbers, one for each local unit. Triple Zero was initially introduced in major metropolitan centers; by the end of the 1980s, coverage was extended nationwide. The three-digit number 000 was chosen for several reasons, one of which was that zero was closest to the fingerstall on Australian rotary dial phones, so it was easy to dial in darkness.

The three-digital number 9-1-1 was considered but was already assigned as the starting numbers for some residents and businesses. Telstra operates the Emergency Call Services and its intended use is for life-threatening or time-critical emergencies. Emergency services and Australia's Communications Regulator preferred the phrase "triple zero" over "triple oh." It was found that the later created confusion of the appropriate number to dial when using alphanumeric keypads. If a caller used the letter "O," the caller would dial "666" rather than "000."

Australia completed its Enhanced 000 service in 1997 converting the Northern Territories to the service. Australia's Enhanced 000 service could send calling line identification (CLI) and service address to the emergency operators.

Europe

Across much of Europe 112 has been adopted as the primary number for emergency services. Many of the emergency calling numbers has been enhanced to transmit the location information to the emergency call center. Like the FCC E9-1-1 in North America, the EU Directive E112

(2003) requires mobile phone networks to provide emergency services with whatever information they have about the location a mobile call was made. Most European countries have another universal number for nonurgent call such as 116 115 in Finland and 114 15 in Sweden. Like the emergency calling number, the nonemergency call numbers are also free of charge. A detail listing of international emergency and nonemergency call numbers is provided in Appendix F—Emergency Call Numbers.

SUMMARY

"A crisis occurs at a breaking point that requires some urgent, usually drastic change" as stated by Rahm Emmanuel. Where a crisis intersected with the inadequate warning has led to significant leaps in crisis communications and mass notification. Out of the fire that led to five fatalities was born the first Universal Emergency Number 9-9-9 adopted in the UK. War and catastrophic disasters have provided the justification governments needed to make significant investments in national emergency communication systems. The quest not to be last, the competitive spirit of U.S. fire chiefs, and a single government official in Alabama led to the birth of the U.S. Universal Emergency Number 9-1-1. Other like systems have sprung up around the globe with most people today covered by a universal emergency calling number for their area. Now that step one is in place, the focus has shifted to leveraging as many of the current communications technologies available in the marketplace and placing them under one umbrella to add dependability. The FCC eventually became the government agency to assume responsibilities for the U.S. national emergency communication program. CONRALED, EBS, EAS, and the program of choice today, IPAWS.

ENDNOTES

1. Hobson, Melody. 2010. Why a crisis can be good for your finances. *Black Enterprise*, November, p. 34.
2. Firenet.org. (2010). 999 History. http://www.fire.org.uk/advice/999history.htm (accessed January 20, 2011).
3. Firenet.org. (2010). 999 History. http://www.fire.org.uk/advice/999history.htm (accessed January 20, 2011).
4. Population Reference Bureau. 2008. *2008 World Population Datasheets*. Washington, DC: Population Reference Bureau.
5. GSMA. 2011. *GSMA Public Policy Annual Review*. London: GSMA.

6. U.S. Department of Commerce Patent and Trademark Office. 1975. *Emergency Reporting System.* http://www.911dispatch.com/911/history/emerg_reporting_3881060.pdf (accessed January 15, 2011).
7. 911Dispatch.com. 2010. *Emergency Reporting System.* http://www.911dispatch.com/911/history/emerg_reporting_3881060.pdf (accessed January 15, 2011).
8. FCC. 2010. http://www.fcc.gov/pshs/services/911-services/enhanced911/Welcome.html (accessed January 15, 2011).
9. 911Dispatch.com. 2010. *Hacking the 911 System.* http://www.911dispatch.com/911/history/hacking911.html (accessed January 19, 2011).
10. Symantec. 2010. Bat.chode.worm.X. http://www.symantec.com/techsupp/vURI.cgi/nav31 (accessed January 20, 2011).
11. U.S. Department of Justice, United States Attorney, Northern District of California. 2004. *Louisiana man arrested for releasing 911worm to WebTV users.* San Francisco, CA.
12. U.S. Bureau of Labor Statistics. 2010. National Emergency Number Association.

CHAPTER 3

Dynamics of Communications and Concepts of Operations

> The Rail Maritime and Transport (RMT)
> Union's Bob Crow said the Tube system
> was "lurching from crisis to crisis."
>
> BBC News, London, February 9, 2011

Communications involves a sender and a receiver. The dynamics of communications can help an individual use the power of many forces to express ideas or transmit information. This can occur with maximum effectiveness within a shared or common space as expressed in Figure 3.1 (Basic model of communications). The richness of the communication option used improves the effectiveness of that communication. The sender and receiver get closer and migrate from documentation only options to modeling options as illustrated in Figure 3.2 (Communications). The more effective the options used by the sender, the more the likelihood the communicated message is understood by the receiver.

Face-to-face communications includes the spoken word and other techniques such as gestures and facial expressions. Face-to-face communications is considered a high form of effective communications. The reason is that the receiver has an opportunity to consider the unspoken words that are offered activating multiple senses—hearing and seeing. Conversely, communications limited to paper has a lower level of richness in communications and effectiveness and uses only one sense—seeing.

Determining if a receiver can read and understand a written message as intended by the sender is driven by

- the legibility of the documentation given the receiver
- the tone of the documentation, as interpreted by the receiver

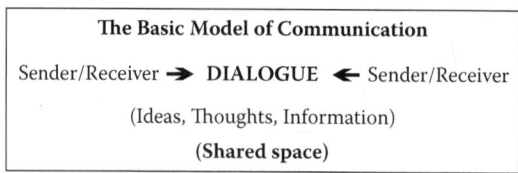

FIGURE 3.1 Basic model of communications.

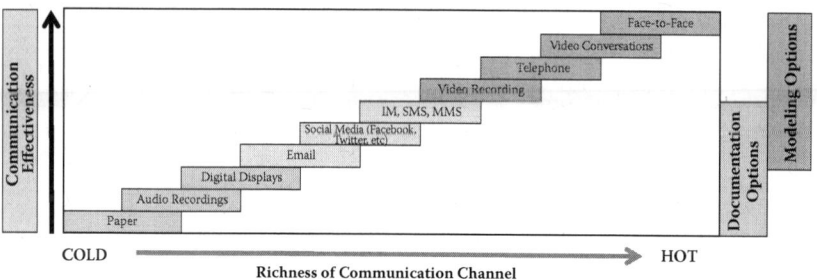

FIGURE 3.2 Communications.

Other forms of documentation options include prerecorded audio messages, digitally displayed messages, and electronic mail. Electronic mail is typically less formal, and dependent upon modeling options. The scarcity of information evokes different improvisations than when there is an abundance of information. The challenge of managing an abundance of information is fundamentally different from the challenge of coping with a scarcity of information.

As depicted in Figure 3.2, from left to right, an individual gains richness, such as physical proximity and real-time two-way communications. Individuals also increase their ability to communicate using more techniques simultaneously, such as gestures, speaking, and facial expressions that affect how well the information is understood by the listener.

EFFECTIVE COMMUNICATIONS

Effective communications is important when speaking in general terms for business, and everyday communications. The time spent communicating, regardless of the situation, is significant. Whether written or verbal, one method or multiple methods, and between or among two or more, communications does make a difference. For example, the influence communications has on the success of a business can be defined as follows:

- 85% of business success is dependent on effective communications and interpersonal skills.
- 75% of the time is spent on nonverbal communications.
- 50% of the time of the average worker is spent communicating.
- 45% of the time spent communicating is on listening.
- 9% of the time spent communicating is on writing.

This could be summarized by stating that communications is one of the most important factors that influence a business' success.

Factors that influence communications go beyond two individuals speaking a common language. Culture, gender, and situation also influence how a message is received. The cultural traditions of Western Europe crossed the ocean to influence the United States. The same holds true near border communities of the Southwestern United States, where strong cultural traditions from Mexico and the rest of Latin America are rooted, and where English is often considered a second language rather than the primary language.

The United States is also a melting pot of cultural traditions from around the world, given the huge influx of immigrants that have brought their own languages, traditions, and cultures. This could mean that what may be acceptable in this country may not be in the country of origin or vice versa. There are other traditions that are strongly held in the country of origin that continue on in the United States. This is true where heterogeneous communities exist. Nationalities that are at war in their countries of origin live as peaceful neighbors in the United States.

Culture, traditions, nationalities, and gender strongly influence everyday communications. Culture assumes a prominent role in effective communications. At the same time, the accelerated pace of technological innovation will significantly alter the traditional paradigm. Communications planning will require more effort and will need to factor in more variables to tailor a message for each targeted audience. This shift will cause organizations to behave differently and force a level of transparency in the midst of a crisis that was calculated in the past.

Effective communications is a two-way process, involving the sender and the receiver who both listen to their messaging as it evolves. Listening entails more than the sense of hearing. *Hearing* is one of the five senses by which noises and tones are received as stimuli using your ears. The ear operates similar to how a microphone detects sound. When applying our ears to hear a message it should not automatically infer that one has chosen to listen to the message. There is a difference between hearing and listening.

To listen to a message means that thoughts, beliefs, and feelings are applied to the message. It takes work by the receiver. Active listening requires the sender to decide first to listen and concentrate on the message sender who is speaking. In face-to-face or video, the speaker can be observed through vocal inflection, style of delivery, body language, and other nonverbal cues. The same applies to a written message although not all factors will apply. For example the style of delivery and the tone used would apply to an e-mail or letter, but vocal inflection and body language may not. Paraphrasing and clarifying questions to confirm that the message as received is as intended by the sender is another step in active listening.

Good active listening skills are important in everyday conversations and in a crisis. Active listeners can

- Hear accurately
- Empathize
- Gather information
- Question assumptions and ideas
- Weigh options
- Find answers
- Show respect
- Build relationships

- Understand
- Draw out ideas and information
- Find answers
- Connect
- Change perspective
- Soothe or heal
- Show appreciation
- Set the stage for something else[1]

Effective listening also has its roadblocks—external and internal. External roadblocks are usually distractions such as noise level, seating, or location. Internal roadblocks cover a range of conditions or reactions by the sender or receiver(s) such as defensiveness, stereotyping, hearing the expected rather than what is said, automatic dismissal, or emotional interference. Roadblocks affect successful communications because it causes differences between the sender and receiver.

Other internal roadblocks that impair successful communication are the differences between the sender and receiver information levels, social systems, sensory channels, communication skills, and inferences.

The sender or the receiver could have significantly more knowledge and information regarding a message than the other. For example, the technical analysts that can draft the technical documentation for others with like knowledge such as the system administrator. The system administrator and technical analyst can most likely draw from the same knowledge to easily understand what is communicated. Both the sender and receiver have a similar level of knowledge regarding the message. Between the technical analysts and the end user of the same product there may be a significant difference in the level of knowledge regarding the product. This could be due to the fact that the product is new or the terminology is too technical for the end user to understand. This gap in communications is an internal roadblock.

The greater the difference in communication skills between the sender and receiver, the higher the probability that communication will be effective. For example, business communications within the organization is typically less challenging that communication externally. The logic is that individuals within the organization have a common knowledge, such as business jargon that is unique to the organization. Use of the same language externally to those who are not familiar with the jargon creates a difference in communication skills. Also, when members within an organization are communicating with people outside the organization, the sender is representing themselves as individuals and as a representative for the organization.

The social systems provide a foundation for interpretation of a message. For example, many 30 years of age and younger use single letters, numerals, typographic symbols and acronyms to represent words or parts of words when texting using mobile phones, a form of *text orthography*. Text orthography is a method defining the correct way of using a writing system to write a language. For example:

r u *o* = are you surprised?
gr8, c u @ 8 b4 dark :-) = Great, see you at 8:00 before dark. Smile

Use of this form of text orthography is appropriate for interpersonal communications yet culturally unacceptable in the classroom for submitting a written assignment. Eye contact is an example to consider as represented in Table 3.1 (Examples of Cultural Norms). If the sender and receiver do not have a like social system, effective communications is more challenging.

Seeing, hearing, touching, tasting, and smelling, the five senses referred to as the *sensory channel*, are the basic channels of communications. As the number of sensory channels increases, the likelihood of successful communications also increases. For example, a discussion regarding a new sugar cookie unveiling. Teleconferencing involves the sense of hearing when the recipe and description of the new sugar cookie is given and how good it tastes. Video teleconference adds seeing, a

TABLE 3.1 Examples of Cultural Norms

Culture	Behavior	Response	Suggested Behavior	Response
Japan	Looking someone in the eye for an extended period	Rude and aggressive	Focus on neck or collar	Desirable
United States	Direct eye contact	Desirable, indicator of confidence and sincerity		Lacks confidence or sincerity

second sensory channel. The receiver of the message can see the new sugar cookie and hear about how good it tastes, the recipe, and a detail description of the cookie. When the same dialogue occurs face-to-face, all five senses can be used. The receiver, can see the new cookie, hear about the recipe, can touch the sugar cookie to determine how it feels, can taste the sugar cookie to determine firsthand if it is as good as stated, and smell the sugar cookie to determine its freshness. This example demonstrates how the success of the communications improves as more senses are used.

Inference, judgment, and generalizations can also create differences between the sender and receiver as well as previous experiences. Inference is the process of coming to a subjective conclusion based on experience without knowing the actual logic and assumptions that were applied to draw the conclusion. This is a concept often used in mediation, enabling parties to identify the root cause of a disagreement by discussion the assumptions and logic used to come to conclusion. For example:

Statement: It is cloudy, dark, and windy today, like on days when it rains.
Conclusion: It is going to rain today.
Assumptions: Today is like other days that were cloudy, dark, and windy.
Logic: Since it rained before under these conditions, it will rain today.

The sender of the message used inference, judgment, and generalizations. Depending upon the level of knowledge by the receiver, they may not question the validity of the statement. The influence of each of the roadblocks discussed has a direct impact on crisis communications.

CRISIS COMMUNICATIONS

Crisis communication is very different from every day communication; people take, process, and react to emergency situations differently. First, it is important to define what a crisis is. A *crisis* is a unique unpleasant occurrence that is characterized by the elements captured in Hermann's model: surprise, threat, and having a short response time. Hermann's model uses adverse international events to conclude that a crisis involves

- threatened high-priority values or a threat to the organization goals
- presents a restricted amount of time in which a decision can be made
- is unexpected or unanticipated by the organization

A surprise can start with a known event such as political protests that escalates to the level of a crisis unexpectedly. The surprise is defined as the discrepancy between the desired or expected, the actual state, and the probability of loss. Also, the thought of suddenly confronting an unanticipated and unfamiliar circumstance.

> **CASE STUDY: POOR CRISIS COMMUNICATIONS—2011 TUNISIAN CRISIS**
>
> Many leaders around the world were surprised by the 2011 Tunisian crisis. Tunisia is the northernmost country in Africa, a former French protectorate. The French did not foresee the events coming as echoed by the French Foreign Minister Michele Alliot-Marie. The Tunisian crisis involved the loss of power of its president, Zine al-Abidine Ben Ali, a longtime ally of France. France began with silence. This silence was perceived as condemning the violent repression of protests and later modified its position. Once the deposed president fled the country, the French president Nicolas Sarkozy's government dropped Ali abruptly. Deposed President Ali found refuge in Saudi Arabia. Alliot-Marie defended the handling of the Tunisian crisis by stating that France had been surprised by the event.

A threat is driven by circumstances that reach beyond the capabilities to respond by an individual or organization. The greater the perceived value of a goal and the probability of a loss associated with the goal the greater the perceived threat.

> **CASE STUDY: POOR CRISIS COMMUNICATIONS AT THE START OF THE 2010 DEEPWATER HORIZON**
>
> When the 2010 Deepwater Horizon fire caused the platform to fall into the Gulf of Mexico and spilled millions of gallons of oil into the prime fishing waters of Louisiana, Mississippi, and Alabama, the crisis threat grew to include most all of the U.S. coastal states including Florida and Texas. The platform was located approximately 50 miles from the Louisiana shoreline. Birds, sea animals, vegetation, and local wildlife were devastated by the spill, threatening the ecosystem of the region. The Deepwater Horizon spill negatively impacted the petroleum industry's image, ultimately leading to a temporary halt to drilling in the Gulf. BP, Transocean, and Halliburton have since spent billions of dollars in litigation, response, and ongoing cleanup efforts that will continue for years to come. Like the *Exxon Valdez* accident in 1989, response to the Deepwater Horizon spill began as a failure in crisis communications.

A crisis also involves a short response time. A crisis has a compressed timeline between the events preceding the situation and the onset of harm that requires an immediate response. There is insufficient time to conduct a systematic analysis.

> **CASE STUDY: SHORT RESPONSE TIME FOR CRISIS COMMUNICATIONS, HAITI 2010 EARTHQUAKE**
>
> Just before 5:00 p.m. ET on January 12, 2010, a 7.0 magnitude earthquake shook Haiti. The epicenter was nearly 10 miles southwest of Port-au-Prince, the capital city. Although not the strongest earthquake in history, it is one of the most destructive natural disasters. The earthquake took by surprise a country with inadequate building codes and left the poorest people in the western hemisphere with buildings reduced to rubble, instantly taking many lives as they fell. The Haitian government estimated that there were nearly 230,000 fatalities, 300,000 injured, and more than 1.5 million displaced. The world had a compress timeline to provide humanitarian assistance, which included temporary shelter, medical, and other basic needs, before survivors suffered further harm. Within a few months, more died and became ill due to poor sanitation and housing and another round of torrential seasonal rainfall.

Each crisis activates crisis communications, a time when people may behave differently than during everyday situations.

Crisis communications is the communications processes, procedures, and methods used early, during and after a crisis. Crisis communications affects the reputations of organizations and agencies and the community these organizations serve or support, including the news media. An effective crisis communications program involves the plan, response, and recovery communications needed for the management of a crisis. This occurs throughout the life cycle of the crisis and the post-review of how the process worked after the crisis has subsided.

Crisis Communications Planning

Crisis communications planning involves first identifying what could constitute a crisis. Within an organization, the planning process incorporates the identification and prioritizations of vulnerabilities. Vulnerabilities are assessed with a high, medium, and low probability of occurrence. Once assessed, communications responsibilities are assigned, supporting materials are developed, and spokespersons and others are trained in how to communicate the desired message. The planning process incorporates formal and clearly defined channels to mitigate a crisis. The process strives to maintain a reputation of leadership and transparency by providing timely and accurate information for the target audiences. The *target audience*, as defined by the Federal Emergency Management Agency (FEMA), is generally everyone who can benefit from the information. Examples of members of target audiences for a college campus shooter event may include

- Jane Doe, 24, freshman, who spends most of her time on Facebook, texting, and chatting with friends when not in class and lives on campus.
- John Simpleton, 76, a continuing education student, who is reliant upon sign language and lip reading as his primary form of communications and lives off-campus.

The definitions of an organization's target audiences should take into consideration what are the typical audience and the extremes for the organization.

The crisis communications planning process begins with the end in mind—achieving goals and objectives. The goals commonly defined with a plan include some or all mentioned in Table 3.2 (Crisis Communications Goals).

Examples of objectives that are commonly found in an organization's crisis communications plan are

- Prepare employees to manage crisis communications.
- Train employees on how to respond with one voice and in a professional manner that reinforces the organizations reputation of leadership and transparency.
- Inform members of the response and recovery community to help shape a consistent response.
- Aid in the public understanding of the event and the value provided by the organization.
- Manage the distribution of critical, often sensitive, information to the media and the target audiences, including the public at large.

TABLE 3.2 Crisis Communications Goals

Accurate	Consistent	Honest	Relevant
Appropriate	Credible	Regular	Timely

Activating the crisis communications plan is the formal start of the crisis response—the first stage. The crisis response is focused on the short-term effort to assess and understand the crisis by an organization's emergency, and response personnel. Strategic decisions and responses are made quickly using available information that is often insufficient. The decisions made are usually targeted on reducing and containing the threat and harm. The time available to communicate is limited during this phase of a crisis.

This stage is defined by compressed timelines that limit the ability to gather and analyze available data. Usually a specific message must contain instructions that result in practical actions by the targeted audiences. Most decision makers handle the response in a manner so that activities can move forward effectively to the next phase of the process. Other decision makers, although rare, working under the pressures of compressed time constraints, find it difficult to make decisions. This is a form of *decisional paralysis*, extreme indecisiveness in the midst of a crisis, as defined by Gouran in 1982.[3]

Determining when a message should be sent and the overall type is critical upon activation of the *crisis communication process*. It is the second stage of crisis communications. In a crisis, achieving as many of the goals of the defined plan as possible is critical. A simple three-phase model, as defined by time intervals, is a good rule of thumb in providing crisis communications and is shown in Table 3.3 (Crisis Communications Phases–Time Intervals).

The third stage of crisis communications planning involves *crisis recovery*. Crisis recovery encompasses the activities completed during the aftermath of a situation that are focused on restoring the organization to normal operations. Most organizations that survive a crisis become a different kind of organization. This is because new or changed policies and processes are developed and implemented to address deficiencies, significant organizational changes, or the availability technologies. During this phase of the process, sensitivity to the target audiences' cultural values and attitudes is critical.

During the recovery phase, when sensitivity to the community's cultural values and attitudes is perhaps most important, a more complex, culturally based model may be more appropriate. Failure to discern

TABLE 3.3 Crisis Communications Phases–Time Intervals

Phase	Time Intervals	Description
Immediate	1–30 minutes	Tactical instructions to occupants
Emergency	15–60 minutes	Emergency announcements
Advisory	1–24 hours	Updates, warnings of potential threats announcements

TABLE 3.4 Communication Issues and Their Impact

Communication Error Type	Average Number of Errors per Communication Error Type
Error of commission (misreading a label)	3/1000
Error of omission (item embedded in procedure)	3/1000
Error of omission (without reminders)	1/100
Error in simple arithmetic (with self-check)	3/100
Personnel on different shift fail to check conditions unless directed by a checklist	1/10
Errors under very high stress when dangerous activities are occurring rapidly	1/4

attitudes, beliefs, values, and rules implicit in different groups could disenfranchise some citizens and harm the community's return to productivity and health. A cultural model is useful because it recognizes community members' shared interest in the community's future.

An organization's crisis communications program is a *continuous improvement process*—stage four. Each phase of the crisis communications model is directed toward sending messages efficiently and directing the target audience toward the desired response. An assessment of what went well and opportunities for improvement is an on-going process. Understanding what went well so those actions can be repeated is important. Equally as important is learning the opportunities for improvement. The four primary issues causing common communications problems are

1. lack of information
2. omission of information that is available
3. misinterpretation of information
4. misrepresentation of information, particularly when under very high stress

The impact of the communication issues increases with severity based on type as defined in Table 3.4 (Communication Issues and Their Impact).

EMERGENCY COMMUNICATIONS CHARACTERISTICS

Next to food, water, and medical aid, information is critical to those that are impacted by an emergency event. Accurate information leads to better decision making and provides reassurance regarding the status of any response or recovery efforts planned or in progress. Ultimately, information can be the difference between life and death. Timely information is also critical in reducing rumors and speculations that quickly fill any

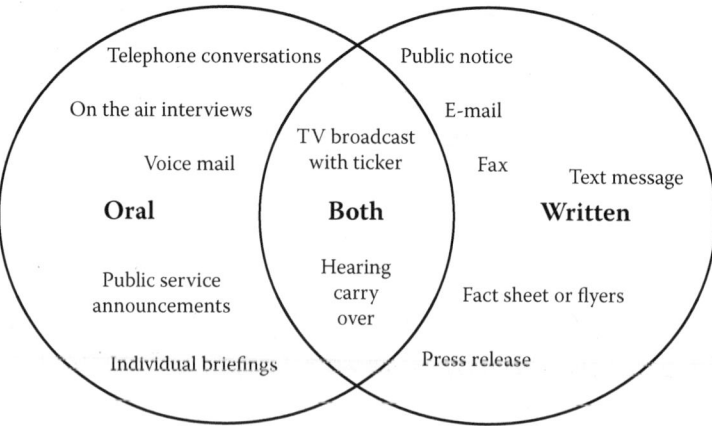

FIGURE 3.3 Three primary communication types.

void of information from trusted or official sources. The failure to deliver timely communications can lead to additional efforts that are needed to counter any misinformation circulated. For example, when using local media outlets, learning their news cycles, deadlines, and appropriate contacts are key components of the crisis communications plan.

The type of message to deliver is also important. If there is a need to increase awareness or provide knowledge, then possibly only an advisory or announcement is given. However, if there is need for the message to elicit a response from the target audience, then an emergency warning message is more appropriate. For example, a tornado warning is imposed if a tornado has been spotted in the area and the target audience must seek shelter immediately. A tornado advisory is given when conditions could produce a tornado in the area and the target audience should be prepared to take shelter immediately. The tornado warning requires a response; the tornado advisory purpose is to increase awareness of a potential threat.

In a crisis there are three primary types of communications: oral, written, and a blend of oral and written communications as illustrated in Figure 3.3 (Three primary communication types). Oral communications include

- Telephone conversations
- On-the-air interviews
- Individual briefings
- Public service announcements

Oral communications is fluid and dynamic, and driven by the sender and the receiver(s). Oral communications involves the verbal and nonverbal components of a communiqué, such as body language, tone, culture, and dialect.

In an emergency, printed communications is preferred. It is more likely to be understood by the audience, particularly as the message is shared with others. Types of print communications include:

- Fax
- E-mail
- Public notice
- Fact sheet or flyer
- Press release
- Text messages

Blended communication methods offer an advantage over the use of only a written or oral message. Combining the strengths of written and oral communications moves the communiqué closer to effective communications by leveraging multiple technologies such as a television broadcast that uses open captioning. This medium of communications enables the receiver to hear and see the message using multiple senses.

Those who are trained in crisis communications have learned to choose the most effective communication tool available to them as defined in the crisis communications plan. The definition of an effective communication tool is that it enables the sender to reach the target audience, the receivers of the message. An effective communication tool enables the sender to give information in a timely way, as needed, and reliably, which enhances the comprehension of the message by the receivers.

How to Recognize When a Message Is Not Being Communicated

How can you tell when your message is not being communicated? An effective crisis communications plan takes this into consideration at each stage of the communication process. Have the target audiences been appropriately defined and have all of them been considered? Often times, crisis communications plans consider the most obvious groups—those that are most like themselves. For example, if planners speak English as their primary language, they may overlook those who are not fluent in English. There are others identified as members of a population with functional and access needs that are members of the target audience. It is unrealistic to become an expert on every culture that you may encounter; it is reasonable for the crisis communications plan to consider the populations that comprise the major parts of your community.

To enhance the crisis communications plan, the planning team should learn the basic customs of ethnic groups within the communities the organization serves and resides in. This may be achieved by reading or attending cultural events in the community where food, art, and entertainment with others can be shared, paying attention to what is said and heard. Another option is to speak with leaders of groups that are a part of your target audience. Many are happy to share key attributes of the group and appreciate the interest in them by your organization. Learn through networking who the resources are before a crisis occurs to help build a relationship so that you are a familiar entity at the time of a crisis.

Continuously scan the environment to assist in determining whether a communiqué has been received and understood by the receivers. Have you seen puzzled looks from the target audience? Almost all cultures have like facial expressions when a message is not understood. If face-to-face, look for changes in body language or tone of voice. For a printed message, are you receiving a high number of clarification questions? Look for signs that the target audience is responding as desired.

CRISIS COMMUNICATIONS IN THE MIDST OF VIOLENT EVENTS

Emergencies Related to Violence by Journalists

Johan Galtung defined 12 points of concern[4] for addressing emergencies related to violence by journalists. Each implicitly suggests remedies that are more explicit. Galtung is a Norwegian mathematician, sociologist, and one of the founders of the discipline of peace and conflict studies. Galtung is a controversial figure, yet recognized for his efforts related to the study of peace. An area of interest for him is how journalism and the media contribute to the violence of the world, and he has formulated ideas on how to act in addressing this problem. He is particularly interested in crisis communications and has identified 12 media flaws that can aggravate misunderstandings:

1. *Decontextualizing violence:* concentrating on the irrational and failing to examine the reasons for unresolved conflicts and polarization.
2. *Dualism:* reducing the number of parties in a conflict to two, when more may be involved. Stories that are limited to internal developments often ignore outside or "external" forces such as foreign governments and transnational companies.
3. *Manicheanism:* depicting one good side and demonizing the other as "evil."
4. *Armageddon:* portraying violence as inevitable, omitting alternatives.
5. *Spotlight individual acts of violence while avoiding structural causes* like poverty, government neglect, and military or police repression.

6. *Confusion:* centering on the conflict arena such as the battlefield or location of violent incidents, excluding the forces and factors that influence the violence.
7. *Excluding and omitting the bereaved*—eliminating the need to explain why there are acts of revenge and spirals of violence.
8. *Failure to explore the causes of escalation* and the impact of media coverage itself.
9. *Failure to explore the goals of outside interventionists,* especially big powers.
10. *Failure to explore peace* proposals and offer images of peaceful outcomes.
11. *Confusing cease-fires and negotiations* with real peace.
12. *Omitting reconciliation:* conflicts tend to reemerge if attention is not paid to efforts to heal fractured societies. When news about attempts to resolve conflicts is absent, fatalism is reinforced. The reinforcement of fatalism helps to engender more violence, when people have no images or information about possible peaceful outcomes and the promise of healing.[5]

Different media outlets or the same outlet within a short period of time can show the same story that has been covered legitimately in different ways—some that promote a conflict while others suggest opportunities for resolution.

Let us use the approaches to reporting on the events in Tunisia as the president lost power as an example. The BBC and CNN reported on the fall of Tunisian President Zine al-Abidine Ben Ali in January 2011. The initial reports reflected shock and awe. This was the first Arab leader to fall due to social pressures and protests that were stimulated using the Internet and social media. There were interviews of a few individual protesters, no interviews of the deposed leader and his family. The focus quickly shifted from the ousting of a leader to the start of a new movement focused on changing Arab leadership across the region through like protests. Al Jazeera, a news media outlet sponsored by Qatar, was able to have the Tunisian ministers give their views before the overthrow and therefore provide more information. This was an opposing viewpoint to the story. Today, many say that Al Jazeera's brand of journalism has changed how Arab media covers events.

> **TIP:** Follow the media's account of events from multiple sources. Most crises have multiple journalists following and investigating. Using as many of these as possible helps to have a more well-rounded understanding of a story.

TABLE 3.5 Four Stages in Preparing a Nation for War[7]

Stage	Description of Activities
The crisis	The reporting of a crisis which negotiations appears unable to resolve. Politicians, while calling for diplomacy, warn of military retaliation. The media reports this as "We're on the brink of war" or "War is inevitable," etc.
The demonization of the enemy's leader	Comparing the leader with Hitler is a good start because of the instant images that Hitler's name provokes.
The demonization of the enemy as individuals	For example, to suggest the enemy is insane.
Atrocities	Making up stories to whip up and strengthen emotional reactions.

Crisis Communications for War Preparation

Investigative journalist Phillip Knightley outlined in *The Guardian,* a British newspaper, the four stages used to prepare a nation for war as defined in Table 3.5 (Four Stages in Preparing a Nation for War).

The application of each stage in preparing a nation for war is demonstrated in the Case Study on the 2000 Desert Shield/Desert Storm conflict.

CASE STUDY: 2000 DESERT SHIELD/DESERT STORM

In the Iraq war (Desert Shield/Desert Storm) of 2000, Saddam Hussein Abd al-Majid al-Tikriti was president of Iraq (from 1996 to 2003). He was a man that, during his presidency, many Arab leaders respected and admired for his aggressive position against foreign intervention. Other Arabs and Western leaders criticized and spoke ill of him as the power behind many deadly attacks on northern Iraq in the late 1980s and for the invasion in 1990 of Kuwait, south of Iraq. U.S. President George W. Bush and British Prime Minister Tony Blair used Saddam's growing unpopularity in the Arab world, claiming he had links to terrorist organizations and had weapons of mass destruction. Saddam was portrayed as a president who was unwilling to negotiate and it was claimed that earlier sanctions on Iraq had not worked. This was used to prepare the nation for war, shape the crisis, and heighten the need to remove him from power. At this point "the crisis" had been defined and Saddam had been demonized as the leader of the enemy. Saddam was also portrayed as an individual who was harsh to his people and even insane. The claim of insanity was not specific; rather the term was used extensively in headlines to define Saddam such as "Hussein Goes Insane" in the *Cornell Review,* 2006. This information was used to demonize him as an individual since the perception was that only an insane person could commit the atrocities he had committed. Saddam was deposed in 2003, found guilty of charges related to the killing of Iraqi Shiites, and was executed by hanging in 2006. Some considered this an atrocity in itself.

Knightley states that some stories that are communicated are known fabrications and outright lies while others may be true. The dilemma is determining what is true and how you can tell that it is true. He concluded that the media insists we trust the information as presented. How you verify the accuracy of the information remains unanswered. Journalists have a growing reputation of dishonesty; the difficulty that honest journalists face is daunting in delivering true and factual information when the consumer of the information has a short memory. This has helped in propelling social media sources as the first place to look for information. Real individuals are providing pictures and first hand accounts of events faster than what many considered as filtered information from traditional media sources.

One difficulty is that the media has little or no memory. War correspondents have short working lives, and there is no tradition or means for passing on their knowledge and experience. The military, on the other hand, is an institution that will continue as long as humankind exists. The military learned a lot from Vietnam, Desert Shield, and Desert Storm, and today plans its media strategy with as much attention as its military strategy.[6]

Propaganda

In 1914, those opposed to World War I (WWI) used mass media for propaganda. Mass media is a segment of the media that is designed to target a large audience. The term *mass media* was coined in the 1920s as radio networks and mass-circulation newspapers and magazines had nationwide audiences in addition to books and manuscripts that have been used for centuries. *Propaganda* is used in conflict to take the battle into the minds of people. Propaganda typically has misleading information using distortions, exaggerations, inaccuracy, subjectivity, and fabrication in order to receive support and to have a sense of legitimacy.

Propaganda is also used in mainstream reporting, leading to partial information and intentional or unintentional systematic contradictions of claims. Techniques that are often used include

- Labeling
- Suppression by omission

- Slighting of content
- Preemptive assumption
- False balancing
- Follow-up avoidance
- Framing
- Attacking and destroying the target
- Face-value transmission

With the consolidation of media ownership increasing, a narrower range of views and discourse arises, although sometimes unintentional. This can lead to heavy bias among the mainstream media outlets. For example, one media outlet in the United States may have followers represented as 75% Republican, 85% male, and 92% white. Followers of another media outlet in the United States may be 12% Independents, 40% Democrats, 30% Republicans, 65% females, and no dominant race. Media outlets routinely conduct polls and then report their findings as fact that is representative of the general population. The problem with these polls is that they typically represent the views of their followers and not the views of the general population. Second, the sample size is usually too small and the conditions under which the poll was conducted may not have been consistently applied. Each variable introduces a bias in the reporting.

Governments from around the world typically have massive amounts of fake and prepackaged news that can be quickly disseminated through the mainstream media. Examples include a video segment that is produced by the government without any mention that the government produced the video. Consider this scenario: A media outlet may elect to use a video of fake reporters covering airport safety that is actually an actor using a false name for the airport security agency or where there has been some acknowledgment of the agency. The news station may rebroadcast them without crediting the source so that the news segment appears real locally. This fake and prepackaged news can reach millions and can benefit the owner and the news outlet, ultimately amounting to a form of propaganda or media manipulation.

SUMMARY

The richness of the communication option used improves the effectiveness of that communication. This is even more evident in a crisis and with activation of your organization's crisis communication plan. Time is limited, decisions are made quickly, and the instructions needed by the target audience must follow soon. How the audience responds to the information will determine what improvements are needed to the plan and the associated training required. Crisis communication is used in response to a crisis, and it is employed to shape the opinions of target audiences in preparing for war. Propaganda by entities and governments can lead to the support of actions that would otherwise be unthinkable. The role of journalists in facilitating the reporting of events and media manipulation of information has lead to violence and has also given rise to new forms of media, social media, and a growing number of challenges in achieving effective communications.

ENDNOTES

1. Hoppe, Michael H. 2006. *Active listening: Improve your ability to listen and lead*. Greensboro, NC: Center for Creative Leadership.
2. Hermann, C. F. 1963. Some consequences of crisis which limit the viability of organizations. *Administrative Science Quarterly* 8:58–80.
3. Gouran, Dennis S. 1982. *Making decisions in groups: Choices and consequences*. Glenview, IL: Scott, Foresman.
4. Schechter, Danny. 2001. *Covering violence: How should media handle conflict?* http://www.mediachannel.org/views/dissector/coveringviolence.shtml (accessed January 2, 2011).
5. Schechter, Danny. 2001. *Covering violence: How should media handle conflict?* http://www.mediachannel.org/views/dissector/coveringviolence.shtml (accessed January 2, 2011).
6. Knightley, Phillip. 2000. Fighting dirty. *The Guardian*. London: The Guardian News & Press Release Office.
7. Knightley, Phillip. 2001. The disinformation campaign. *The Guardian*. London: The Guardian News & Press Release Office.

CHAPTER 4

Challenges to Effective Communications

> Effective communication is important in everyday life, and is critical during a disaster .... Clear, unambiguous messages are essential.
>
> FEMA, 2002[1]

Reacting with the appropriate response in a timely manner is the most challenging element of crisis communications. This is given that behavior precedes communication. In an emergency, delayed or inappropriate responses can cause late reactions by those needing the information. It can also mean embarrassment, prolonged visibility, and potential litigation for the sender. Helping decision makers understand the impact of an inappropriate or a poorly crafted crisis response before a crisis occurs is a strategic move that a public information practitioner can offer before an event occurs. This is part of a proactive approach to effective crisis communications. How communications is managed in a crisis can become a career-defining moment for decision makers and an opportunity to promote the strength and goodwill of an organization or ruin its image.

An effective crisis management program strives to meet community standards and expectations. This translates into understanding all the potential target audiences as defined in Figure 4.1 (Communication interrelationships). The program should also consider the need for each audience to

- manage the causes of the crisis
- aid victims and others directly affected by the crisis
- communicate with internal audiences, primarily employees to solicit their support

FIGURE 4.1 Communication interrelationships.

- inform those indirectly affected
- deal with external communications focusing on timely and accurate information

The key challenge is having the resources available to achieve the key activities defined as quickly as possible.

Many organizations will spend only a brief period on crisis management. From these meetings, these organizations find that progress is impeded, given a few typical assumptions. These assumptions are more prevalent when an organization has never faced a crisis. The typical assumptions for placing crisis communication planning as a low priority are

- crises cannot be planned for
- crises will not happen to us
- we are safe, secure, and well run
- most crises resolve themselves
- crises turn out not to be important
- crises are just a cost of operations
- planning for crises is a luxury we can't afford

What organizations generally learn is that these assumptions are fallacies and that the cost associated with the failure to plan is more costly than a tested plan.

These costs can be defined as the *costs of quality*, the costs associated with preventing, detecting, and dealing with defects. The costs of quality does not refer to the costs of using a higher grade product such as Grade AAA eggs instead of Grade A eggs or using 720 pi (the number of scan lines on a screen) instead of 1,080 pi for a high-definition television. Instead, the costs of quality refer to all costs incurred to prevent defects or problems that result from defects. Costs of quality encompass all activities in an organization, from initial research and development through customer service. Total quality cost can be quite high unless management gives this area special attention.

These assumptions can lead to

- Limited surge capacity when it is needed most
- Limited testing and exercising before an event
- Shortfalls existing in how to respond to a major event
- Higher potential for negative press
- Higher potential for major unplanned expenditures
- Higher potential for noncompliance with laws and regulations and increased auditing

The consequences of failed communications are many. The total costs of quality can be quite high unless management gives this area special attention.

THE COST OF QUALITY

Generally, the most effective way to manage crisis communication challenges is to avoid having defects in the process from the start. It is less costly to prevent a problem from ever occurring than it is to find and correct the problem after it has occurred. Crisis communications planning is a *prevention cost*. Prevention costs are activities that aim to reduce the number of defects. Organizations that plan and have available to them multiple techniques to prevent defects such as controls, training, exercises, and improvement plans are in a better position to deliver effective communications.

Like the practice found in the manufacturing of a product, any defective components or processes should be caught as early as possible in the production process. The same holds true for crisis communications. Identification of defects in the message before it is disseminated to the targeted audiences is a form of *appraisal or inspection costs*. Having a second person review a message before it is communicated is an appraisal or inspection of the message to ensure that the targeted audiences understand it is an opportunity to identify a defect before the message is delivered. This process may not eliminate all errors in messaging but it can minimize the likelihood that it will occur. An army of reviewers is costly and can delay the delivery of an effective message yet a failure to have any checks is more likely to more costly.

Failures in effective communication fall into two areas: internal and external. *Internal failures* related to effective crisis communications are the failure of the message and its delivery conforming to the specifications outlined in the crisis communication plan. Internal failures are those that are identified before the message is delivered. Examples of internal failures include not identifying who should create or deliver a message, obtaining the approvals before delivering a message using the stated process, or a message is written without the target audience in mind. The costs associated with these internal failures include

- rejection of the message by the decision makers
- scrapping and rewriting the message
- the message approval process taking longer than it should, causing a period of undesired silence

The more effective the appraisal process, the more was increased the chance of identifying defects internally and avoiding external failures that could be devastating.

External failures are those messages that have been sent that result in issues, perceived or real by the receivers of the messages. Examples of external failures are messages written in U.K. English for a U.S. audience, messages that are delayed, creating an air of silence, vague messages that lack enough factual information to act upon, and using an unauthorized or the wrong individual to deliver a message (such as using an untrained volunteer rather than a trained employee). External failures can lead to

- the target audience taking the wrong action
- the need for damage control due to loss of confidence from the target audience
- damage to the organization or incident command team's reputation and image
- liability
- unplanned costs

CASE STUDY: CRISIS COMMUNICATION CHALLENGE—HOW TIGER WOODS HANDLED THE CAR CRASH QUESTION

SITUATION:

Tiger Woods, from the start of his crisis, demonstratd a lack of knowledge or chose to ignore the basic tenets of crisis communications. He began with silence over an extended period. The silence was followed by a vague response to an expectant world of social media. These failures in communications led to speculation, rumors, and a tarnished image. The consequences included cancelled contracts from sponsors and unknown personal losses.

ISSUES:

Internal failures —message delivered without the target audience in mind.

External failures—message delayed, creating an air of suspicious silence followed by a vague message lacking enough factual information.

OUTCOME:

The loss of his brand's image and his personal reputation. There was also a need for further damage control.

SOLUTION:

- Strike a balance between transparency and privacy—a demand in the world of social media today.
- Adhere to basic crisis communication processes of delivering timely information.

This has become a classic case study of ineffective crisis communications.

Some decision makers have the attitude that their legal department must review and analyze every message before it is disseminate to minimize any legal liabilities. Unless the legal experts and public relations are at the same table constructing the communiqué and can quickly come to agreement, the message, most likely, will be delayed. It may also be void of information containing instructions the target audience needs to move forward and take the appropriate action. When this occurs, people go to other sources for the information they seek, such as social media and rumors from untrusted sources. The source may be an untrusted source from the organization's perspective yet a trusted source for the audience wanting more information. This approach results in high external failure costs, and impaired goodwill for the organization. External communication failures give rise to hidden costs and damage to an organization or crisis team's image and brand.

Examples of four types of cost of quality concerns are provided in Table 4.1 (Cost of Quality Types).

The need to issue time-sensitive communications at the onset, during, and after a crisis, is the operational foundation of crisis management.[1,2] In summary, the effectives of ineffective crisis communications for an organization can be devastating. Having a crisis communications

TABLE 4.1 Cost of Quality Types

Prevention	Internal Failures
• Crisis communications planning • Preselection of authorized parties • Training/orientation • Exercises • Drills • Supervision of prevention activities • Quality data gathering, analysis, and reporting • Quality improvement plans measured • Cross-functional teams used	• Scrapping the communiqué by approver • Selecting the wrong individual to create or deliver message • Rework due to errors or omissions • Failure to obtain approvals before delivering message • Message written without the target audience in mind • Retesting/approval of reworked products • Taking too long to get the message out • Analysis of the cause of defects in the communiqué
Inspect/Appraise Activities	**External Failures**
• Test and inspection of incoming materials • Test and inspection of in-process communiqués • Final testing and inspection of message • Supervision of testing and inspection activities • Continuous improvement of review and approval processes • Crisis communication plan use • Field testing and appraisal	• Message drafted in a manner that is difficult to understand by the target audience • Delayed message or no message, creating an air of unwanted silence • Vague messages lacking enough factual information to act upon • Using an unauthorized or the wrong person to deliver a message • Target audience taking the wrong action • Damage control due to loss of confidence by the target audience • Damage to organization or incident command reputation/image • Liability arising from defective messaging • Unplanned costs arising from quality problems • People go to other sources for the information they seek, often untrusted sources • Impaired goodwill for the organization

plan is a key resource in the toolkit an organization use. The failure of not having a program that is documented, tested, and shared with employees could lead to

- Liability for executive management and the board of directors due to the lack of adequate protection and care of its resources, including its employees and the community

- Random and conflicting decision making by various managers who have good intent yet are misguided. This causes delays or disseminating errors, or omitting needed information
- Loss of life or serious injuries due to slow delivery of instructions or critical life-safety information
- Unfavorable reactions by stakeholders that could lead to loss of business or significant remediation costs
- Incorrect or delay messaging that can result in lost revenue, audits, or penalties
- Negative media reactions based on rumors, innuendo and the absence of fact-based information

COMMUNICATIONS FOR PEOPLE WITH DISABILITY AND OTHERS WITH FUNCTIONAL AND ACCESS NEEDS

Populations with disabilities and others with functional and access needs are subgroups having unique needs that may not be fully addressed using only traditional approaches. These subgroups include those who are

- physically disabled
- mentally disabled
- cognitively impaired
- hearing impaired (deaf and hard of hearing)
- visually impaired (blind, unable to see at night or different colors or shapes), and mobility challenged

Other subgroups that may qualify as having access needs, particularly in terms of crisis communications, include those who are

- non-English speaking (or not fluent in English)
- culturally or geographically isolated
- homeless
- medically or chemically dependent
- children
- elderly

Each subgroup, with few exceptions, has the same right to services and information as the general population.

In 2010, the Center for Personal Assistance Services reported that 38,487,000 Americans, approximately 13.5% of the U.S. population, have one or more disabilities.[3] FEMA in 2010 stated that 12% of the population, age 65 and above, has a "self-care disability" and 18.9% have a "go-outside disability." These percentages in either category should not be added together since people with a "self-care disability" are often a subset of the "go-outside disability" subgroup.

Many laws mandate the integration and equal opportunity for people with disabilities including:

- The Stafford Act and Post-Katrina Emergency Management Reform Act (PKEMRA)
- The Americans with Disabilities Act of 1990 (ADA)
- Rehabilitation Act of 1973 (RA)
- Federal civil rights laws
- State counterparts to the laws mentioned

People with disabilities must be given information that is comparable in content and detail to that given to the general public. It must also be accessible, understandable, and timely. This includes the need for auxiliary aids and services to ensure effective communication. These resources may include pen and paper, sign-language interpreters via on-site or video, and interpretation aids for people who are deaf, deaf-blind, hard of hearing, or have speech impairments. People who are blind, deaf-blind, or have low vision or cognitive disabilities may need large print information or people to assist with reading and completing forms.

Meeting the needs of children and adults with disabilities and others with functional and access needs is a key component of any crisis communication and mass notification plan. This segment of our population is often overlooked. During a disaster, these populations are more susceptible to poor outcomes if their unique needs have not been previously considered. Technology continues to advance. We are gaining a better understanding the nature of their communication requirements and the technologies available to service their communication needs in a crisis. Any crisis communication plan should be inclusive of these population subgroups. The unique needs of each subgroup should be considered individually when developing communication strategies.

A short list of communication devices using methods that can deliver timely messages that can be understood are

- Hearing aids
- Text Telephone (TTY)/Telecommunications Device for the Deaf (TDD) phones
- Cap Tel phones (for captioning)
- Computer-assisted real-time translation
- Hearing aid batteries of different sizes (including batteries for cochlear implants)
- Synthesizers used with PCs for text-to-speech
- Screen readers

- Screen magnification programs
- Scanning systems for low-vision users[4]

Where possible, emergency managers should already have contracts or memorandums of agreement in place with the vendors or agencies that can provide the equipment and services needed if activated. Any agreement should include testing to ensure that resources have a good understanding of what has been agreed to and whether it will meet the core requirements of the population to be serviced.

Primary Groups

From a crisis communication perspective, communication disorders introduce their own set of challenges. Communication disorders are estimated to affect between 5% to 10% of the U.S. population and cost up to $185 billion per year, almost equal to the gross national product of 2.5% to 3% in 2008. Data suggest from numerous studies that people with communication disorders are more likely to be more economically disadvantaged than those with less severe disabilities. People with severe speech disabilities are more often found to be unemployed or in a lower economic class than people with hearing loss or other disabilities.[5]

Speech Disabilities

More than 2.5 million Americans experience speech disabilities making it difficult to be understood by others. Many are unable to speak at all, and others have physical disabilities that prevent expressive communication techniques such as using hand signs, writing, or typing. According to Robert J. Ruben, M.D., people with severe speech disabilities are more often found to be unemployed or in a lower economic class than people with hearing loss or other disabilities.[5] Reaching this target audience early in the crisis communications process is critical. This population typically needs more time to prepare.

For the speech impaired, presenting information slowly, using simple language, and speaking in short sentences, are ideal. Providing a qualified sign language or oral interpreter, augmentative communication devices, posting messages in a central location, or having notepads, pens, and pencils available are also acceptable strategies. Mechanical devices such as the *electronic larynx*, a speech aid that now offers full digital control of volume, tone, and pitch, gives the user an increased variety of speaking voices.

The field of study dedicated to addressing the communication and related needs of individuals with significant impairments that restricts communications is called *augmentative and alternative communication*

(AAC). The goal of AAC is to assist individuals in adopting and using the most effective communication methods for their unique circumstances by applying research, providing education, and offering clinical service delivery. AAC involves supplementing or replacing natural speech and writing using aided or unaided symbols and an associated transmission device.

Mentally Disabled, Developmentally Disabled, and the Cognitively Impaired

According to the World Health Organization (WHO), mental and behavioral disorders are estimated to account for 12% of the global burden of disease. Approximately 25% of any population will suffer from neuropsychiatric conditions during their lifetime.[6]

Developmental disability refers to a broad range of conditions that interfere with the ability of an individual to function in everyday activities. Developmental disabilities include spina bifida, autism, cerebral palsy, Prader–Willi syndrome, and mental retardation. Many who are classified as developmental disability are on multiple medications and treatments, have communication challenges, and may require different strategies for communications.

Being *cognitively impaired* (sometimes known as having an "unsound mind") is the inability to perceive all relevant facts related to a situation, condition, and treatment or actions to make an intelligent decision. This is regardless of whether the inability is temporary, has existed for an extended period, or occurs intermittently. Examples of persons who may be considered cognitively impaired or mentally disabled include some individuals with dementia, schizophrenia, delirium, mental retardation, bipolar disorder, and stroke.

Those who are developmentally disabled or cognitively impaired need additional safeguards to protect their welfare, particularly those who may be vulnerable to coercion or undue influence. Individuals with psychiatric, cognitive, or developmental disorders may have a limited capacity to understand the information presented and may not be able to make a reasoned decision about the appropriate action to take. Any crisis communications that are expected to include the cognitively impaired must address how determinations will be made to protect the affected.

There are two broad groups of cognitively impaired or mentally disabled persons who are recognized as either institutionalized or noninstitutionalized. Those who are recognized as *institutionalized* are nonvoluntary or dependent resident in an "institution" such as a hospital or group home. *Noninstitutionalized* persons are individuals who are free living. Whether institutionalized or noninstitutionalized, these persons may not be competent to give informed consent. When unable to give informed consent, the disabled may have a *legally authorized representative* (LAR), an individual, judicial, or other body authorized by law to consent on behalf of the disabled. In many cases, it is important to know who the LAR is for the disabled within the community you service. A starting point may be a state or local agency responsible for disabled persons.

A service offered across the United States is *2-1-1*. Calling 2-1-1 is free and provides confidential information and referral for help. Services include assistance with

- food
- housing
- employment
- health care
- low-income tax preparation
- counseling
- general relief services from a comprehensive database of social services

The 2-1-1 services are available 24 hours a day, seven days a week, in many languages. This number also serves as a primary point of contact during times of disaster. Disaster assistance includes registry in a database before a disaster occurs for travel, sheltering, and other special needs. For example, in Texas, registry in the 2-1-1 database helps emergency management teams ensure they have transportation addressed for those who are without their own transportation.

Transportation services include free pick-up and drop-off at a shelter before a hurricane strikes their area. Transportation planning takes into consideration whether the individual will be traveling with a pet, or require evacuation by air, ambulance, or bus. The 2-1-1 service also ensures that the receiving shelter is prepared, and their planning includes the return of the evacuee to the original residence.

Telecommunications Relay Services (TRS) is a telephone service that enables a person having hearing or speech disabilities to place and receive local and long distance telephone calls at no cost. Just as you can call 9-1-1 for an emergency and 4-1-1 for information, you can dial 7-1-1 to connect to certain forms of TRS anywhere in the United States.

TRS uses a communications assistant (CA) and an operator. A CA is a live agent who relays messages in real-time between individuals with hearing and speech disabilities and other persons. The party can initiate the call with the appropriate devices or through 7-1-1. The CA places an outbound traditional voice call to the receiver of the message. Once a call is initiated, the CA serves as a link for the call, relaying the text of the calling party in voice to the called party, and converting to text what the called party voices back to the calling party.

There are several forms of TRS:

- *Captioned Telephone Service*—Uses a text screen to display captions of what the other party to the conversation is saying. A captioned telephone allows the user on one line to speak to the called party, and to simultaneously listen to the other party and read captions of what the other party is saying.
- *Hearing Carry-Over (HCO)*—For individuals who have limited or no speech capability to make or receive telephone calls, HCO relay is available with use of a teletype-writing device for the deaf (TTYs/TTDs). With an HCO relay with a specially designed telephone that has a text display, the HCO user can listen to the voice of the other person on the call. The HCO user can type a response to a CA, who voices the response to the other person on the call as depicted in Figure 4.2 (HCO overview).
- *Internet Protocol (IP) Relay Service*—A text-based form of TRS that uses the Internet for calls between a person with a hearing or speech disability and the CA. Otherwise, the call is generally handled just like a TTY-based TRS call.

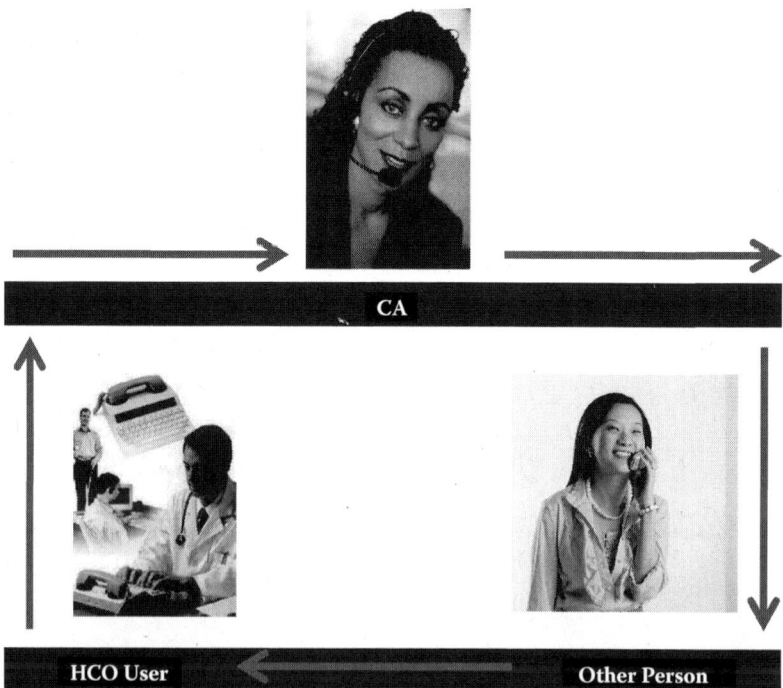

FIGURE 4.2   HCO overview.

- *IP-Captioned Telephone Service*—One of the newer TRS, combines elements of captioned telephone service and IP Relay that does not require any specialized equipment. IP captioned telephone service uses the Internet to provide the link and captions between the caller with a hearing disability and the CA. It allows the user to simultaneously both listen to, and read the text of, what the other party in a telephone conversation is saying.
- *Shared Non-English Language Relay Service*—Spanish and French speakers in the United States have increased substantially in numbers. Spanish-to-Spanish and French-to-French, among other languages can use traditional TRS services that are funded by federal resources.
- *Speech-to-Speech (STS) Relay Service*—This form of TRS is used by a person with a speech disability and does not require any specialized equipment. A CA who is specially trained in understanding a variety of speech disorders repeats what the caller says in a manner that makes the caller's words clear and understandable to the called party.
- *Text-to-Voice TTY-Based TRS*—"Traditional" TRS. A person with a hearing or speech disability uses a special text telephone, called a TTY, to call the CA at the relay center. TTYs have a keyboard and allow people to type their telephone conversations. The text is read on a display screen and/or a paper printout. A TTY user calls a TRS relay center and types the number of the person he or she wishes to call. The CA at the relay center then makes a voice telephone call to the other party to the call, and relays the call back and forth between the parties by speaking what a text user types, and typing what a voice telephone user speaks.
- *Video Relay Service (VRS)*—This Internet-based form of TRS allows persons whose primary language is American Sign Language (ASL) to communicate with the CA in ASL using video conferencing equipment. The CA speaks what is signed to the called party, and signs the called party's response back to the caller. VRS allows conversations to flow in near real time and in a faster and more natural manner than text-based TRS.
- *Voice Carry-Over (VCO)*—Like HCO, VCO relay requires the use of TTY and a telephone with a text display. VCO is available for individuals with hearing loss significant enough to prevent them from hearing and understanding conversations over the telephone yet can speak clearly. A VCO user speaks directly to the other person on the call, with a CA typing what the other caller speaks for the VCO user to read.[7]

Communication devices available for use include a single message communication device such as the Advanced Multimedia Device, Inc. (AMDi) Partner Plus. As skills are acquired, more can be added. The device starts by using picture cards and objects, and enables a person to record and then have the user play back a single message using a light-touch membrane panel. Numerous low-tech tools are available that are picture exchange communication systems (PECS) with recording abilities such as the Logan ProxTalker. The ProxTalker also has a voice-output.

Other assistive technology devices that are focused on communications using picture communication symbol (PCS) software are the Dynavox Boardmaker family of products. They offer individuals with significant speech, language, and learning disabilities the Intellitools/ IntelliKey. These are programmable alternative keyboards enabling users with physical, visual, or cognitive disabilities to easily type, enter numbers, and navigate on-screen. Any of these tools can be used with prescripted messages and associated processes as needed.

The population able to live independent lives at home is growing. Another resource commonly used by this segment of our population is *service animals*. A service animal is any guide dog, signal dog, or other animal trained to provide assistance to an individual with a disability. The service animal can perform specific activities as trained such as

- informing the owner when they are past due in taking medication
- signaling a change in a medical or mental condition
- aiding in stress reduction caused by certain environments

It is quite important for an organization to understand the unique communication needs for the mentally or developmentally disabled and the cognitively impaired. It is important to provide decision makers and those responsible for crisis communications with training to enhance their understanding of mental disabilities and their symptoms. This process will help to develop basic crisis communications skills to use in creating messages.

General Guidance
- Provide identification
- Project a calm and competent demeanor
- Allow extra time for the individual to process the information that has been shared
- Show respect
- Be empathetic and reassuring
- Provide accurate information using short, simple and concise sentences

- State one instruction at a time—some may only understand one word at a time
- Speak slowly using your normal speaking voice
- Practice evacuation and shelter-in-place drills for large buildings

Shaping the Message

- Many who are mentally or developmentally disabled, or cognitively impaired trust figures of authority (e.g., police or doctor), and will go to them for assistance.
- Use familiar emergency experiences, such as a fire drill
- Use symbols or pictures such as siren
- Enact instructions, such as showing people leaving home or going to school
- Be repetitive—repeat the same concept using different words
- Be patient.

Media

- Use
  - 2-1-1
  - Community social services and faith-based organizations
  - Group homes and institutions that provide shelter for this subgroup
  - Government support agencies, such as centers for the disabled
- For the speech impaired use
  - Sign-language interpreters
  - Teletype-writing Device for the Deaf (TTYs/TTDs)
  - Hearing carry-over (HCO)
  - Telecommunications Relay Services (TRS) 7-1-1

Hearing Impaired

Speech is defined as a basic form of communications used between people to share information, convey thoughts and ideas, and maintain social contacts. The *hearing-impaired*, those deaf or hard of hearing, are faced with the challenge of communicating in a hearing-capable world. This barrier can be debilitating in an emergency, leaving the individual with an inability to understand the message or take the appropriate actions. To overcome these obstacles significant advancements in existing technology have occurred.

According to the Hearing Industries Association (HIA), key trends for the U.S. industry for the foreseeable future will be the following:

- The Veterans Affairs (VA) will be a trailblazer in adopting new hearing technologies
- Rapid adoption of wireless aids, loop systems, and peripherals
- A hearing aid tax credit
- Remote programming and tele-audiology improvements

In 2010, the VA was responsible for approximately 18% of all hearing aids dispensed in the United States.

Many returning veterans have suffered hearing loss and there is a rapidly increasing aging veterans population where hearing loss and supportive technologies are needed to maintain basic quality of life.

Since the 1990s, hearing aids have been transformed from large analogue devices to small digital devices. Hearing aids are smaller, can fit invisibly inside the ear, and are more precise with automatic adjustable settings to accommodate most any type of listening environment, reducing background noise and improving sound quality. The exception to this strategy is power aids for those with severe-to-profound hearing losses. Despite these improvements, hearing aid product lines have traditionally used a one-model-fits all strategy.

Today's products are reaching a broader audience with products available at different price levels offering good-better-best product lines, kid-friendly, extended bandwidths, and FM/DAI capabilities. A new generation of "smart" hearing instruments using algorithms with a broad range of capabilities can replicate the sound-shaping features of the pinna. These instruments can automatically select the best possible listening settings based on the auditory environment. Many hearing aids have wideband with frequency ranges of 8 to 10 kHz using speech cues and hi-fi sound for both children and adults. With wireless technologies, the devices can be linked with cell phones, MP3 players, and other hearing aids.

Television viewing can be enjoyed by the deaf and hearing impaired through the aid of *captions*. Captions are on-screen descriptions that are synchronized with the video to provide the dialogue, identity of speakers, and describe any relevant sound to the audio in text format. There are two types of captions—*open* and *closed*. *Open captions* are always on and available to all viewers. *Closed caption* is available if selected by the viewer. Like open caption, closed caption use has been expanded for use by those learning to read, learning to speak a non-native language in an environment where audio is muted or difficult to hear, and for those who prefer to read a transcript along with the program audio.

Amplified telephones are available in corded and cordless, analog and digital modes that offer greater clarity and background noise suppression. The advantage of the fully digital amplified cordless phones is that they can adapt to an analog or digital high speed for greater sound quality. Many of the devices include features such as talkback that confirms the buttons that have been pressed and a talking caller ID that announces who is calling. Many of these units also have handsets or hands-free amplified message playback.

Other supportive technologies include smartpens and smartphones that can amplify their audio outputs. The smartpens are able to record audio while writing. The audio and data captured by the smartpen can be replayed simultaneously so that an individual who is hearing impaired can read what was stated. Many smartphones have built-in features and third-party accessories available to aid the hearing-impaired to use these devices for communications. Most have push-delivery technology that allows text-based messages, including Short Message Service (SMS) and Multimedia Messaging Service (MMS) text messages, and instant messaging (IM), and then notify the phone user that the message has arrived.

Notification of messages, e-mails, and inbound phone calls can be given visually, or by extended vibration or audible tones. Smartphones offer open and closed captioned multimedia content that can typically be customized. Most smartphones are hearing-aid compatible, and have sound-isolating headsets that offer clear communications while taking calls and improved sound quality when listening to music or watching videos.

The growing reliance on the Internet to service the hearing impaired is profound and expected to continue. A study was conducted in an audiologic laboratory located in Bern, Switzerland, on 21 adult hearing-impaired patients. The purpose of the test was to determine whether Internet telephony with a broadband transmission (0.1–8 kHz) of speech improves speech perception in comparison to conventional telephony (0.3–3.5 kHz) in hearing-impaired and normal-hearing adults. The conclusion drawn from this study is that Internet telephony provides significantly improved speech perception to hearing impaired with minor modifications.[8]

Despite the advances in technology, many prefer the availability of real-time speech transcription anywhere, anytime. The service represents a potentially life-changing opportunity for the hearing-impaired to improve their communication capability. This solution has its challenges, however. There is a shortage of trained transcribers, and the cost for the services is high compared to the capability of many within this population to pay for the service.

Another technology available for crisis communications are *Automated Speech Recognition systems* (ASRs). ASR is a technology that allows users of information systems to speak their entries rather than punching numbers on a keypad. ASR is used primarily to provide information and to forward telephone calls. Basic ASR systems recognize single-word entries such as yes-or-no responses and spoken numerals. An example of ARS are the systems commonly used by customer service organizations for routing a caller to the appropriate analyst through automation.

The primary benefit of an enhanced ASR system is that the ASR system will shorten the menu navigation process entry or response to a query, such as requesting driving directions or a telephone number for a business. This process is shortened by reducing the number of decision points and the number of instructions a user receives and comprehends. The weakness of this solution is the limited recognition accuracy, particularly for some languages that are difficult to interpret, for colloquial speech, or where compound words are made from two languages.[9] Cellular phones with weak connections can cause the system to misinterpret the input; the systems can also be expensive.

Media
- 2-1-1
- Amplified phones
- Captioning (open/closed) (an FCC requirement)
- Cell phone and text messaging
- Electronic messaging
- Emergency e-mail
- Listening systems
- Public service announcements
- Sign language interpreters
- Telecommunications Relay Services (TRS)
- 7-1-1
- TTY/TTD
- Voice carry-over (VCO)
- Wireless network alerts.
- Use television *crawler* or *ticker*, a line of text that scroll at the bottom of the television screen such as those used for news headlines and severe weather warnings.

When consider the needs of the hearing impaired, the fourth most commonly used language in the United States is American Sign Language (ASL). It is a language that uses signs made with the hands and includes other body movements such as facial expressions and body posture. In an emergency, ASL should be included in any live or videotaped emergency communication. The ASL alphabet is shown in Figure 4.3 (ASL alphabet).

Visually Impaired

More than 10 million Americans are either blind or visually impaired. *Visually impaired* include the blind, and those unable to see at night or to distinguish different colors or shapes. More men than women are blind or visually impaired. Blind and visually impaired individuals face

FIGURE 4.3 ASL alphabet.

a huge language barrier since they are unable to physically read standard newsprint, Internet information, or books. Word of mouth remains the most common and useful method of communication for the blind and visually impaired. Braille continues to be the standard form of reading and writing used by those who are visually impaired in virtually every language around the globe.

General Guidance
- Announce your presence when entering an area
- Speak naturally and directly to the individual
- Do not shout
- Offer assistance but let individuals explain their need
- Communicate written information orally.

Shaping the Message
If visual impairment is limited to reading in large print you may consider

- If captions are used, sentences should be simple.
- Use visual aids such as maps and pictures.

Media
- Use radio and sirens
- Ensure that websites can be accessed using screen readers and screen pointers
- Use electronic and information technology accessible to people with disabilities including those defined by the U.S. Rehabilitation Act, Section 508
- 2-1-1
- Audiocassettes, CDs, DVDs, USB sticks and other devices that can store audio
- Audio-described videos
- Braille
- Cell phone
- Disks
- Electronic messaging
- Large print
- Telecommunications Relay Services (TRS) 7-1-1
- Voice carry-over (VCO)

A number of smart devices are available on the market today for the visually impaired. Many of the features of smartpens and smartphones that are used by the hearing impaired can also be used by the visually impaired such as extended vibrations and audible notification for incoming phone calls, text messages, e-mail, or news alerts. Fonts can be customized so that they appear larger or the style can be changed for easier viewing as specified by the user. The keys, trackball, or trackpad, can have unique event sounds, tones, or audible clicks applied to each so that the user can confirm the key they are pressing or the event that is occurring, such as a low battery or full charge, or if the device is connected to an accessory.

Color is very important for those who may have difficulty viewing displays, but who are not completely blind. Features that can be of value include font colors that can be changed, reverse contrast of the display colors from dark on light to light on dark, a bright LCD display, browser zoom to magnify web content, or the use of a grayscale, which converts all colors to shades of gray. Many phones offer assignable ringtones so that a unique ringtone can be applied to each caller or sender of an e-mail using the Contact List so the user need not view the display to know who is calling or sending a message. Most smartphones allow a user to speak into the phone to place a call using voice dialing by name, phone number, or speed dial shortcut key. Many smartphones have tactilely discernible keyboards and identifiable keyboard "nib" for using the keys without looking at the phone.

Mobility Challenged

Individuals with mobile impairments have some level of dependency on others who cannot always be there. Those who are mobility challenged are those who use mechanical devices to move from one place to another within their residence, workplace, or places they visit. Mechanical devices include wheelchairs, walkers, walking canes, and crutches. It also includes those whose residence may have been modified by installing handrails, widening doors, or installing handicapped-accessible bathrooms and entry ramps. It is important to ensure access to communications is available in the event of an emergency.

There are a number of two-way communication devices that can help, such as an alerting device that can be worn as a pendant around the neck. The user can press a button on the device or call from their telephone. When using the telephone, an agent answers the call asking what services are needed. If a button is pressed on the pendant and the agent calls and does not receive a response, the agent will immediately notify emergency responders for assistance using the same protocol followed by 9-1-1 operators. This service can be extended beyond the home or workplace to enable cross-country communication using wireless communications much like a wireless cell phone.

Other Groups

Non-English Speaking

In much of the Western world, including the United States, English is the dominant or only language spoken and used. This means millions who speak a language in addition to or instead of English with limited access to effective communication resources in a crisis. Many of those who do speak English indicate that their ability is poor or nearly nonexistent since they speak other languages more fluently. Many who do not speak the local language may not respond well to law enforcement or others with whom they have little in common. Some non-English speakers may have come from other countries where different views exist, possibly due to cultural differences, regarding emergency communications, preparedness, and response. This often requires that first responders are dependent upon English-speaking family members to accurately relay the information.

General recommendations suggest that, prior to an event, individuals with foreign language proficiency should be identified to communicate messages to the media and community leaders. In shaping the message, establish an agreement with a translation service and develop scripts to be translated into the primary languages used by the community leaders and public information personnel prior to the event.

To deliver messages to this subgroup post all communiqués electronically, using electronic media outlets and social networking sites that may be frequented by members of this subgroup. Request to have emergency contact information for local community centers, community newspapers and local ethnic media outlets.

Limited English Proficiency (LEP)

Limited English proficiency (LEP) individuals are those whose primary spoken language is not English. An individual's *primary language* is the language in which the individual is most effectively able to communicate. These individuals have a limited ability to read, write, speak, or understand English. Some LEP individuals are in the process of learning English; however, they are not yet proficient. Other LEP individuals have acquired basic English language skills to communicate their name, address, etc., but are unable to communicate detailed information such as medical condition, the status of others around them, etc., in English.

Any emergency communication system should take into consideration the need to take reasonable steps to provide meaningful access to LEP and non-English speaking individuals. Language barriers can put people and property at risk, and can prevent LEP individuals from understanding what is being asked of them or to receive meaningful access to needed services and information.

To develop a plan for addressing the needs of this subgroup, a number of resources are available, such as language access documentation available at www.LEP.gov and the Policy Guidance Document for LEP offered by the U.S. Department of Justice (DOJ) at www.justice.gov. The framework includes a four-factor analysis for LEP individuals incorporating:

- Number or proportion encountered by an organization, including any variations in population
- Frequency of contact
- Nature and importance of the different types of encounters an organization has with LEP and non-English speaking individuals
- Organizational resources available and the costs associated with providing language services

This framework enables an organization to prioritize types of language services available as needed.

Options available when provisioning language services include both oral and written language services. Oral language services include *direct communications* and *interpretation*. Direct communications is monolingual communication in a language other than English between a qualified *bilingual employee* or representation and an LEP individual. Direct communications with LEP individuals using qualified bilingual staff who are fluent in English and the language of the LEP person is most effective. The qualified bilingual staff should have their fluency assessed in both languages and receive ongoing training.

Another oral service option is *interpretation*. Using a qualified interpretation service is important when qualified bilingual employees are not available to communicate with the LEP person in their native language. Consider using both contracted in-person and telephonic interpreters. It is important to note that interpreters need ongoing training in interpreting using various modes of interpretation, for example, simultaneous, consecutive, and sight. In an emergency, family members, neighbors, friends, bystanders, and children can be used for interpretation, although this is the least desirable approach. Using children should be considered the last resort. Sensitive or negative information may be traumatic to the child and have long lasting unintended consequences. For example, having a child tell her non-English speaking father he has cancer. When communicating information regarding victims, investigation, negotiations or other sensitive situation an uninterested third party is highly recommended, such as a volunteer or paid contractor.

Whether to use an in-person or telephonic interpreter depends upon the circumstances. If interaction with an LEP individual will be lengthy or could result in significant potential consequences, then in-person interpretation is recommended. Telephonic interpreters are more appropriate for brief encounters and where a qualified in-person interpreter is unavailable.

A directory with contact information should be maintained of all qualified bilingual employees and interpreters. The directory should contain all non-English languages spoken or written, and contact information including any radio equipment.

Written communications for frequently encountered LEP groups should be *translated* into the language(s) spoken within this group. These communications should take into consideration

- the time needed for translation
- when to access these translations
- how to request addition translations

Quality control measures should be implemented for all written communications to ensure the message has been translated with the proper tone and meaning in the LEP's language. Retain this service throughout an event to ensure that updates are communicated timely and accurately.

Wherever possible, 9-1-1 communications with LEP individuals should be conducted in the language of the LEP caller. Bilingual staff should be prepared to assist upon short notice and backup resources should be placed on notice. Dispatch operators taking calls should learn to say "Please hold" in the more commonly used non-English languages within the jurisdiction to locate bilingual resources. As a backup to the in-person interpreter, a detail contact list of telephonic interpreters should be readily accessible by the dispatch operation. A 9-1-1 call center should track calls made by LEP individuals, recording the date and time of the call and the language spoken by the LEP caller.[10]

Culturally or Geographically Isolated

Culture is defined as a full range of human behaviors related to values, beliefs, and practices shared by a group of people. The attributes of a distinct and shared culture can be found in small groups of people, a region, nation, or even an organization. Each person belongs to multiple cultures, for example, a Canadian who speaks French as the primary language, lives in a rural area of northern Canada, believes in conserving energy, and values saving all earnings. This is an example of an individual belonging to multiple cultures and geographically isolated.

An individual having an opportunity to come to Mississippi from Sudan to experience American culture and values will still be culturally more Sudanese and may have difficulty adapting, possibly due to language barriers and, say, lack of familiarity with the norms of agricultural life in the State of Mississippi. If he is the only one from his family in the United States and there are no Sudanese in his area, fitting into social situations could be difficult for him. This is an example of an individual who is culturally isolated.

In the world today, thanks to applications like Google Earth, there are few inhabited areas that have not been discovered. Like those who live a nomadic existence moving around tent sites at U.S. state parks, and those who go to great lengths to avoid modern civilization. There are also individuals who have modern technology in their homes, but they live miles away from their closest neighbor. There are people who spend their free time hiking in mountains far from the trappings of modern civilization. These are examples of individuals who are geographically isolated.

Cultural differences and geographical and cultural isolation affect communication styles and the methods used in a crisis. Often, persons who are culturally or geographically isolated also have limited use of technology. For example, cellular phone service may not reach into high mountainous terrain. Do these individual use a landline or satellite phone service? Power outages may last longer where fewer people live since restoration usually begins where the greatest concentration of resources and people reside. Do they have adequate backup power to service their televisions, radios, telephones? Do they use any of these? To reach those who are isolated often requires individuals trusted by the community to knock on doors. Another approach is having an intermediary to relay a message to a subgroup with limited access due to cultural differences.

The Incarcerated Those who are in jail are members of this subgroup. Inmates have limited legal capability to make major decisions for themselves, and are dependent upon the jail administrators for their day-to-day care before, during, and after an event. During Katrina, there was not a plan in place to address the incarcerated at the time of the crisis. As a result one of two solutions was implemented. There were those who were released into the general population and left to manage their own needs, and using the honor system, were expected to return to the jail on their own. Quite naturally, many never returned nor have been found. Others commited new crimes in other communities. A few were not released and were trapped in their cells with no resources. A lesson learned by communities across the United States is to have a plan in place for this subgroup that includes the special transportation and shelter needs throughout the process. Communications are often restricted to that offered by jail administrations to the inmates and possibly that which is seen on television or shared among the inmates themselves.

Homeless

Those without a fixed, regular, and adequate nighttime residence, and those living in emergency or temporary living arrangements are classified as homeless. The homeless population includes those who live on the streets, camp in the woods, and sleep in their vehicles. Many suffer with physical and developmental conditions that are disabling, with mental illness, or substance abuse problems. This subgroup creates unique communication challenges in a crisis when workers are attempting to notify them of danger and need to direct them to a safe location. Like the elderly and children, communication technologies that can be used for other vulnerable populations also apply to this group.

For those who reside in shelters and for migrant populations, responding to a message by taking the appropriate action is not always possible. Many are without access to transportation, and are transient since their location may change frequently, and some may have privacy concerns and fear of discrimination. The migrant subgroup, in particular, may have limited English proficiency and work closely with language and cultural-media outlets. Within a single shelter, there may be multiple cultural and socioeconomic backgrounds coexisting.

In preparing a message, the cultural aspect of the target audience should be considered. The homeless have healing and coping concerns that may include a loss, grief, or death, or are related to seeking help that the general population does not face. Those in the shelter may require translation into the language used by the receiver rather than the primary language used by the general population.

The communication medium used should be predetermined through the planning process. Establishing relationships with community outreach groups and shelter administration workers to help inform and prepare this subgroup for a crisis before it occurs needs to cover both emergency evacuation and shelter-in-place procedures. Involve cultural leaders and brokers in the planning process that promotes the development of formal and informal communication networks. Additional mediums include

- Drafting approved messages or securing a translation service to recreate the message using the predominant languages of the subgroup prior to an event.
- Identify staff with foreign language proficiency to communicate messages to small groups via the media and community leaders.

Medically or Chemically Dependent

Individuals who are medically or chemically dependent are a protected class. They may be unable to make sound decisions or, by law, make major decisions for themselves while in this condition. This subgroup includes individuals in specialty care facilities such as hospitals, oncology clinic, and methadone clinics, and those who are under the influence of a chemical such as alcohol or illegal drugs—even those who were under anesthesia less than 72 hours prior to the emergency or whose judgment is impaired while recovering from surgery.

This subgroup is fractured, having multiple needs and representing many difference socioeconomic groups. The mediums used to communicate with this segment of the population must be broad enough. Development of emergency communicates should take into consideration the behavioral health component of the message so that the stress that may be created by the situation does not impair the health or recovery process.

The primary communication mediums to use in a crisis are electronically posted messages that are tagged for screen readers on electronic media outlets and social networking sites that can be accessed by the general population. It is wise to convey messages through clinical staff and local support agencies whenever possible and other trusted sources that have been identified.

Children

Children are classified as a vulnerable population because they are dependent upon others to provide an accurate interpretation of a communiqué and to provide for their needs, whether in an emergency or for everyday necessities. This is a segment of the population that is often overlooked.

Although many children can read and comprehend a message, they often lack the resources or the authority to take action independently. For example, a school can issue an immediate evacuation order and ask parents to come pick up their child. A parent may ask a neighbor to pick up the child if she is unable but unless prior arrangements have been made, the school cannot release a child to the neighbor. This presents unique challenges in an emergency for a school, day care, or other venues where children may be found without a parent or guardian. The adult responsibility is transferred to school administrators, faculty, and employees. It is important to know what agencies are used when a parent or guardian is not available in an emergency, for example, law enforcement or government sponsored social service agencies.

There are many technologies today that target the needs of children, including those who are disabled, to independently communicate such as viewing and interacting on electronic tablets and pads. However, electronic and text messages can cause panic and confusion.

Elderly

The elderly are those who are 65 years of age or older, a growing segment of the world's population. In the United States, baby boomers have begun to enter this age range. Many live independently in naturally retiring communities (communities 30 or more years old) and without the need of any community or specialized services. Many who are mobile will claim short-term residency in different areas of the country for a variety of reasons. There are *snowbird* (seasonal resident) populations that will move from colder climates to warmer drier climates during the winter months. Others will move from their city of longtime residence to areas where the cost of living is more affordable or access to desired services is closer. Yet others may move away from populated urban areas to less populated suburban or rural areas for a more relaxed environment.

Those of particular concern are those 85 years of age and older who continue to live alone or occupy their own personal residences and those who require aids for hearing, vision, and mobility. In shaping the message for the elderly population, use of logical sequencing when creating messages is helpful. Logical sequencing focuses on the flow of the message from beginning to end on a single topic before going to another topic.

Mediums:
- Call and request or register for 2-1-1 services before an event and update your information as it changes.
- Repeat written messages in broadcast format.
- Tag for screen readers electronically posted messages, including those posted on websites.
- Use electronic media outlets and social networking sites that reach the general and targeted population.
- Use large print (14 points or larger) for print messages.

COMMUNICATIONS METHODS FOR POPULATIONS WITH DISABILITIES AND OTHERS WITH FUNCTIONAL AND ACCESS NEEDS

The changing realities of communications must consider the essential services needed by the portion of the population with disabilities and others with functional and access needs. This segment of our population represents those who are physically or mentally disabled, medically or chemically dependent, elderly, children, homeless, non-English speaking, and LEP. Recent disasters provide data on the overlooked needs recorded for Hurricane Katrina and Rita or earthquake survivors within this population. Some who survived the hurricane later died because their needs had not been addressed by the traditional response agencies that were often ill-equipped to respond to these needs. Emergency communications needs must become a part of the basic response.

The challenges to effectively communicate with all affected by a crisis goes beyond reaching the general population able to speak the dominant language without special aid. Effective crisis communications takes into consideration the needs of people with disabilities and others with functional and access needs affected by the crisis. For example, a one-half page message to warn people to evacuate is ineffective if text messaging or television tickers are used. An extended list of solutions is provided in Table 4.2 (Summary of Communication Mediums for Populations with Functional and Access Needs). Understanding the following Five Ws and one H and applying them to any crisis facilitates helping those in need:

- *Who* are the target audiences?
- *What* is the message you want to convey?
- *How* will you deliver the message?
- *Why* it is important that the receiver stops to understand the message?
- *Where* people will be to receive the message?
- *When* they will receive the message?

Key points for effective communications that is inclusive of people with disabilities and others with functional and access needs are as follows:

- Use culturally appropriate messages and the appropriate format in developing a trusted message.
- Use trusted community representatives, natural leaders, and service agencies who are able to reach subgroups and communicate the "what" or "why." Agencies include senior centers, faith-based, or cultural organizations.
- Collaborate and coordinate efforts to provide training to emergency preparedness planners and service delivery personnel in both the public and private sector, and discuss how to best reach targeted audiences
- Create messages using a fourth- to sixth-grade reading level. Check the readability level, using tools such as the Microsoft Word proofing function that provides a readability score as a part of spelling and grammar checking.
- Use actionable information, such as "Check our website" or "Go to the parking lot," etc.
- Use universal symbols and pictures to convey instructions and other information.
- Identify resources before an event that are consistent with the needs of a group.
- Secure a translation service before an event to aid in crafting messages for use with technology aides for people with disabilities and others with functional and access needs, such as a CA for VCO and HCO or conversion to Braille.
- Identify resources before an event that can translate messages in multiple languages that are culturally appropriate.
- Structure posted electronic messages so that they can be accessed using screen readers.

SUMMARY

Consideration for population subgroups is no longer an option but a mandate from the citizens of developed countries. Mapping a strategy that is inclusive of the many and various subgroups is a requirement of an effective communication plan. This is the changing reality of emergency management, crisis communications, and mass notification planning. Effective communications must be timely, accurate and inclusive.

Given the potential for a future catastrophic disaster is high anywhere around the globe, it is imperative that the new norm includes protocols and plans for delivering emergency communications services to people

TABLE 4.2 Summary of Communication Mediums for Populations with Functional and Access Needs

Media	Audience											
	Physically Disabled (Including Speech Impaired)	Mentally Disabled	Cognitively Impaired	Hearing Impaired	Visually Impaired	Mobility Challenged	Non-English Speaking	Culturally/Geographically Isolated	Homeless	Medically or Chemically Dependent	Children	Elderly
2-1-1	X	X	X	X	X	X	X	X	X	X	X	X
Amplified phones	X	X	X	X		X						
Audiocassettes, CDs, DVDs, and other devices that can store audio	X	X	X	X		X	X					
Audio-described videos		X	X		X		X				X	X
Braille					X							
Captioning (open/closed)	X	X	X	X								
Cell/satellite phone	X	X	X		X	X	X	X	X		X	X
Electronic media outlets/social networking sites	X			X	X	X	X	X				X

Challenges to Effective Communications

Method										
Electronic messaging	X		X	X	X	X	X			X
Emergency e-mail					X					
Hearing carry-over (HCO)	X				X					
Large print	X		X							
Listening systems	X					X			X	
Print messages, use large print (14 points or larger)	X		X		X					
Public service announcements	X		X	X	X	X				
Radio					X					
Sign language interpreters						X				
Siren	X		X			X		X		X
Telecommunications relay services (TRS) 711	X			X	X		X	X		
Television/video display "crawler" or "ticker"	X		X	X		X	X	X		X
Text messaging (SMS)	X		X		X	X		X		
TTY/TTD	X		X		X					
Voice carry-over (VCO)			X							
Website that can be accessed using screen readers and screen pointers	X		X		X	X	X		X	X
Wireless network alerts	X					X				X
Written preparedness material	X		X		X	X	X	X	X	X

with accessibility, cultural, and language needs. The challenge of emergency management professionals is to integrate the skills and knowledge of community resources into emergency services plans and strategies to connect this underserved population with local government. This connection enhances the response and recovery efforts for vulnerable populations.

ENDNOTES

1. FEMA. (2002). Summer Storms. Communications Challenges. http://www.training.fema.gov/EMIWeb/downloads/is111_Unit%205.pdf (accessed February 10, 2011).
2. Hamilton, Dennis. (March 2010). In-crisis decision making: Communicate or expect the worst. http://www.continuitycentral.com/feature0757.html (accessed July 7, 2011).
3. University of California, San Francisco Center for Personal Assistance Services, *Disability data for US from the 2009 American Community Survey* (2010). http://www.pascenter.org/state_based_stats/disability_stats/acs_prevalence.php?state=us (accessed April 29, 2011).
4. FEMA. (2010). *Guidance on planning for integration of functional needs support services in general population shelters.* Pg 20. Washington, DC.
5. Ruben, Robert J. (2009). Defining the survival of the fittest: Communication disorders in the 21st century. *The Laryngoscope*, 110(2): 241.
6. World Health Organization. (2001). *WHO: The world health report 2001—Mental health: New understanding, new hope. 2001.* http://www.who.int/whr/2001/chapter1/en/index.html. (accessed February 3, 2011).
7. FCC. (2010). *FCC consumer facts—Telecommunications relay services.* http://www.fcc.gov/cgb/consumerfacts/trs.html (Accessed February 5, 2011).
8. Mantokoudis, G., Kompis, M., Dubach, P., and Caversaccio, M. (2010). How Internet telephony could improve communication for hearing-impaired individuals, *Otology and Neurotology*, 31(7): 1014–1021.
9. Gupta, N. K., Dantu, R., Schulzrinne, H., Goulart, A., and Magnussen, W. (2010). Next Generation 9-1-1: Architecture and challenges in realizing an IP-multimedia-based emergency service, *Journal of Homeland Security and Emergency Management*, 7(1). http://www.bepress.com/cgi/viewcontent.cgi?context=jhsem&article=1657&date=&mt=MTI5Njk3MDE5NQ==&access_ok_form=Continue (accessed February 6, 2011).

10. U.S. Department of Justice. (2005). *Sample for discussion purposes planning tool: Considerations for creation of a language assistance police and implementation plan for addressing limited English proficiency in law enforcement agency.* http://www.lep.gov/resources/Law_Enforcement_Planning_Tool.htm. (accessed February 10, 2011).

CHAPTER 5

Changing Realities in Emergency Communications

> "The World Bank says food prices are at 'dangerous levels' and have pushed 44 million more people into poverty since last June ...." "President Mahmoud Ahmadinejad and the opposition protests seen in Iranian cities on Monday ...." "Security forces in the Yemeni capital, Sanaa, have used tear gas and batons to disperse thousands of anti-government protesters calling for President Ali Abdullah Saleh to stand down ...." "The cholera outbreak that has killed 3,600 people in Haiti since October 2010 has not been suppressed ...."
>
> BBC headlines, *February 15, 2011, 6:00 p.m. CT*

The next emergency is out there. There will be small isolated incidents such as a house fire to catastrophic disasters that affect multiple countries. If history is any indicator of the future, the future will present situations this generation has never experienced. The trend is toward more frequent and more expensive events occurring due to the socioeconomic and climatic changes occurring around the globe. Over the last 60 years, since 1950 there have only been 2 years—1952 and 1958—where there were no recorded natural catastrophes.[1] At the same time, the world is trending toward mega-cities. This translates into more people living and working in smaller spaces closer together. This same population is increasingly dependent upon technology and power to facilitate the fast pace of living.

Ten years ago, only a few in the United States could have envisioned a hurricane like Katrina that would shut down a large metropolitan area, leaving most of its residents with uninhabitable homes and hospitals. The 2000 census reported 484,674 residents in New Orleans, Louisiana. A year after Katrina, the population had declined nearly 54%

to 223,388.[2] It is estimated that New Orleans will remain at nearly one-half of its pre-Katrina population due to the long-term impact Katrina has had on the region. Neither the government at any level nor the public were prepared for the long-term recovery needs of the region that continue today. Hurricane Katrina remains the most expensive natural catastrophe to date in terms of damage and losses.[3]

The *reinsurance business* is about insurance companies selling insurance to other insurance companies. They continuously scan for *geo-intelligence*. Geo-intelligence is used to develop precise estimates of the threats posed by hazards and their ramifications. Long-term survival of these companies and the organizations they support is dependent upon knowing the calculated costs of current and future risks.

In 2009, there were 860 recorded catastrophic events, 110 more than were reported in 2008 and 90 more than the 10-year average. The loss was $50 billion, with 17 events exceeding $1 billion.[4] Of this, the total losses the insurance industry incurred were $22 billion with the balance of these losses incurred by government and businesses. Some cost recovery was received through public and private donations. Earthquakes causing extremely high economic losses have been three of the four most expensive catastrophes since 1950. Earthquakes are geophysical by nature and often trigger multiple events, such as tsunamis, major building and infrastructure damages, fatalities and injuries.

2009 WORLDWIDE CATASTROPHES STATISTICS

- Natural catastrophes
  - Type:
    - 93% were caused by atmospheric conditions
    - 7% earthquakes and volcanic eruptions
  - Where:
    - 35% (300) occurred on the American continent
    - 34% occurred on the Asian continent
    - 15% appeared in Europe
    - 8% (70) each occurred in Africa and Australia
  - Fatalities:
    - 11,000 fatalities in 2009 due to natural catastrophes
      - Far below the 30-year average of 57,000 per year
      - 1,200 fatalities were caused by the Indonesian Sumatra earthquake (deadliest event)
      - 2,000 fatalities happened in Asia from a series of severe typhoons
      - 500 fatalities were recorded in Australia due to wildlands fires and extreme heat waves[5]

TABLE 5.1 Deadliest Events Recorded

Year	Event	Country	Fatalities
1970	Tropical cyclone and floods	Bangladesh	300,000
1976	Earthquake	China	242,000
2010	Earthquake	Haiti	225,000
2004	Earthquake Tsunami	Indonesia, Sri Lanka Thailand, India	220,000
1991	Tropical cyclone	Bangladesh	139,000
2005	Earthquake	Pakistan, India, Afghanistan	88,000

TABLE 5.2 Costliest Events for the Overall Economy and the Insurance Industry

Year	Event	Country	Overall Economy Losses ($ m)*	Insurance Industry Insured Losses ($ m)*
2005	Hurricane Katrina	USA	125,000	62,000
1995	Earthquake	Japan	100,000	
2008	Earthquake	China	85,000	
1994	Earthquake	USA	44,000	
2008	Hurricane Ike	USA, Caribbean	38,000	18,500
1992	Hurricane Andrew	USA, Bahamas		17,000
1994	Earthquake	USA		15,300
2004	Hurricane Ivan	USA, Caribbean		13,800
2011	Earthquake	Japan	235,000**	TBD

Note: * Denotes the original values.
** Unofficial estimated.

The deadliest events recorded are captured in Table 5.1 (Deadliest Events Recorded). These data are followed by the costliest events for the overall economy and the insurance industry in Table 5.2 (Costliest Events for the Overall Economy and the Insurance Industry).

Munich Re, Arch Re, Kuwait Re, and Swiss Re, to name a few, are reinsurance companies—insurance companies that provide insurance for other insurance companies. *Reinsurance* is an approach used by a primary insurer to protect against unforeseen or extraordinary losses, to increase individual insurers' capacity, to share liability when losses overwhelm the primary insurer's resources, and to help insurers stabilize their businesses in the face of the wide swings in profit and losses. Munich Re, a reinsurer, gathers data, calculates, and publishes the *National Hazard Total Risk Index*. The National Hazard Risk Index is

TABLE 5.3 Formula for GDP

Formula for GDP
GDP = Consumer Spending + Investment + Government Spending + (Exports–Imports Value)

a review of the top 50 mega-city trends and the hazards each are exposed to, subsequent risk, and cost-to-value ratio. Their data indicates that the trend of risks and losses from climate-related natural hazards are rising, averaging $100 billion per annum in the last decade alone.[6]

The uninsured losses that Munich Re has covered are growing. By 2050, the disaster losses are predicted to exceed the world's gross domestic product (GDP). *GDP* is the total market value of all goods and services produced in a country in a given year, equal to total consumer, investment and government spending, plus the value of exports, minus the value of imports as defined in Table 5.3 (Formula for GDP).

In 1900, approximately 2% of world's population lived in urban areas and cities; today 50% now live in urban areas and cities. Urbanization is growing. Global population growth affects the local population. Immigration, jobs transfer, restless youth pressures, and *echo immigrants* (those displaced due to lack of resources in their local area) are relocating to urban areas to live and have the basics and geopolitical influence.

Munich Re has stated that peak oil supplies were achieved in 1970; concurrently an oil crisis occurred at the same time. By 2010, peak oil discoveries had been reached with production forecasted to peak in a few years given current knowledge and technology. Energy consumption continues to increase as developing and developed countries' needs increase. Validation of this hypothesis will occur as statistical data indicates a slowdown in production. As production slows and alternative fuels are unable to close the gap, this will cause an increasing number of issues. Today, it is a business and political issue but over time it may become an emergency management issue. There will be countries competing for diminishing output, and pressure will build.

Another growing concern is that the economies of developed countries are facing huge budget woes that are reflective of the increased cost to maintain the base. State and local communities have faced bankruptcy such as the State of California and one of its cities, San Diego. California is looking for new ways to generate revenue include legalizing some medical marijuana initiatives to help pay for law enforcement. The federal government is exploring new taxes on motorists—a tax based on the number of miles driven rather than gasoline taxes. The U.S. government is having to deal with budget deficits and rapid changes in the marketplace.

Portugal and possibly Spain may seek financial aid after Greece and Ireland requested bailouts. These requests have negatively affected Europe's monetary unit, the euro. Investors are asking whether these economies are predictors of what may be coming for other European countries.

DISASTER TYPES PREDICTED OVER THE NEXT 25 YEARS

Earthquakes

A frightening and destructive phenomenon of nature is a severe earthquake and its terrible aftereffects. An earthquake is a sudden movement of the tectonic plates of the earth, caused by the sudden release of strain that has accumulated over a long time. If the earthquake occurs in a populated area, it may cause many deaths, injuries, extensive property damage, and economic losses. According to the U.S. Geological Survey, Alaska is the most earthquake-prone state in the United States and one of the most seismically active regions in the world. Alaska experiences a magnitude 7 earthquake almost every year, and a magnitude 8 or greater earthquake on an average of every 14 years.

A major portion of the world's earthquakes are around the rim of the Pacific Ocean ("Ring of Fire"). This area is referred to by seismologists as the circum-Pacific belt, and is the most probable location for an earthquake. An earthquake can occur anywhere since no region is entirely free of earthquakes. The largest recorded earthquake in the world was a magnitude 9.5 in Chile on May 22, 1960. Earthquakes are one of the most costly natural hazards faced in the United States, posing a significant risk to 75 million Americans in 39 states. Earthquakes can trigger secondary events; building fires and utility and transportation disruptions, tsunamis, landslides, and volcanic eruptions.

On March 11, 2011, Tōhoku earthquake, officially named the Great East Japan Earthquake, was an undersea mega-thrust earthquake off the coast of Japan. It was a magnitude 9.0, making it one of the five most powerful recorded on earth to date. The earthquake triggered destructive tsunami waves of nearly 38 meters (125 ft.) and caused a number of major nuclear accidents at the Fukushima I Nuclear Power Plant. The cost is expected to exceed $235 billion making it one of the costliest disasters ever recorded.[7]

Neither the U.S. Geological Survey (USGS) nor any other scientists have ever predicted a major earthquake. They do not know how, and they do not expect to know how anytime in the near future. The probabilities can be calculated for potential future earthquakes using scientific data. The exact day, time, or location remains a mystery. The odds of an earthquake striking Northern California increases with time. In April 2008 USGS compiled a report on Bay Area earthquake probabilities. In this report it states it is predicted that there is a 50% chance of a magnitude 7.0 or greater earthquake on the San Francisco segment of the San Andreas Fault in the next 45 years, and a 75% chance during the next 80 years. Such an earthquake would lead to 5,800 deaths according to the USGS. In Southern California along the same fault, there is an 80% to 90% chance of a 7.0 or greater hitting Los Angeles within the next 20 years. Such an earthquake would lead to 18,000 deaths in Los Angeles area, according to USGS.[22]

Although there are claims of the ability to forecast large earthquakes accurately and many days in advance, forecasting earthquakes remains a science in development. An early warning of earthquakes is available in California's Coachella Valley around Palm Springs—the first installed and operational earthquake early warning system. There are 12 locations in operation in 2011 with 120 sites planned by 2013. This early warning system is meant to detect an earthquake and give people 15 to 20 seconds advance notice. This is time enough to get under a table, and tell others in a classroom to duck, cover, and hold on, or in the case of a fire station, get the engines outside of the building before the ground starts shaking. Every other natural hazard has warnings systems in place to warn the public; earthquakes do not.

Since ancient times, many believe that animals are able to sense the approach of earthquakes. Stories are told of cattle becoming uneasy and hard to manage, dogs fearful and continuously howling, snakes leaving the area, fish appearing in large numbers in areas where they were normally scarce, and birds ceasing to sing and leaving the trees.

Before, animal prediction of earthquakes was considered superstitious folklore, but it is now studied by scientists and parapsychologists, who refer to the phenomenon as *anpsi*, the psi faculty in animals—in other words, an animal's psychic abilities and an animal's ability to find their owners in places they have never been ("psi trailing"). Other explanations of how animals can predict earthquakes and volcanic disruptions are the following:

- Animals are aware of super- and sub-sonic frequencies; they hear the initial sound waves of an earthquake that are inaudible to humans.
- Animals perceive electromagnetic field variations.
- Earthquakes produce an intensification of positive ions in the atmosphere, acting on the nervous system of creatures, that is not detected by humans.

Earthquakes occur every day, one occurred somewhere in the world today. Earthquakes can occur most anywhere, such as the 5.8 magnitude quake that occurred August 23, 2011 in Virginia, that was felt as far north as New York City and central Maine, and as far south as central Georgia. As urban growth continues around faults, more injuries will occur due to the existence of more structures. Predicting whether an earthquake will occur is easy; determine exactly when and the magnitude is where science still has a job to do.

Volcanic Disruptions

Many believe that volcanic disruptions only impact those people living in the vicinity of volcanoes. In 1985, nearly 23,000 people died from mudflows ("lahars") from the eruption of the Nevado del Ruiz volcano in Colombia. The primary effects of volcanism are lava flows, pyroclastic activities, and poisonous gas emissions. Lava flows can travel as fast as 64 km/hr; however, most are slower, allowing adequate time for people to move out the way. Lava flows lead to more damage to property and anything in its path than to people. To control the path of lava flows, attempts have been made to redirect lava flows by bombing flow fronts and spraying them with water to cool (and thus slow or dam) the flow.

Pyroclastic activities, such as *pyroclastic flows*, cause death by suffocation and burning everything in its path. Pyroclastic flows are high-speed avalanches of hot ash, rock fragments, and gas that move down the sides of a volcano during explosive eruptions or when the steep edge of a dome breaks apart and collapses. Pyroclastic flows can reach 1,500°F and move 100–150 miles per hour. These are common with violent volcanic disruptions, and are the most dangerous aspect of volcanism. The flows can travel so quickly that most humans are unable to escape.

Also stemming from these disruptions are *mudflows*, a mixture of water and sediment that blanket an area like snow, and yet more destructive are *tephra falls*. Tephra falls is a term used to define fragments of volcanic rock and lava that are blasted into the air by explosions or carried upward by hot gases in eruption columns or lava fountains. Tephra deposits are twice as dense as snow and do not melt like snow. They destroy vegetation and the livestock that eat vegetation covered by the ash. Where tephra deposits exist, agricultural activity may take years to return to normal, a secondary or tertiary effect of volcanic disruptions.

Volcanic disruptions emit poisonous gases. The poisonous gases include carbon dioxide (CO_2), hydrogen chloride (HCl), hydrogen fluoride (HF), and hydrogen sulfide (H2S). Lake Monoun, in Cameroon, a crater lake in this African country, is where 37 died from CO_2 gas escaping from the bottom of the lake in 1984. Two years later, 1986, CO_2 gas escaping from Lake Nyos in Cameroon left 1,700 people and 3,000 cattle dead.

The atmospheric effects from large quantities of tephra, volcanic gases, and volcanic ash that is injected into the atmosphere can cause the temperatures to be cooler for several years after a large eruption. There are approximately 600 active volcanoes on Earth; 50 to 60 of these volcanoes erupt each year. Recovery efforts take years.

Predicting a volcanic eruption cannot be determined today based on a single event. Each volcano behaves differently. Volcanoes usually give warnings that they will erupt. USGS scientists have developed a forecasting system to alert public officials and the general public of these warnings. Many events must be monitored, and collectively the data gathered is used to predict an eruption.

There are also volcanoes that erupt without any advance warnings or patterns identified or available for prediction. For example, it was a magnitude 4.2 earthquake recorded on March 20, 1980, that provided scientists with the first early warning signs that Mount St. Helens might erupt. The number of earthquakes recorded increased dramatically after the 20th. March 27, the suspicions of scientists were confirmed when Mount St. Helens began a sequence of explosions that continued intermittently until May 14. This was the first eruption of Mount St. Helens since 1857 (see Figure 5.1 Mount St. Helens May 18, 1980 eruption).

The larger and most explosive volcanic eruptions spew tens to hundreds of cubic kilometers of magma onto the Earth's surface. When the magma is removed from beneath a volcano, the ground subsides or collapses into the emptied space to form a depression called a *caldera*. Some calderas are more than 25 kilometers in diameter and several kilometers deep. Calderas are also referred to as *super volcanoes*.

The Yellowstone volcano in Yellowstone National Park is among the largest super volcanoes on Earth. It last erupted 640,000 years ago. An eruption cycle for this volcano is 600,000 years, meaning this volcano is overdue. An eruption is predicted to be catastrophic, as much as 2,500 times greater than the Mount St. Helens eruption in 1980. A volcanic winter is predicted to follow as a plume of ash would reach 20 miles into the atmosphere leaving the western half of the United States covered with ash. Other super volcanoes include:

- *Lake Toba, Sumatra, Indonesia*—considered Yellowstone's "big" sister
- *Long Valley, California*—Second only to Yellowstone in North America
- *Lake Taupo, New Zealand*—filled with water and one of the most beautiful landscapes in the world
- *Valles Caldera, New Mexico*—a sleeping monster in the heart of New Mexico
- *Aira, Japan*—One of the most recently troubling calderas in the world
- *Siberian Traps*—the largest lava flow in Earth's history, occurring 251 million years ago[8]

FIGURE 5.1 Mount St. Helens May 18, 1980 eruption. (From U.S. Geological Survey USGS.)

The impact of a volcanic eruption on airspace can be predicted once an eruption has occurred. Volcanic ash can affect people and equipment hundreds of miles from the volcano.

The airline industry is getting better prepared since aircraft ash can become a growing problem. Air traffic within the region is halted until the ash has settled. National plans are needed to address the long-term grounding of aircraft with consideration given to the financial impact on stakeholders as well as arrangements for repatriation of travelers.

Tsunamis

Anything that rapidly displaces a large volume of water can cause a *tsunami*. Tsunamis are usually caused by underwater earthquakes, yet landslides, volcanic eruptions, calving icebergs, and, rarely, meteorite impacts can generate tsunamis. These types of events can cause large disturbances in the surface of the ocean, and as gravity pulls the water down, the tsunami is born. Major tsunamis occur on average once per decade. Historical data provided by USGS indicate that 59% of the world's tsunamis occur in the Pacific Ocean, 25% in the Mediterranean Sea, 12% in the Atlantic Ocean, and 4% in the Indian Ocean.

The wave speed of a tsunami is controlled by water depth. For example, at an ocean depth of 6,000 meters (3.7 miles) unnoticed tsunami waves can travel at the speed of a commercial jet, over 800 km per hour (500 miles per hour), according to National Oceanic and Atmospheric Administration (NOAA). The shallower the waters, the slower a tsunamis will travel. As the tsunami approaches coastal waters, the wave heights increase dramatically. The danger can last several hours or with larger tsunami, for a few days.

Tsunami notices are provided to the community using a three-level system—warnings, watches (advance alerts), and advisories. *Tsunami warnings* are issued when there is an imminent threat of a tsunami from a large undersea earthquake or following confirmation that a potentially destructive tsunami is under way. The initial warnings may be based only on seismic information to provide the earliest warning possible. When a warning is issued, area population is advised to evacuate low-lying coastal areas and move marine vessels to deep water. Warnings are updated at least hourly, as conditions change, or to end the warning. *Tsunami watches* are issued when seismic information indicates that a destructive tsunami is under way, however, the data has not yet been confirmed. *Tsunami advisories* are issued to coastal populations within areas not currently in a warning or watch status yet a tsunami warning has been issued for another region in the same ocean.

Since earthquakes, meteor strikes, and avalanches cannot be predicted, knowing when a tsunami will happen before any of these events occur is not feasible today. Once a large potentially tsunamigenic earthquake does occur, forecasting the arrival times and wave heights of a tsunami is possible using computer modeling.[9] Tools used by regional tsunami warning centers (TWC) around the world—primarily, the United States, Japan, and Africa—to disseminate official messages include

- Global Telecommunication System (GTS)
- Aeronautical Fixed Telecommunication Network (AFTN)
- Advanced Weather Interactive Processing System (AWIPS)
- Emergency managers Weather Information Network (EMWIN)
- Weather Wire
- NOAA Weather Radio
- RANET

- USGS Earthquake Notification Service (ENS)
- e-mail
- fax
- telex
- RSS feeds

Atmospheric Conditions

Climate variability, including more frequent and damaging extreme events, are anticipated in the years to come. Climate fluctuations and associated risks for many regions are large. The fluctuations are expected to disproportionately affect those areas already faced with climate sensitive development challenges such as rapid and unplanned urbanization, resource scarcity, epidemic stresses, and land degradation. What is expected is already characterized by issues faced today.

There are stories of military operations manipulating the weather as a force multiplier. Manipulation of the climate using chemicals is a form of *climate warfare*, a weapon of mass destruction that is capable of destabilizing economies and agricultural and ecological systems globally. It can be directed against enemy countries or "friendly nations" without their knowledge. Weather manipulation is an excellent preemptive weapon. During the Vietnam War, cloud-seeding techniques were used to extend the monsoon season that blocked enemy supply routes. Disruptions in weather manipulation can lead to a greater dependency on food aid and imported grain staples. An international treaty banning weather modification as a weapon of war was signed in 1976.

There are also stories that climate change will cause greater volatility in weather patterns. Whether one believes in climate change or not, the weather is changed through technology every time a car is started or a lightbulb is connected to the power grid or when a dairy farm has too many cows concentrated in a small area. While corporations and governments bear the largest responsibility for the problem, billions of individual actions are contributors.

Changes in weather have been occurring throughout Earth's history. El Niño and La Niña determine the severity and frequency of severe flooding and hurricanes or drought and heat. *La Niña* is characterized by unusually cold ocean temperatures in the Equatorial Pacific, compared to *El Niño*, which is characterized by unusually warm ocean temperatures. El Niño is a disruption of the ocean-atmosphere system in the Tropical Pacific. These weather patterns have important consequences for weather and climate around the globe. El Niño leads to increased rainfall in the southern United States and Peru, causing destructive flooding, and drought in the West Pacific, sometimes associated with devastating brush fires in Australia. El Niño and La Niña are opposite phases of the El Niño–Southern Oscillation (ENSO) cycle, with La Niña sometimes referred to as the cold phase of ENSO and El Niño as the warm phase of ENSO.

At higher latitudes, El Niño and La Niña are among a number of factors that influence climate. The impact of El Niño and La Niña at higher latitudes are becoming more apparent in wintertime. In the continental United States, during El Niño years, temperatures in the winter are warmer than normal in the North Central states, and cooler than normal in the Southeast and the Southwest. During a La Niña year, winter temperatures are warmer than normal in the Southeast and cooler than normal in the Northwest.[10]

Sub-Sahara Africa has a population of more than 800 million, most of whom are dependent on rain-fed agriculture. On average, 25% of the region's GDP originates from agriculture, and 70% of the livestock and workforce is in the rural sector. The dependence on rain-fed agriculture makes African economies particularly susceptible to climate variability. Households in the rural areas have limited assets to withstand climatic-related shocks. The impact that climate variability can have on such agrarian economies is captured throughout history such as in the case of Ethiopia where economic growth and food imports closely track variations in rainfall.[11]

The Asia and Pacific region has a population of more than 4 billion people, including two-thirds of the world's poor, and over 60% of the undernourished people in the developing world. The climate of the region is highly variable, with frequent droughts, floods, and typhoons, which affect livelihoods, food security, and health. Many people remain heavily dependent upon rain fed agriculture; concurrently, demands for water for urban and other uses are increasing rapidly. In India, for example, more than 60% of agriculture is rain fed. A lackluster monsoon season can affect food prices and livelihoods and a failed monsoon season can seriously affect the economy.

The Latin America and Caribbean region has been called a "Land of Contrasts and Disparities" and contains some areas with the largest year-to-year climate variability in the world. Such variability and the expected longer-term changes in climate are putting significant pressures on agricultural production, water resources management, and human health.

Superstorms

Tokyo is considered the largest mega-city communication hub in the world, followed by San Francisco and Los Angeles/Long Beach. Interestingly, Washington, DC, where national policy is driven for the United States, is ranked a distant 13th in the world as a mega-city communication hub.

The risks associated with a major power outage on the U.S. West Coast from an extremely and unusually destructive *superstorm* are not fully appreciated by those on the East Coast. Not only individuals and businesses would be affected on the East Coast but Wall Street, the Pentagon, the White House, and Congress, among many core operations that move the United States forward.

> **CASE STUDY: CALIFORNIA FACES THE RISK OF A MASSIVE SUPERSTORM**
>
> Scientists and experts from the USGS say that California faces the risk of a massive superstorm that could flood a quarter of the state's homes and cause $300 billion to $400 billion in damage, four or five times greater than the damage from a major earthquake. Superstorms have occurred in the past on an average of every 100 to 200 years; the threat of a cataclysmic California storm has been dormant for the past 150 years. In 1861–1862, a 300-mile stretch of the Central Valley was flooded, causing the relocation of the capital from Sacramento to San Francisco. Governor Leland Stanford had to take a rowboat to his own inauguration, according to reports published. Larger storms occurred in 212, 440, 603, 1029, 1418, and 1605, according to geological evidence. The risk is increasing as temperatures rise in the atmosphere, which makes weather patterns more volatile. Modeling showed that a superstorm could last for more than 40 days and dump 10 feet of water on the state. The storm would be aggravated by an "atmospheric river" that would move water at the same rate as 50 Mississippi Rivers discharging water into the Gulf of Mexico, according Daisy Nguyen of the Associated Press on January 14, 2011. Flooding, wind speeds reaching 125 miles per hour, and landslides could affect one-fourth of all housing in the state.[23]

Solar Flares

The High-frequency Active Auroral Research Program (HAARP) is a program studying the properties and behavior of the ionosphere. It is a program jointly managed by the U.S. Air Force, the Federal Aviation Administration (FAA), and a number of other government agencies. HAARP is focused on being able to understand and use the ionosphere to enhance communications and surveillance systems for both civilian and defense purposes. Today it is possible to produce computer simulations of ionospheric processes. The development of computer visualization enables one to see the enormous variability and turbulence that occurs in the ionosphere during a major solar geomagnetic storm and the affect it can have on radio communication and navigation systems.

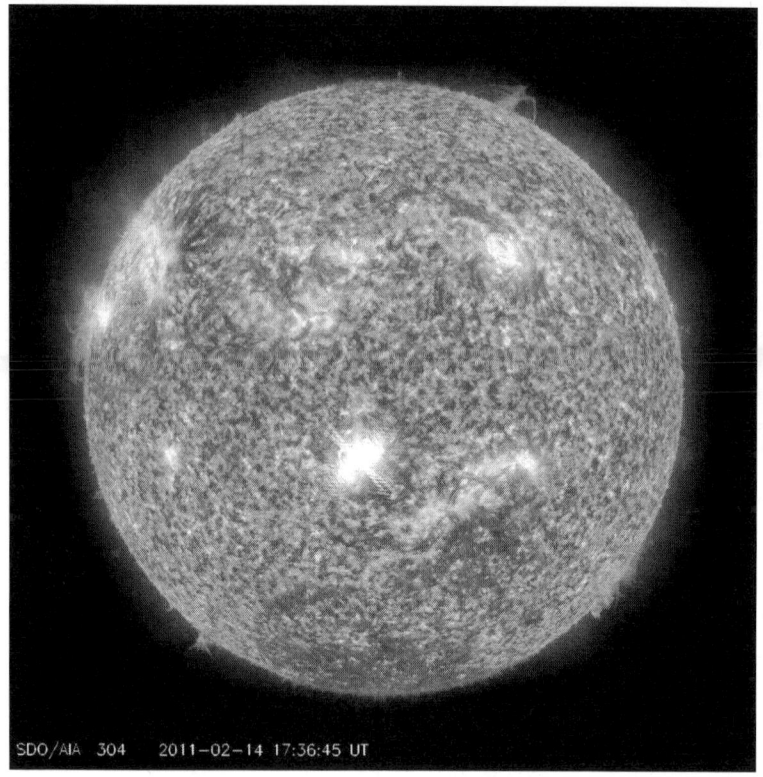

FIGURE 5.2 Solar flare image. (From National Aeronautics and Space Administration [NASA]. www.nasa.gov/mission_pages/sunearth/news/News021411-xclass.html)

The sun has a dominant effect on the ionosphere, and solar events such as solar flares or coronal mass ejections can lead to worldwide communication "blackouts" on short wave bands. A *solar flare* is an explosion on the Sun that occurs when energy stored in twisted magnetic fields, usually above sunspots, is released suddenly. Solar flares generate a burst of radiation across the electromagnetic spectrum, from radio waves to X-rays and gamma rays. The largest solar flare in 4 years was emitted from the sun on February 14, 2011, as pictured in Figure 5.2 (Solar flare image). NASA's Solar Dynamics Observatory (SDO) tracks solar flares. This was a powerful class X2.2 flare, the strongest so far in the Solar Cycle 24.

Astronomers have five categories of flares: A, B, C, M, and X. The *X-class flares* are huge, a major event that can trigger planet-wide radio blackouts and long-lasting radiation storms. Solar flares are caused by a sudden release of magnetic energy, and they emit light, ultraviolet waves, and X-rays. There is often a disruption of radio communications shortly after major flares. When the charged particles reach the earth, they can create a geomagnetic storm that can influence the planet in several ways, including disrupting satellite communications and power grids. The particles also interact with the earth's magnetosphere to create spectacular auroras known as the Northern and Southern Lights. These flares usually reach the earth within 19 to 37 hours after the eruption.

M-class flares are medium sized and can cause brief radio blackouts that affect the Earth's polar regions. Minor radiation storms sometimes follow an M-class flare. *C-class flares* are small, with few noticeable consequences on Earth. The amount of energy a solar flare releases is the equivalent of millions of 100-megaton hydrogen bombs exploding simultaneously. An *A0* solar flare is the weakest that can be categorized; a *B0* solar flare is ten times stronger than a A0 flare. A *C0* is ten times stronger than a *B0*, and 100 times stronger than an *A0*, etc. The strongest category is X9, which occurs once in 500 years. The frequency of flares coincide with the sun's 11-year cycle. When the solar cycle is at a minimum, active regions are small and rare, and few solar flares are detected. As the solar-cycle progresses, an increase in the number of solar flares occurs as the Sun approaches the maximum part of its cycle.

The February 14, 2011, eruption was strong but not as big as the November 2003 eruption, when the NOAA Space Environment Center released warnings so that electrical utilities, airlines, and spacecraft managers were able to take preventive action to minimize disruption of service—and the economy. This is probably the first of many more x-class flares to occur over the next 2 to 4 years as we reach the peak of Cycle 24.

Comets, Asteroids, and Meteor Strikes

Sometime in the future, most astronomers agree, the Earth will be struck by a large meteor or asteroid. The effects could be catastrophic—coastlines drastically altered with a number of coastal metropolitan areas submerged under water; new seaports would be created in areas that are inland today and significant changes would affect climate patterns. On September 27, 2003, a meteor hit the Earth and fragmented into several pieces, which injured three people in the Orissa region of eastern India.

Sandia National Laboratories' scientists conducted a computer simulation of a 1 kilometer comet (or asteroid) striking in the open ocean. Based on the simulation, the comet and 300 to 500 cubic kilometers of ocean water would be vaporized from the energy generated at impact. This energy would be the equivalent of 300 gigatons of TNT—10 times the explosive power of all the nuclear weapons in existence in the 1960s at the height of the Cold War. The impact would send hundreds of cubic miles of superheated water vapor, melted rock, and other debris into the upper atmosphere and space. Much of the debris would then rain over the world for the next several hours forming a high global cloud. The shock wave from the impact would level much of the area around the site for hundreds of miles with the heat incinerating cities and forests. The global cloud would then lower temperatures worldwide, and a global "nuclear winter" would follow.[12]

SHORTAGES OF KEY RESOURCES

Water Shortages

Water shortages are the next big global problem. Fresh ground water is less than the water in the oceans, ice caps, and other saltwater sources. Hoover Dam in Lake Mead is an example of the problem that is looming. Hoover Dam was the largest hydroelectric project in US. history. In 2010 it experienced an extended low water level that could force the closure of the project. The collision between rising energy demands and diminishing reserves of fresh water is cause for concern.

The world is warming and with it have come shifts in weather patterns. Severe winter weather with subfreezing temperatures, snow, and ice are experienced farther south at greater frequency and intensity. Extremely hot and dry summers have followed, and changed rainfall patterns have left areas such as portions of California with extended drought conditions. In 2011, Texas experienced record drought conditions leading to burn bans in 99% of the state and mandatory water restriction.

Extreme swings in weather cycles may be more problematic than long-term climate direction, for example, a dry year followed by a wet year and long droughts followed by flooding. Seven of the 10 most destructive storms occurred within a two-year period, in 2004 and 2005. The risks are increasing.

Cities and urban areas are growing at a time where there is less funding available to address an aging infrastructure, increased crowding, inadequate mass transportation, and growing at-risk populations. The rapid advancements in technology present unknown hazards and risks. Government agencies with dwindling resources are tasked with the burden of mitigation planning when the private sector owns more than 80% of all available resources. More than half the jobs in existence today did not exist in 1999.

To give an idea of the exponential increase in costs when servicing a disaster, we compare the Los Angeles and San Francisco area in the mid-1800s to early 1900s and today. In 1857, Los Angeles County had a population of 9,000 people; today, the area has 10 million people. The value of the land in Los Angeles County in 1857 was $750,000 in 2010 it is valued at more than $950 trillion. The 1906 San Francisco earthquake yielded $524 million in losses; today, the same earthquake would cause economic losses of $100 billion to $200 billion.

Additional risks are introduced with the increased costs of construction and home and office contents, more expensive and complicated than ever. The trend is just-in-time inventory and consolidation of warehousing, ultimately putting more products and materials in fewer places. Just-in-time logistics are automated and fully mechanized with a greater reliance on Radio-Frequency Identifications (RFIDs), bar codes, lasers, and belts. Logistical systems are complex and require specialized procedures to restart.

As world economies are faced with challenges, planning for a major catastrophe becomes increasingly difficult. The challenges include financial recovery, rebuilding, and addressing the need of the poor and those affected by previous disasters, each placing an enormous load on services. The world is moving toward a period of global resource scarcity and unrest.

From 1900 to 1979 an average of seven or fewer catastrophes were recorded annually. Since 1979, this number continues to increase. Common characteristics of disasters today include

- Insufficient communication
- Responders becoming victims
- An unknown degree of related problems
- Confusion and disorder among public and private institutions
- Disruption of the social fabric
- Long-lasting economic impacts
- Loss of economic and social synergy

Man-made events such as rising prices and natural events such as powerful storms, severe droughts and floods, and other unexpected events are likely to create mayhem with the fabric of global society, producing chaos and political unrest. Global consumption patterns are beginning to challenge the planet's natural resource limits.

A report entitled *Water: Sustainable Management of a Scarce Resource,* authored by Dr. Farouk El-Baz, director of the Center of Remote Sensing at Boston University, and others warned that without fundamental changes in policies and practices, the water supply situation would get worse. Water shortages would have drastic social, political, and economic ramifications.

Nearly two-thirds of the water supply sources for the Arab world originate outside the region. Thirteen Arab countries are among the world's 19 most water-scarce nations, and per capita water availability in eight countries is below 200 cubic meters annually, less than half the amount designated as a severe water scarcity. By 2015, only Iraq and Sudan will pass the water scarcity test. The Arab region is one of the driest in the world. More than 70% of the land is arid, and rainfall is sparse and poorly distributed. Climate change will intensify the situation.

The era of cheap water is ending as the global population grows, and food and water demands increase. Products requiring water for manufacturing and processing are confronted with a new natural limit on something that was once considered virtually infinite. Peter Gleick, president of the Pacific Institute in California, refers to this phenomenon as *peak ecological water*. The Pacific Institute is a leading authority on global freshwater resources.

Consider how much water is used to make a single everyday product. A glass of orange juice needs 850 liters of fresh water to produce, an average hamburger needs 2,400 liters, 1 kilogram of paper requires 125 liters of water to process, excluding the water needed to grow the tree, and the manufacture of a kilogram of microchips uses 16,000 liters for cleaning to remove chemicals. Billions of people live today without basic water services.

Water will increasingly be the cause of violence and even war since water is a basic condition for life. West Africa; the Ganges–Brahmaputra river system in Nepal, Bangladesh, and India; and Peru over the next decade will have a significant risk of violent conflict over water. In China, water usage and pollution have overtaxed freshwater resources and could begin to threaten the stability of the region.

A study conducted by David Zhang, a geographer at the University of Hong Kong, and published in the United States by National Academy of Sciences journal *Science*, analyzed 8,000 wars over 500 years and concluded that water shortages played a far greater role as a catalyst for war than previously thought. Conditions today are an indication that resource shortages are becoming a major cause of conflicts that emergency managers must prepare for. The costs of *potable water*, water that is fit for consumption by humans and other animals, will increase as the methods needed to purify water become more expensive. Despite technology, the world is using more water than the ecosystem can sustain.

According to United Nations (UN) estimates, in 2010 more than one-third of the world's population was suffering from water shortages; by 2020 water use is expected to increase by 40% from current levels, and by 2025 two out of three people could be living under conditions of *water stress*. Water stress occurs when the demand for water exceeds the available amount for a given period or when poor quality restricts its use.

Food Shortages

By 2020, the UN projects that there will be 50 million *environmental refugees* who will travel into the global north, fleeing food shortages sparked by climate change, as shared by University of California–Los Angeles School of Public Health Dr. Christina Tirado in February 2011. Environmental refugees are people who are unable to have a secure livelihood in their homelands where drought, soil erosion, desertification, deforestation, and other environmental issues threaten their homelands. These conditions are occurring simultaneously with population pressures and profound poverty issues.

The migration from the south to the north is already being experienced in Southern Europe as sharp increases in what was a slow but steady flow of migrants from Africa. Many of the migrants risk their lives to cross the Strait of Gibraltar into Spain from Morocco or sail in makeshift vessels to Italy from Libya and Tunisia. The flow grew to a flood after a month of protests in Tunisia in early 2011, set off by food shortages and widespread unemployment and poverty that brought down the government of Zine El Abidine Ben Ali.

It is predicted that the migration north will become more common as food shortages, food security, and food safety issues increase. Geopolitical instability and unsustainable conditions are also predicted. Whether it is in the Middle East, Latin America, or in the Americas where politics, religion, and other matters cause the poor to focus on survival, it is the increasing inability of people to eat and feed their families that sparks migration.

The World Health Organization estimates that 2.2 million deaths in developing countries are caused each year by food- and waterborne diseases.[13] Food shortages, food security, and food safety are a result of rapid increases in fuel costs and extreme weather conditions. Flooding, drought, freezing, heat-related crop damage, barren land, and the increasing cost of fuels used for planting, harvesting, and getting items to market contribute to food- and waterborne diseases.[14] An increasing frequency of disruption or reduction in agricultural production will breed food insecurity. In coastal communities that experience disturbances to its ecosystems, sea levels rising or contaminated marine habitats may diminish the availability of sea foods.

Power Shortages—Blackouts and Brownouts

Blackouts and brownouts are a time for waiting until the lights are back on. Humans are dependent upon electricity. Electricity is getting harder to produce. In Switzerland, the population has grown 20% over the last generation while electricity consumption has risen 60% according to the Swiss Federal Office for Energy (BFE) Authority. The primary causes of rapidly increasing energy consumption are

- more gadgets
- growing population
- electricity becoming a substitute for fossil fuel energy
- many devices that are never completely turned off even though they are not in use, such as computers, coffeemakers, televisions, game systems, continue to consume massive amounts of energy in stand-by mode

Like many growing countries, Switzerland is heading for a big shortage that could lead to brownouts and blackouts by 2020.

Blackouts and brownouts have occurred in the United States and other countries and are expected to continue as demand increases and energy generation fails to keep pace. The United States has 57,000 megawatts of new resources identified while 141,000 megawatts will be needed by 2020, a shortfall of 84,000 megawatts. This gap is the equivalent of 160 large power plants. China had 380 million kilowatts in 2003 and expects to have 950 million kilowatts installed by 2020. In underdeveloped countries, nearly 300 million homes are deprived of adequate lights. Blackouts and brownouts have changed lifestyles, schedules, and spending habits like the advent of electricity did to its generation.

The UK, in 2009, faced the prospect of widespread power cuts for the first time since the 1970s, according to projections by the Department of Energy and Climate Change in London. The shortage of supplies was forecasted to hit the equivalent of as many as 16 million families for at least one hour during the year. The report highlights the first shortfall in 2017. The "energy unserved" level reaches 3,000-megawatt hours per year, the equivalent of the whole of the Nottingham area being without electricity for a day.

Based on a government report titled "Low Carbon Transition Plan" (July 2009) Shadow Energy Secretary Greg Clark states that by 2025, the situation worsens with the shortfall reaching 7,000-megawatt hours per year, the equivalent to an hour-long power cut for half of the UK. The scale of the blackouts could be three times worse than government predictions. He said some of the modeling used was "optimistic" as it assumes little or no change in electricity demand up until 2020. The last time the UK experienced regular power cuts because of shortages of supply was in the early 1970s, when a miners' strike caused coal restrictions. The country was forced to do everyday tasks by candlelight and a 3-day week was imposed on all but essential services to try to conserve electricity.

HEALTH DISASTERS

The impact of health disasters creates a different challenge. If you are not a victim, you will know someone who is. Pandemic and other health catastrophes will overwhelm authorities such as the cholera outbreak in Haiti after the 2010 earthquake.

A report published by the Inter-Governmental Panel on Climate Change (IPCC) states that changes in climate have already led to increased disease levels, the geographical movement of certain infectious diseases, and increased heat-related deaths. Current climate trends indicate that a range of environmental consequences such as heat waves and other natural disasters having health consequences will increase. This translates to increased malnutrition, airborne disease, diarrheal disease, dengue fever, and cardiorespiratory disease.

Natural disasters, considered *direct impact threats*, can lead to significant health consequences, but man-made events that arise from social, demographic, and economic disruptions, thought of as *indirect impact threats*, can also lead to negative health consequences. Indirect impact threats include air pollution, biological changes such as changes in crop yields, and ecosystem changes, such as pests like bed bugs and fire ants migrating to new areas.[15] Research on health risks indicate that by 2020 that there will be increased incidences of heat stress and heat-related illnesses, allergic diseases, food poisoning, increased trauma from drought and natural disasters, and increased demand for aid from neighboring countries affect by climate change.

Aging Infrastructure

Pipeline

There are 2.4 million miles of natural gas pipelines in the United States, enough pipelines to wrap the Earth nearly 96 times (see Figure 5.3 Oil and gas refinery pipelines). These pipelines are usually buried in rural areas and crisscross through urban area near homes, schools, and businesses, typically forgotten by the public until an incident occurs. There are incidents daily with someone in the hospital or dead nearly every 9 to 10 days, according to Carl Weimer, executive director of Pipeline Safety Trust. The large transmission pipelines—like the one in San Bruno, California, that exploded in September 2010—do not cause many of the problems today. The 300,000 miles of the large transmission pipelines in the United States are newer and made of thick steel.

The bigger concern is the small distribution lines that bring gas close to homes. There are 2.1 million miles of these pipelines, made from a wide variety of materials such as wood dating back to the late 1800s

FIGURE 5.3 Oil and gas refinery pipelines. (Courtesy of Shutterstock.com.)

according to the American Gas Association. Today, cast iron pipes are an increased concern in wetter climates where the pipe has a shorter life span. Newer pipeline systems are made of more durable plastic or steel.

> ### Case Study: Massive Gas Pipeline Explosion, San Bruno, California, September 2010
>
> Around dinnertime on this date, in San Bruno, a suburban area south of San Francisco, a 30-inch gas main ruptured and obliterated a neighborhood. In an instant, without warning, the explosion occurred leaving 8 dead, 50 injured, 38 homes destroyed, and more than 50 homes damaged by the inferno. For the survivors, there began a psychological rebuild in addition to rebuilding their homes. For example, one mother relives the nightmare daily. She panicked when she first heard of the explosion since she knew her 20-year-old daughter was there visiting her boyfriend. He suffered second- and third-degree burns; a visit from the county coroner's office ultimately confirmed that her daughter was among the deceased.
>
> Preliminary reports from federal investigators indicate that defective welds in a 60-year-old pipeline were the primary cause. The pipeline owner maintained poor records on the system. Although the pipelines had aged, if they had been built and maintained properly, most likely the explosion would not have occurred. The total bill for the explosion could exceed $760 million.

FIGURE 5.4 A typical water treatment facility. (Courtesy of Shutterstock.com.)

Waste Treatment Facilities

The majority of United States publicly owned treatment works were constructed during the 1970s and 1980s (see Figure 5.4 A typical water treatment facility). The need to rehabilitate or upgrade treatment is expanding, and the need to leverage innovative treatment technologies is critical. Today, there are 240,000 water main breaks per year in the United States. Water main breaks lead to low water-pressure that can increase incidents of diarrhea, based on a 2005 UK study. As these systems near the end of their service life, the numbers of breaks substantially increase. Large utility breaks in the Midwest increased from 250 to 2,200 per year over a 19-year period. In 2003, there were 1,190 water main breaks reported in Baltimore, Maryland, an average of three to four per day.

The estimated water lost from water distribution systems is 1.7 trillion gallons per year at a national cost of $2.6 billion per year according to the USGS. For example, according to Houston, Texas, Mayor Annise Parker, "In August 2011 the city of Houston was dealing with dwindling water supply due to drought, heat, and its system of water pipes bursting at a rate of 700 each day, up from the usual rate of 200 per day at this time of year."

Wastewater Sewer Lines

There are 600,000 miles of wastewater sewer lines in the United States. A network of pipes, pumping stations, and other equipment make up the wastewater collection systems. These collection systems transport storm, irrigation and wash water, sanitary and industrial wastewater, and infiltration and inflow. Many of the early sewers installed before the 1900s remain in service. As urban areas grow, the need for wastewater and storm water management is necessary to protect human health. Current technologies are inadequate for addressing emerging issues.

Roads, Bridges, and Transit Systems

Many roads, bridges, and transit systems are old and falling into disrepair or need replacement.

Based on recent assessment of data by the U.S. Federal Highway Administration, only half of the nation's major roads are in good condition. In high-traffic urban areas 25% of roads are in poor condition with some major urban centers having more than 60% of roads in poor condition. In 2007, these urban roads carried two-thirds of the nation's vehicle traffic.

The average service life of a bridge is 50 years yet in the United States, the average bridge is 43 years old. One out of four bridges is too narrow to handle today's traffic or in need of repair. Half of the nation's transit buses and rail cars have exceeded their service life or will do so within the next 6 years. The U.S. population, travel, and economic activities are increasing, leading to more wear and tear on the nation's roads, bridges, and public transit systems. The U.S. population grew by 23% between 1990 and 2010. During the same time, the number of miles driven nearly doubled to 41%, and mass transit increased by 41%.[24]

> **CASE STUDY: 2007 MINNEAPOLIS I-35W BRIDGE COLLAPSE DURING RUSH HOUR**
>
> Wednesday, August 1, 2007, in Minneapolis, Minnesota, a section of the I-35W Bridge, known as the St. Anthony Falls Bridge, that spanned the Mississippi River catastrophically collapsed during rush hour. The bridge carried more than 140,000 vehicles a day. Approximately 1,000 feet of the deck truss fell into the river as depicted in Figure 5.5 (Minneapolis I-35W bridge). Dozens of cars and their occupants were plunged into the river. Thirteen died in the incident, 145 were injured, and some risked their lives to save others. This disaster disrupted transportation for the region. The loss of the bridge cost $400,000 per day. The National Transportation Safety Board concluded that a design flaw was the main cause of the collapse. Gusset plates and large steel sheets that connected the old bridge's beams were cracked and corroded. Another contributor to the collapse was the lack of redundancy in its structural support system. The new bridge took 11 months to complete and reopened September 18, 2008, over a year later. The new bridge was built for a 100-year life span. It was 80 feet wider, with four additional lanes and an added shoulder, and made of high performance concrete. Approximately 300 sensors on and inside the bridge's concrete will transmit data for analysis of the bridge's movement and road and hazard conditions recorded.[16]

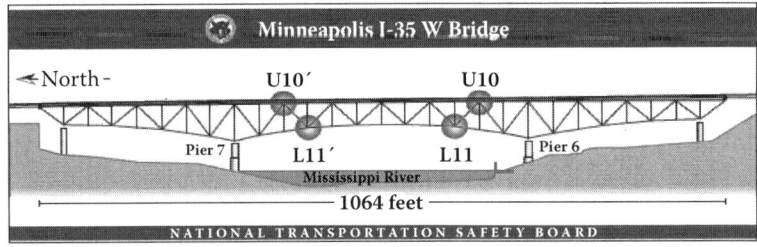

FIGURE 5.5   Minneapolis I-35W bridge.

CYBER ATTACKS AND CYBER WARFARE

Cyber warfare is a growing concern. Some of the most secure websites such as Mastercard and Visa were down the week of December 9, 2010, due to a *distributed denial-of-service* (DDoS) attack. A *DDoS* is an attack where many compromised systems attack a single target, thereby causing a denial of service or denial of access to the target. DDoS attacks were conducted by hackers who were upset by the WikiLeaks founder being jailed in the UK while awaiting the judge's decision on whether he would be extradited to Sweden. There are threats remaining that a "poison pill" will be released if he is harmed.

The newer generation of attacks includes *advanced persistent threat (APT)*. APT are associated with cyber attack espionage assaults. APT is a malware that has taken advantage of weaknesses in targeted networks of an organization with the intent of stealing data, especially intellectual property. Not always are these attacked "advanced" or sophisticate; they may be exploiting known vulnerabilities on systems with poor or relaxed patch management practices. APTs are persistent rather than a one-time event and can be stealthy.

APTs are often linked to *cyber-espionage* or *cyber-sabotage* attacks against organizations that are involved in national security or significant global economic activities. Everything connected to the Internet could be subject to attack. Google in 2010 was a victim of network-based intellectual property theft whose origins were reportedly in China. Those behind APT attacks are often sponsored or directed by a nation-state having very different goals than the average cyber criminal. These types of APT attacks are focused on obtaining intelligence for a military, political, or economic advantage.

APTs also target social media sites as they are increasingly used within an organization.

Open source espionage is the process of gathering information that is readily available posted information on social media sites such as Facebook or LinkedIn. *Spear-phishing* is often used by cyber criminals to gather data from these sites. It is also a method for distributing malware to intended victims. A recent example was the fake White House e-mail Christmas postcard luring government victims to download ZeuS, a malware malicious code used to steal documents.

Governments are modernizing their military operations to enhance their ability to fight high-tech wars. This modernization effort is known as *informationization*—to fight a local war under conditions where control of an adversary's information flow is established and to maintain dominance in the battlespace using comprehensive computer network exploitation (CNE). The program is focused on developing a networked architecture capable of coordinating military operations on land, in air, at sea, in space, and across the electromagnetic spectrum. From China and Russia to countries in Latin America, Europe, and North America, many nations are modernizing their militaries to fight high-tech wars.

A strategy of informatization is the coordinated use of computer network operations (CNO), electronic warfare (EW), and kinetic strikes designed to hit an enemy's networked information systems, creating "blind spots" that different forces could exploit at predetermined times or as the tactical situation warrants. Attacks on vital targets such as an adversary's intelligence, surveillance, and reconnaissance (ISR) and communication systems will be the responsibility of EW and counterspace forces using sophisticated jamming systems and anti-satellite (ASAT) weapons. Computer network attacks and exploitation units will conduct attacks on an adversary's data and networks. Simultaneous application of electronic warfare and computer network operations against an adversary's command, control, communications, computers, intelligence, surveillance, and reconnaissance (C4ISR) networks and other essential information systems appears to be the foundation of cyber warfare. This strategy will be employed starting in the early phases of a conflict, targeting a preemptive strike against an enemy's information systems and C4ISR systems.

Individuals who support illegal hacking activities, *black hat* programmers, from around the world are developing tools to exploit vulnerabilities in software that vendors have not yet discovered. The exploitation of these vulnerabilities is known as *zero day exploit* or *0-day*. This community of programmers is not bound by physical geographic boundaries. Proving that black hat programmers have government affiliation is difficult.

Unlike hack attacks against department stores and credit card companies offering monetary value to cybercriminals such as bank account information or credit card numbers, the targets have no inherent monetary value. Rather the exfiltration is on information regarding a nation-state defense industry, space program, selected civilian high technology industries, and other critical resources so that foreign military planners can form an intelligence picture of a country's defense networks, logistics, and related military capabilities that could be exploited during a crisis. In a conflict, a country could attack another country by targeting nodes on a military's nonclassified networks and unclassified defense and civilian contractor logistics networks of the enemy and its allies.[17]

These sophisticated attacks could impair telecommunication systems from coast to coast, even those that are considered less vulnerable. At times of conflicts, governments are known to block and censor Internet activities and intercept or eavesdrop on telephone conversations.

Effective emergency communications uses both high tech solutions such as satellites and the Internet and low-tech solutions such as ham radio operators, pen and paper, and word of mouth. In the age of electronic warfare, it is critical to have multiple low technologies, no technology solutions available, and a team to disseminate emergency communiqués should the high technology solutions become compromised.

WAR/CONFLICT

Factors within the wars of today include

- accidental or intentional missile launches
- geopolitical instability
- civil unrest and wars
- resource shortages and power struggle over remaining resources
- leaders grabbing for power or fighting to hold on to power against their own people
- disputes over territorial boundaries

One or several of these factors can start a conflict or escalate a conflict to a war. Political leaders are increasingly faced with making life and death decisions while navigating through a maze that constantly changes, and the intelligence used is often incomplete, incorrect, or if accurate, poorly timed. Allies and the countries themselves have conflicting or competing priorities. After the recent uprising in Libya, some countries, although opposed to the Libyan leader Mu'ammar Gadhafi's approach, are hesitate to apply much pressure because they are dependent upon his oil. By August 2011 the United States and other NATO allies asked him to step down. By October 20, he was dead. Killed at the hands of his own people.

> **CASE STUDY: LIBYA REFUGEE AID OPERATION INTENSIFIES**
>
> Between February 20 and March 5, 2011 more than 100,000 refugees fled from Libya as the uprising against Libyan leader Mu'ammar Gadhafi continued. The refugees crossed into the Tunisian border according to the Tunisian civil defense agency. European countries and the United States helped Tunisia to mobilize for receiving and repatriating the tide of additional refugees fleeing the unrest. The United States worked with the International Organization for Migration (IOM) to repatriate its foreign nationals—Egyptians, Africans, and South and Southeast Asians who fled Libya. The normal flow was approximately 10,000 new arrivals daily. Many refugees were waiting close to the border on the Libyan side in nearby towns, knowing that the border was jammed. Most Egyptians, Americans, and many Europeans had been evacuated. Bangladeshis, Somalis, Ghanaians, and Vietnamese refugees were crossing the border on foot, all to campsites along the border. Some Egyptians are now in camps unable to return to Egypt or Libya.
>
> Countries from around the world assisted in this effort in addition to the United States. Egypt used frigates, France used the helicopter carrier *Mistral*, and Italy provided a navy patrol boat carrying tents, blankets, and water purification kits to Benghazi.
>
> Meanwhile the UN demanded urgent access for humanitarian organizations to the civilian population that had been shelled by Gadhafi tanks. Reported are untold numbers of injured and dying.
>
> The UN Security Council unanimously condemned the violence in Libya, stating that they supported accountability for the perpetrators, and that they stand with the Libyan people. On October 20, 2011, President Obama confirmed Gadhafi's death. October 20 is marked as the day that Libya issued the ultimate rebuke of Gadhafi's dictatorship and his final act—dying as a fugitive.[18]

- Changes are occurring across the Middle East by the people of the region who are seeking a better life. The price of food, scarce resources such as clean drinking water, high unemployment, low wages for the working poor, an oppressive leadership, and the Internet are key factors that have sparked the change. As stated by President Barack Obama, "[they]just want to be able to live like human beings ... it is the most basic of aspirations that is driving this change ..." in his press release on February 23, 2011. Although most of these factors have been present for generations, access to the Internet is a fairly new phenomenon for the general public in this region. The Internet, and in particular social media and free international telephone services such as Skype, have open the doors to a world beyond geographical borders where youth learn what life is like elsewhere and strive to have the same within their own borders. To turn off, censor, or restrict access to the Internet only increases the curiosity of people who want to know what it is that others do not want them to know. The rule of effective communications is to understand that eventually it will all come out.

Although countries use intelligence to help determine where to place resources, the intelligence is sometimes flawed or missing critical components. Communicating a plan to refugees of a conflict of where to go for assistance depends upon effective communication tools. Reaching those who are geographically isolated and working through cultural norms makes effective communications complex. This means coordinating among nations and groups who can help bridge the divide so that people respond appropriately. Conflicts and wars require disaster planning during peacetime, and these plans are fine-tuned during times of war.

DISASTER PLANNING

As experience with Hurricane Katrina and with the 2010 Deepwater Horizon oil spill have shown, no incident command system can protect an affected community from leadership that is unable to handle the problem. Local resources may have knowledge but little capacity to respond alone, creating a void. Disasters of the future will push the boundaries of government and private sector resources. From geopolitical instability and war to resource shortage and health disasters, disaster planning of the future will require ample advance planning. The failure to plan is a plan for disaster. Communications is critical and can be the glue that keeps a disaster from become a catastrophe.

Catastrophes require different procedures at a time when communications is limited, power and transportation may operate below required levels, external responders lack local knowledge, and demands exceed available resources. This void causes a dependency on state and federal resources that do not possess the local knowledge. This gap in knowledge and resources among resources lead to delays, miscommunications, and finger-pointing while victims become pawns among the agencies involved. Although the private sector has the resources to help, often their concerns for reimbursement and litigation will contribute to the many problems a catastrophe present.

Other risk that must be addressed during planning is the aging U.S. population and infrastructure, more specialists than generalists, and a great reliance on technology when it may not be available. Regardless of the risks, the costs of failure are increasing. Risks and losses from climate-related natural hazards continue to rise, averaging $100 billion per annum in the 1990s. Losses are incurred from tangible and intangible sources—a competitive and unforgiving environment, contemporary expectations, concentration of values, and increased replacement values. Not repairing the infrastructure since funds are not available is one example of the kind of political impact that can occur by not addressing issues before a catastrophe happens. The effects of many catastrophes could be reduced if disasters were planned for rather than considered a one-of-a-kind event that an organization never expected. To receive federal assistance or insurance, bureaucracies must show that they can be adaptive for non-routine events. If they use existing strategies to keep from outlaying funds, providing assistance will assume a rigid and detached process. In a catastrophe, this approach causes matters to become worse and incapable of rapid customized responses to meet the regional and individual needs of the target audiences. In a day where the competitive advantage one organization may have over another is customer service, customer service by agencies is often lacking in providing services in a catastrophe.

Mega-cities have complex social systems and disaster services that vary, based on size and complexity, in efficiency in response to a disaster. These systems and services include prediction, warning, and evacuation systems that are dependent upon sophisticated technology, bureaucracies, and a cooperative public and private sector. Disaster preparedness competes with day-to-day priorities, yet the failure to plan and prepared is known to increase loss of people and property. Responding and recovering takes longer. The problems of disaster management are complex and evolving with a need for simultaneous solutions. Today Chernobyl is known as the dead city after the nuclear reactor suddenly erupted on April 26, 1986. Planning for the possiblity of a nuclear reactor meltdown was inadequate because the thought was the probability was low risk.

Disaster management is faced with a number of emerging issues of complex and connected hazards. For example, massive storms can shut down air traffic at Chicago's O'Hare airport that affects global air traffic, causing delays and cancelled flights. This means you have stranded passengers trapped in airports and filling hotels awaiting the next flight. In the last few years, there has been an evolving legal definition of disaster. Using the closed airport example, an airplane in the United States can no longer sit on the tarmac for an indefinite period, they must return to the gate and provide passengers with compensation, something unheard of before 2009.

An increasing number of people are dissatisfied with communities' disaster response. After Hurricane Ike in 2008, many that were left homeless were unhappy with the extended stay in shelters and the level of security offered. Many communities have plans for a disaster response and short-term recovery, yet fail to develop and test long-term recovery and mitigation plans, making them ineffective when needed. Standards are increasing, such as the requirement for smoke detectors in attics and integrating fire alarm systems with mass notification systems, as outlined by the National Fire Protection Associaton (NFPA), and the need for shelters to accommodate the general population. More standards are coming.

The failure to plan is of growing concern in an environment where more emergencies occur. For example, Donna Prince sued Mike Waters, Fire Control Coordinator, County of Onondaga, New York, for failure to implement National Incident Management System (NIMS) Incident Command System (ICS) on behalf of her husband. (See *Donna Prince L. v. Waters, CA 0701233 [4th Dept 2-1-2008] 2008 NY Slip Op 00879, 48 A.D.3d 1137, 850 N.Y.S.2d 803]*. Donna Prince L., Individually and as the parent and the Natural Guardian of Philip Lawrence L., an Infant, and as Administratrix of the Estate of *Timothy John L., Deceased, Plantiff-Appelant, v. Mike Waters, as the Fire Control Coordinator of the County of Onondaga, and County of Onondaga, Defendants-Respondents. CA 07-01233*. Appellate Division of the Supreme Court of New York, Fourth Department. Decided on February 1, 2008.)

Among the greatest dangers many communities face are discounting hazards. The longer the time span between emergency events, the more likely that the reality a disaster presents is forgotten and memories fade. Many organizations think in terms that hazards that have been identified will not be "that bad," should they actually occur. Communities will trust in preparedness because they have a plan, yet the plan has not been tested. Failures in ignoring, misjudging, or mishandling disasters have come in varied forms throughout history, as with Rome, Chernobyl, New Orleans, the fireproof Chicago New Iroquois Theater, and the export price of oil in the 1970s. The failure to plan is the same as planning to fail.

It is human nature to attempt to solve the problems we can handle, putting off those that are harder to address. Conversely, planning does not reduce the impact to zero, due to limited, if any, control of the environment, legislation, or growth and development. The key to effective planning is to go beyond the red lights and sirens (short-term perspective) and consider real worst-case scenarios that can be used to increase internal preparation. Planning should incorporate the adoption of new tools, for example, social media, expanding the knowledge base, and simplify wherever possible to reduce stress. The human brain operates differently when stressed.

Whether an emergency event occurs tomorrow or 20 years from now everyone should be prepared. Preparedness is gathering supplies for your personal needs for at least 72 hours. The new recommendations have increased to five to seven days.

The planning process, whether for a multinational organization or your home, must include training and exercising the plan. Begin by training people with what they already have and use and what is available. People should also be trained on how to improvise when life is at stake. In the 1980s and 1990s the primary threats were related to severe weather, earthquakes, and fires. The realities of today include the threats of previous decades and new ones such as cyber warfare, water shortages on a mass scale, and the expectations that someone will "rescue me" if I am not prepared. Since threats vary widely, it becomes quite difficult to be prepared for every type of emergency. Having the basics and learning to improvise when life is at stake could make a critical difference in how long it takes to recover from an event.

Case Study: Improvise When Life Is at Stake

Carmen Castro of the Associated Press reported on February 15, 2011, that an 84-year-old man survived five days with his car in a ditch in the Arizona desert. Henry Morello made a wrong turn while driving home February 7 from Cave Creek, a suburb in the desert near Interstate 17 north of Phoenix. He made a U-turn and went into a ditch. His cell phone battery soon went dead, proving useless as rescuers looked for him. He was able to get out of the car but could not get far and returned. He took a piece of chrome from his car and placed it on the roof, hoping someone would see the reflection. He drank windshield wiper fluid, as he became thirsty, used car mats to stay warm in 30-degree nighttime temperatures, and read a car manual to pass time. Finally, hikers noticed him inside and knocked on his window. Doctors at John C. Lincoln Hospital say he arrived in good condition.

Stress levels are higher for those in a disaster who do not know what to do next. In addition to exercises, taking advantage of Community Emergency Response Training (CERT) programs is another avenue to help community members prepared. CERT programs teach first aid and some include cardiopulmonary resuscitation (CPR). The 2009 Italy earthquake is an example of where civil protection responders (CERT equivalent) worked side-by-side with fire and police collectively to help more victims faster immediately after the earthquake. Ideally, individuals and organizations should prepare themselves to operate on their own for a minimum of 72 hours, preferably five to seven days and have adequate basic supplies to use during this time. Supplies should include medication, water, food, clothing, cash, and identification.

The New Mobility

In the late 1990s, the transition from analog to digital cellular created an explosion in wireless services. As with most technologies, it began as an expensive service for those who wanted to be on the leading edge of technology. The wireless industry has since grown into a global offering with more than four billion subscribers, overtaking wire-line connections. For many developing countries, the introduction of wireless services was the first foray into phones and accessing the Internet for its people. We are now into the fourth generation of networks, referred to as 4G, pairing mobility with broadband, enabling multimedia smart devices to select voice capabilities or not to connect to the wide-area cellular network.

The new mobility is about using networks of the future in creative ways at unimaginable scale and depth. Service providers will be managing billions of individual connections rather than counting the millions of subscribers they may have. The new mobility is about using any device or object with or without active electronics to transmit a wireless signal directly with other devices or by using the multiple local and personal area networks in the spectrum, and interconnect directly with the many different devices in the home, workplace, vehicle, or public area.

What does this "new mobility" mean for the future of an organizations; more broadly, Intranet and internal communications? As dependency on technology for day-to-day operations and in a crisis increases, organizations must overcome the challenge of granting its employees access to their Intranets that may be limited due to network bandwidth. This is especially true for geographically dispersed locations such as offshore oil rigs and organizations with multiple locations such as banks, multinationals, and universities.

One approach explored by organizations is the use of hosted cloud solutions to bring the Intranet to employees, and others who are launching mobile Intranet platforms that replicate the desktop user's experience. In some cases, to make this tactic feasible, organizations may use developers to build applications customized to their internal communications platforms, deploying these apps on top of their current Intranets to deliver an integrated digital mobile experience and access to mission critical applications typically housed on the Intranet.

Another alternative is to use social media sites like Facebook, which began as a platform limited to Harvard students, to build closed or partially open (to vendors, business partners, and providers of data and other information). Internal communications platforms can provide mobile and desktop users access to this content within a social-networking framework. This model is in development and today has only limited use.

Organizations can leverage the Intranet as a self-service repository for company-specific business processes, tools, information, and other nonstatic content. This information can be accessible using mobile and desk-bound employees via the cloud or customized applications. The goal in a mobile society is to join the age of the superconnected, migrating data from the desktop to an organization's Intranet site that can be accessed anywhere, anytime.

More organizations are planning on investing more in mobility solutions in the years to come. Lauren Brousell of *CIO* magazine stated in her article "Mobile on Your Mind" in December 2010 that business strategy is pushing the need to invest more according to 43% of the Information Technology (IT) professionals surveyed in late 2010. After surveying 276 IT professions, her study concluded that in 2011 organizations planned to purchase enterprise mobility solutions as follows:

- On average, organizations plan to invest around $870,000 in enterprise mobility solutions in 2011.
- The primary targets for mobile-technology investments are:
  - 71%   Devices (smartphones, tablets, etc.)
  - 70%   Security and data management software
  - 63%   Applications
  - 62%   Wireless services

Business strategies increasingly include emergency preparedness for securing mission critical data and securely accessing this data anywhere, anytime. This has led to the increased need for smaller mobile devices, security applications, and enabling wireless services for a cross section of employees.

The new mobility takes into consideration more people with unique or special needs. The FCC assigned the phone number 2-1-1 for community information and referral services nationwide for everyday needs and in times of crisis. In many areas where the service is deployed, individuals can register in their communities, predefining needs in the event of emergency. For example, in preparation for Hurricane Ike, the State of Texas State Operations Center used the registry information maintained by 2-1-1 to develop mass evacuation plans and activation of those place 10 days before landfall. The state knew who needed to be evacuated by air and the specific medical equipment that would be needed for transport and sheltering those who are without transportation. Plans also included those with a pet dog, cat, or bird that they want to stay with them throughout the evacuation and sheltering process.

Communications in the Future

A broad scenario for communications is to begin with how data will be stored and accessed. In 10 years, the thought is that wireless users could store all kinds of data about their lives in the cloud and authorize various algorithms and computing systems to analyze it and communicate it later—the "Invisible Internet" that intuitively links disparate personalized information across databases and devices. And just as wireless services become an integral part of the Internet, the wireless network comes to resemble the Internet itself.

For example, it can remind a person of names and addresses that are today captured in contact lists on cell phones, in e-mails, or the yellow and white pages, little tidbits of miscellaneous data for a trivial game, or facts needed for an essay. With cloud computing, it may be capable to do this without the need for personalized mobile devices such as the phones and tablets used today.

By 2020, an individual could go to a neighbor's home or a hotel room across the country and use a wireless device connecting to a television to send a message or make a call using a password, key fob, or biometrics to access the desired data. Retrieving the data on the caller could include more than their name and contact information; it could also include other unique information about the person being called, from a favorite color to grandma's favorite dish.

A *password* is a secret string of characters, phrases, or word(s) that is used to gain access to information. A *key fob* is a keyless remote entry device or a security token, a small hardware device with built-in authentication mechanisms. *Biometrics* is a characteristic and a process. Biometric is a measurable biological and behavioral characteristic that can be used for automated recognition such as fingerprints, palm, earlobe or iris. The biometric process encompasses automated methods for recognizing an individual based on measurable biological and behavioral characteristics, such as the FBI's Integrated Automated Fingerprint Identification System (IAFIS) or palm readers used for granting authorized users through an otherwise locked door. Biometric technologies have become effective tools that are being used by government agencies to support national and homeland security, justice and eGovernment missions, and by the private sector to enable new business practices that improve efficiencies and competitiveness for American industries.

New mobility in the years to come is having the answers to everything readily available with the information being more mobile and ubiquitous—available anywhere, anytime. An example of how data is intuitive is a job interview with Jennifer Jones of HP about two hours from now by an applicant we will call Ken.

> Ken's day begins with his hot cup of coffee awaiting him on the counter with the coffeemaker saying "Good morning, Ken, you have five minutes to leave and here is your coffee." Ken grabs his jacket and gets into his car that has already started. His mobile phone is a part of his

wristwatch. As soon as Ken gets in the car, his mobile device begins talking to other specialty devices in the car such as the navigation system and car radio. The car, having the information on Ken's next stop, and with Ken seatbelted, closes the car door and navigates itself through traffic. The devices begin sharing information. The radio asks "What would you like to hear?" After checking his calendar, address book, and other information, the system checks Jennifer Jones' last 10 postings on Facebook (social media) and cites to Ken the latest quarterly earnings results and history of HP using information available in the cloud. The navigation system asks him, "Do you want me to calculate and use the fastest route?" While driving, all the technology works together to prepare Ken for the interview. During Ken's commute to his interview, the navigation system comes on and says, "There is an accident ahead. You may be late for your interview. Do you want to call Jennifer and let her know?" Using a Bluetooth connection to the audio system, the mobile device starts the call to Jennifer. The data, devices, and the cloud all work together in a synchronized, integrated manner.

This scenario could be completed today if more time were available. Ken would need to conduct his research before leaving, when he can stop during his commute, or if he could arrive earlier. He could input the destination information in the navigation system once he entered his car. The devices are a means to access content and information. Devices of the future will be active nodes on the Internet, collecting data about the device and their owners and sharing the data with other devices and the network.

As we move further into the future, a rich repository in information will be in the cloud and people will become independent of devices like smartphones and tablets. Rather than carrying multiple devices, such as a smartphone, camera, tablet, and e-reader, an individual could instead use a device in anyone's home, work, or public space. To access the information in the cloud the user of a device authenticates them on the device. The cloud will be intuitive based on the pattern analysis of individuals. The greatest hurdle to overcome in adopting these innovations is the need to address personal privacy, and manufacturers from Apple to Veracruz, from televisions to smartphones, will also need to find new markets for revenue or disintegrate.

When pattern analysis for individuals is adopted, the same will be needed for emergency communications. Automation and pattern analysis can determine and deliver emergency communiqués in a manner that reaches the target audience when needed. The system will need the intelligence and mapping tools, location data, and algorithms to know the time of day, day of week, location, and business versus personal communications to enable the best form of communications from within the cloud. The need for low technology options remain: pen and paper, and word of mouth, where the dependency on power is less critical. Do you need to send the message when one is away on vacation and away from the area of concern or do you need to reach the individual regardless of where they are in the world due to an event?

As people use the Internet they are reaching the point where they expect the Internet to be always available using the devices they have on them, for example, smartphone, tablet, e-reader, laptop, and even the television and desktop. They are also looking to retain the confidentiality and privacy of their data, be it for an organization or personal.

The Internet is slowly becoming the "Invisible Internet" associated with the concept of the "Internet of Things." This is where everyday objects are connected wirelessly—from the refrigerator and television to the smartphone and vehicle. Every object has the intelligence to make decisions for itself—such as a carton of milk or can of vegetables has the ability to communicate what it is and its expiration date. Collectively, this data can create a form of ambient intelligence, allowing them to self-organize as a group. The Internet will migrate from an activity performed in front of a screen, a virtual space, to people interactions in a real space, a meet space where thousands of personal and public objects interact with one another.

Global Positioning System (GPS) and cellular triangulation sense location; accelerometers and digital compasses sense movement and direction; and digital cameras can see for the devices are available today. The challenge is processing, interpreting, and combining the data. Devices become chameleons, adapting to whichever radio standard and frequency would be optimal at any given time or place.[19]

Satellites

An increasingly critical component of emergency communications solutions are satellites having a nonterrestrial-based platform. Access to reliable and redundant communications is critical. The satellite industry and solutions cover a broad range of services that are effective for emergency situations—communications, GPS, navigation, remote sensing, and broadband. There are hundreds of geostationary communication satellites in orbit today that are both government and commercially owned and operated. Communication solutions include wireless networks, messaging, telephony, mobile satellite phones, Internet backbone, and *VSATs* (Very Small Aperture Terminals).

GPS and navigation applications include position location, timing, land or sea rescue, and mapping. Remote sensing solutions can be used for utilities (oil, gas, and water pipelines, and electrical distribution monitoring), transportation monitoring (rail, vehicle location, and status management), infrastructure planning and safety, forest fire prevention, urban planning, flood and storm watches, and air pollution management.

Broadband solutions offer services such as telemedicine, tele-education, and videoconferencing. In times of emergency, entertainment solutions can be converted to emergency use such as direct-to-consumer television, broadcasting and cable relay, and digital audio radio service (DARS).

Mobile satellite services and devices offer anytime, anywhere telecommunications that is considered reliable. Satellite services know no geography. Their communications limitation is where devices are located in areas surrounded by heavy metals and concrete, such as under bridges or in the basements of some commercial buildings. Satellite services are effective in rural areas where available communications options are often limited.

The next generation of fixed satellite services (FSS) and mobile satellite services (MSS) will provide greater use of automation to deliver more power and enhanced spot beam capabilities. Next-generation ground solutions will also take advantage of automation and miniaturization of component in devices to make them smaller and develop hybrid networks that can offer satellite, cellular, Wi-Fi, and other services under a single provider.

Through triangulation of calls, the location of emergency calls is possible. Satellite is also the backbone of most of national TV and radio broadcasting. Search-and-rescue operations also benefit from the use of satellite communications. Maritime operations can send distress calls. The military can use the navigation and GPS systems for unmanned aerial vehicles, cruise missiles, and precision munitions for war fighting.[20]

For emergency purposes, satellites have the advantage of limited dependency on terrestrial networks that can be subject to physical damage and stress due to high demand during an emergency. Satellites are effective devices for emergency responders in the field, particularly when communication devices are placed in rugged hardware made to withstand extreme conditions.

THE NEW MOBILITY PUSHES THE FEDERAL COMMUNICATIONS COMMISSION (FCC) CHANGES TO 9-1-1 SERVICES

The FCC has begun to revolutionize the United States' 9-1-1 services for consumers and first responders by focusing on the new mobility, the

Next Generation 911 (NG911). NG911 will enable the public to obtain emergency assistance by means of advanced communications technologies that go beyond traditional voice-centric devices. NG911 will use text messaging, e-mail, video, and photos from mobile and landline broadband services. In 2010, there were more than 270 million wireless consumers nationwide and nearly 70% of all 9-1-1 calls were made from mobile handheld devices.

Today's 9-1-1 systems support voice-centric communications and are not designed to transfer and receive text messaging, videos, or photos. Many Americans, particularly those with disabilities, rely on text messaging as their primary means of communication. Sharing of timely and relevant videos and photos would provide first responders with on-the-ground information to help assess and address emergencies realtime. For example, these technologies could help report crimes as they are happening, giving law enforcement officials an increased advantage when responding. Prior to the implementation of NG 911, several factors must be addressed, namely

- Technical feasibility and limitations of text messaging video streaming and photos
- Consumer privacy issues, particularly the sharing of personal electronic medical data
- Technical and policy standards
- Consumer education and awarenes
- Intergovernmental coordination
- Coordination with public safety agencies[21]

SUMMARY

Working collaborative is a core competence needed in an emergency among law enforcement, fire, and mutual aid partners. This collaboration, coupled with coordinated emergency drills and planning by local, state, and federal agencies, is key to being prepared for the changing realities of disasters. When a disaster strikes, multiple entities must quickly and effectively respond. First response must be able to talk to each other, regardless of the agency they work for. The inability of some agencies to effectively and seamlessly communicate across boundaries and jurisdiction can create delays, are perceived barriers to action, and question who's involved in the control of communications systems and services that in any way impair response to major incidents. Communications interoperability is critical and is mandated by the federal government.

ENDNOTES

1. Munich Re. (2010). *TOPICS GEO. Natural catastrophes 2010* (US version). http://www.munichre.com/publications/302-06295_en.pdf (accessed April 15, 2011).
2. U.S. Census Bureau. (2010). *Quickfacts by state.* http://quickfacts.census.gov/qfd/states/22/2255000.html (accessed January 5, 2011).
3. Munich Re. (2010). *TOPICS GEO. Natural catastrophes 2010* (US version). http://www.munichre.com/publications/302-06295_en.pdf (accessed April 15, 2011).
4. Munich Re. (2010). *TOPICS GEO. Natural catastrophes 2010* (US version). http://www.munichre.com/publications/302-06295_en.pdf (accessed April 15, 2011).
5. Munich Re. (2010). *TOPICS GEO. Natural catastrophes 2010* (US version). http://www.munichre.com/publications/302-06295_en.pdf (accessed April 15, 2011).
6. Harmeling, Sven. 2008. *Global climate risk index.* Berlin: Germanwatch e.V.B
7. The World Bank. (2011). *The recent earthquake and tsunami in Japan: Implications for East Asia.* http://siteresources.worldbank.org/INTEAPHALFYEARLYUPDATE/Resources/550192-1300567391916/EAP_Update_March2011_japan.pdf?cid=EXTEAPMonth1 (accessed April 30, 2011).
8. O'Hanlon, Larry. (2011). *SuperVolcano. What's under Yellowstone.* Discovery Channel. http://dsc.discovery.com/convergence/supervolcano/under/under.html (accessed March 3, 2011).
9. http://nctr.pmel.noaa.gov/tsunami-forecast.html
10. NOAA. (2010). *El Niño.* http://www.noaa.gov (accessed February 20, 2011).
11. The International Research Institute for Climate and Society. (2010). http://portal.iri.columbia.edu/portal/server.pt?open=512&objID=430&parentname=CommunityPage&parentid=3&mode=2&in_hi_userid=2&cached=true (accessed February 23, 2011).
12. Sandia National Laboratories, German, John, Crawford, David, and Breckenridge, Arthurine. (1998). News Release. *Real (not reel) deep impacts: Sandia scientists predict what an asteroid strike would look like, really.* http://www.sandia.gov/media/comethit.htm (accessed May 2, 2011).
13. World Health Organization. (2011). *Water related diseases.* http://www.who.int/water_sanitation_health/diseases/diarrhoea/en/ (accessed February 28, 2011).
14. World Health Organization. (2010). *Communicable diseases and severe food shortage.* Geneva, Switzerland: WHO, Disease Control in Humanitarian Emergencies Department of Global Alert and Response.

15. Research Australia's http://www.aph.gov.au/library/pubs/ClimateChange/effects/social/health/health.htm).
16. Williams, Terry. (2008). SB-08-02 *NTSB urges bridge owners to perform load capacity calculations before modifications; I-35W investigation continues*. Washington, DC: National Transportation Safety Board. http://ntsb.gov/pressrel/2008/080115.html (accessed May 5, 2011).
17. DeWeese, Steve (October 9, 2009). US-China economic and security review commission report on the capability of the people's Republic of China to conduct cyber warfare and computer network exploitation. Prepared for The McLean, VA: Northrop Grumman Corporation c/o US-China Economic and Security Review Commission.
18. Obama, B. (2011). Obama hails death of Muammar Gaddafi as foreign policy success. *The Guardian.* http://www.guardian.co.uk/world/2011/oct/20/obama-hails-death-gaddafi (accessed November 7, 2011).
19. Fitchard, Kevin. (2009). *Wireless 2025: A look at wireless in the year 2025.* http://connectedplanetonline.com/wireless/news/wireless-future-year-2025-0409/ Excerpted with permission. All rights reserved. http://license.icopyright.net/7.5528-24468 (accessed April 29, 2011).
20. Cavossa, David, *Satellites as Critical Infrastructure*, http://www.sia.org/industry_overview/SatellitesasCriticalInfrastructure.ppt (accessed April 24, 2011).
21. Donovan, Patrick. (2010). FCC Takes First Step to Help Revolutionize America's 9-1-1 Services for Consumers, First Responders... Rapid Sharing of Videos, Photos and Data to Improve Emergency Response. *FCC 10-200.* Washington, DC: FCC's Public Safety and Homeland Security Bureau.
22. U.S. Geological Survey. U.S. Department of the Interior. (2009). *2008 Bay Area earthquake probabilities.* http://earthquake.usgs.gov/regional/nca/ucerf/ (accessed September 19, 2011).
23. U.S. Geological Survey. U.S. Department of the Interior (2011). *Overview of the ARrkStorm scenario.* Multihazards Demonstration Project. http://pubs.usgs.gov/of/2010/1312/ (accessed September 19, 2011).
24. U.S. Department of Transportation. Federal Highway Administration. (2009). *2008 status of the nation's highways, bridges, and transit: Conditions and performance.* Report to Congress. Washington, DC. http://www.fhwa.dot.gov/policy/2008cpr/index.htm (accessed September 19, 2011).

CHAPTER 6

Emergency Communications Framework

> The emergency communications technical "... framework should include coverage requirements, network architecture, security, robustness and resiliency ..."
>
> The FCC, January 25, 2011

Any organization or community can expect to experience emergencies. These events could threaten the health and safety of an organization, community, and property. An event could necessitate the implementation of protective actions for the population at risk. Hurricanes Katrina and Rita were events that caused significant harm and damage. Numerous studies were conducted after these storms questioning the adequacy of U.S. disaster response capabilities. The areas defined as having the greatest concern were:

- Situational assessment and awareness
- Emergency communications
- Evacuations
- Search and rescue
- Logistics
- Mass care and sheltering

Communications play a critical role in emergency operations. A reliable communications system is essential to obtain information on emergencies Reliable systems are needed to direct and control resources responding to those situations. Equipment and computer-based systems should be available to provide the communications necessary for emergency operations.

As we know, power outages and disaster may disrupt computing systems, telecommunications, television systems, and radios, which carry warning messages and provide instructions. There are occasions when a disaster strikes without warning, and the public information system cannot react fast enough to properly inform the community. The emergency communications framework should incorporate information about the communications equipment and capabilities available during emergency operations for an entity to manage during these times.

The United States is nowhere near where it needs to be in terms of first responder communications, according to the Federal Communications Commission. The FCC chairman Julius Genachowski stated that steps have been taken to launch the framework for the new emergency communications network. This action is to ensure that the nation's public safety broadband network is interoperable nationwide as it sets aside spectrum and create a network. The process begins with designating the *Long Term Evolution* (LTE) mobile communications standard as the broadband platform for the network. This action is followed by creating a technical framework that encompasses coverage requirements, network architecture, security, robustness, and resiliency.

The cost of the interoperable network is estimated to be between $12 billion and $16 billion over the next 10 years. This network is based on recommendations from the 9/11 Commission to create an interoperable public safety network. The key to getting started is funding. One approach under consideration is to reauction the D block of spectrum for a public–private partnership. This would create an interoperable emergency communications network for commercial use that in an emergency would be turned over to first responders. The standard would also enable mobile broadband for sending video, photos, data real-time, and on-site scanning and diagnostics.[1]

During periods of an emergency, information regarding protective actions to minimize loss of life and property should be provided to the community. The framework should contain information on the means, organization, and processes to use to provide the appropriate information and instructions to first responders and the public during emergencies. Those involved in developing, maintaining, or using the framework and systems should be provided with disaster-related education. This should come well in advance of emergencies to reduce the likelihood of personnel and the public placing them in harm's way, necessitating an additional emergency response.

Those who apply the emergency communications framework within their organization should consider and review assumptions that would be applicable to them. Assumptions can be mapped to the interrelationships

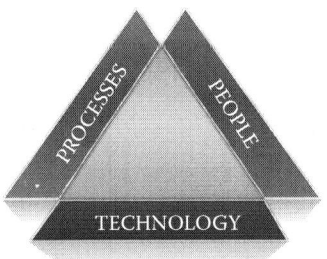

FIGURE 6.1 Emergency Communications Framework interrelationships.

among people, processes, and technology as illustrated in Figure 6.1 (Emergency Communications Framework interrelationships).

- *People*
    - Media interest could go beyond the capacity of available public information staff.
    - Organizations should know their primary sources of emergency information.
    - Some people who are threatened by a hazard may ignore, not hear, or not understand the warnings issued.
    - Many people will not participate in or retain pre-emergency education until they are affected.
- *Processes*
    - Adequate communications are vital for effective and efficient warning, response and recovery operations.
    - Warnings are delivered in a timely manner to the community threatened by impending or active emergencies.
    - Make provisions to provide warnings to special-needs groups.
    - Have an effective program that combines education and emergency information. Special emphasis must be placed on the delivery of information during emergencies and disasters.
    - Local media will be requested to disseminate warnings and public information during emergencies.
    - The predefined public information officer (PIO) will take the lead role in public information issues. The PIO will be supported by other PIOs for an organization, as needed.
- *Technology*
    - A particular hazard or incident may neutralize current communications.
    - Additional communications equipment and services will be required the longer the incident remains active.
    - Outside agencies and the general population may supplement emergency operations.

- The local National Oceanic and Atmospheric Administration (NOAA) Weather Radio station will broadcast weather watches and all-hazards warnings issued by the National Weather Service (NWS). Weather radios are activated when such messages are broadcast.
- Local radio and television stations will broadcast Emergency Alert System (EAS) messages when appropriate.

PUTTING THE FRAMEWORK INTO OPERATIONS

Communications play a critical role in emergency operations. Extensive communication networks and facilities are needed the longer the incident remains active or extends beyond the scope of a single entity or a single shift. Effective communications goes further than alerting people to the presence of danger. Emergency communications needs to guide affected persons to safety, warn of hazards and areas to avoid, advise to shelter in place, and return normal activities when it is clear. This process applies to natural disaster, fire, environmental accident, workplace violence, or terrorist threat. Regardless of the hazard type, reliance on a single communication system means that a percentage of persons will not be reached in a timely manner. When the capabilities of multiple communication systems are properly coordinated, response activities become more effective and efficient. More communication methods used means you can reach more people.

Building the framework begins with defining the systems that an entity could used for emergency communication. Begin with "no tech" options of pen and paper, word of mouth, or a bullhorn. No technology, also referred to as *no tech*, is any communication system that does not require a power source. Despite best efforts, it is a given that at some point technology will fail. It is essential to have solutions that do not require technology as a part of the toolkit.

Communication Technologies—No Tech		
• Pen and paper/posters	• Bullhorn	• Gestures
• Word of mouth/face-to-face	• Signing	• Picture exchange

Low- and medium-technology options, or *low-/medium-tech* communication solutions are any communication systems that require a source of power, are very easy to program, or require some level of training to adequately program and maintain the device. It is important that the crisis communications toolkit contain solutions from each category of communications. Low-/medium-technology options include the following:

Communication Technologies—Low/Medium Tech		
• Landlines • Internet • Paging • Fire alarm pull station • Loudspeakers • Media notification	• Texting • Company e-mail • Facsimile (Fax) • Public address (PA) systems • Intercoms • Phone trees/telephony	• Voicemail • Ham radios • Switchboard greetings • Two-way radios (200, 400 MHz)

Technology solutions can cover a broad range of technologies. High technology, or *high tech*, is any communication system that requires a power source and extensive training to competently program and maintain the device. Most organizations have access to one or more solutions within their organization, such as their own website, that are classified as high tech. Other examples of high-tech solutions are

Communication Technologies—High Technology	
• Company website • Communications-on-Wheels (COW) • Computer Aided Dispatch (CAD) • Internet • High-frequency (HF) radio • Networked-attached computers, projectors, etc. • Digital Displays (LEDs, changeable message signs, etc.)	• Voice over Internet Protocol (VoIP) • Social networking (Twitter, Facebook, etc.) • Augmentative/Alternative Communication • Satellite telephones • Two-way radios (700, 800 MHz) • Voice Output Communication Aids (VOCAs) • Smartphones

Existing communications technologies available within an entity serve as the initial and basic communications to be used for emergency operations. Most buildings have fire alarms or smoke detectors. Most organizations have e-mail and a phone system. The next step is to prioritize the use of each technology by type of notification.

For example, the fire alarm system, PA system, and text messaging are tools recommended for use in a fire. The fire alarm and PA system notify persons to take immediate action during the response phase. Text messaging could provide another method to communicate the need to evacuate during a fire for those with a mobile device with them. Once occupants have cleared the building, a follow-up communication would be to advise occupants and other affected parties on the status of the building fire or if the alarm was false. Subsequent messages would offer additional instructions.

Other methods of communications to consider during this phase of the emergency are e-mail, switchboard greetings, and the company website. The follow-up messages would be classified as advisories and information messages. As the situation moves from response to recovery, more informational messages are relayed giving occupants more detailed information and instructions—an alternate place to go to, when the site will return to normal operations, etc. Other communications may be introduced. The company website, e-mail, text messaging, switchboard greetings, posters, signage may be used as the method of communications for ongoing updates on the situation.

Hazards with a medium to high level of probability are threats that an entity should start with in building its communication plan. The plan should include consideration for:

- What technologies will be maintain
- Procedures used for communicating that are used for field operations during emergency operations
- What technologies are available to keep communications going between field operations and Emergency Operations Center (EOC).

It is critical that EOC communications be up at all times, so that they can stay informed of their operations at all times, by whatever means available. This means the framework should consider dependability, backup, and redundancy requirements.

The day-to-day emergency communication capabilities may be insufficient to meet the increased communications needs for an emergency. The plan should include resources that are outside the management of an entity such as local or state agencies, amateur radio operators, and business/industry radio systems. Communication technologies that are typically beyond the management of an organization are

Other Communication Technologies		
• Highway alert signs • Government Alert Systems (Amber, Silver, Traffic) • Dynamic Message Signs (DMS)	• Intra/Inter Computer Aided Dispatch (CAD) • Anonymous Wireless Address Matching Systems • Regional Incident Management Systems • Call boxes	• Doppler Radar Imaging • Automated Flood Warning Systems • Voice messaging attached to fire alarm system

SENDING MESSAGES

Effective emergency communications guide affected persons to safety, warn of hazards, inform of areas to avoid, and state whether to evacuate or shelter in place. A follow-up message should be provided stating when it is clear to return to normal activities. Activation of emergency communications often begins with advisories, that are escalated to alerts and warning. The primary objective of a warning system is to notify the affected community of emergencies and disseminate timely, accurate warnings and instructions to the population at risk from the threat or occurrence of an emergency. Rapid dissemination and delivery of warning information and instructions may provide time affected persons to take action to protect themselves and their property.

Roles and responsibilities should be preidentified and documented, and the parties responsible for communications trained. A key function mentioned in the framework is the *local warning point (LWP)*. The LWP is the focal point of the warning function who is authorized to send a warning, and select where and when the warning will be sent, given the available information. For example, the LWP for a college or university may be the Public Safety Director within the Dispatch Center.

The LWP can receive warning of an actual or potential emergency from a variety of sources. Sources may include employees; an organization's executives; the local office of emergency management; outside agencies; local, state, and federal agencies; local officials; businesses; industry; the news media; or the public. The systems by which the LWP may receive a warning vary from entity to entity—from a telephone call or a dispatcher monitoring a camera feed. The key to the process is the LWP verifying that the warning information using the information is available and then disseminating the necessary information to specified officials, internal departments, and affected persons.

The LWP may be authorized to activate local warning systems and warn affected persons immediately of time-sensitive warnings, such as issuing a "lockdown" when an active shooter situation is in progress. In other situations, an official or executive must first approve activation of the warning system and determine the appropriate instructions to accompany the warning before the message is disseminated to the affected individuals. A third scenario is where an EOC assumes responsibility for creating warning messages and public instructions that are sent to LWPs for a given area or provided to the media for dissemination. The receiver should have some predefined validation or handshake process in effect to confirm the identity of the sender and the authenticity of the message before the message is redistributed.

Emergency communication systems, as a best practice, should be reserved for emergency preparedness, advisories, warnings and alerts, evacuations, sheltering, and testing of the system to ensure soundness. Emergency communication systems can also be used to increase public awareness about potential hazards and how people should prepare for them. Emergency communication systems should not be used for other purposes such as marketing or building contact lists and profile information for other nonrelated systems or selling.

In most instances, at least two messages are needed. The initial message advising of a situation or potential situation. The second message is a closing statement advising the affected parties that the situation is under control or has returned to normal. *"All clear"* is a common phrase used by first responders to indicate that a situation is under control. Multiple messages may be sent to keep the affected parties informed on the progress of events and to help control rumors. The pulse of the public should be measured throughout this process to determine the level of effectiveness the emergency communiqués are having. Receipt of advisories, alerts, and warnings of actual emergencies or the threat of emergency information can be redistributed at their discretion.

RECEIVING WARNINGS FROM TRUSTED SOURCES

The National Warning System (NAWAS)

NAWAS is a 24-hour nationwide, dedicated, multiple-line telephone warning system to disseminate civil emergency warnings from federal agencies to State Operations Centers (SOCs). This system is operated by the U.S. Department of Homeland Security's (DHS) Federal Emergency Management Agency (FEMA) and controlled from the FEMA Operations Center (FOC) in Washington, DC, and the FEMA Alternate Operations Center (FAOC) in Olney, Maryland.

The FOC is a central 24/7 operation that collects, analyzes, and disseminates time-sensitive information. The information is used to advise key decision makers in a region or regions of warnings on domestic and worldwide events. The FOC also facilitates the information flow between FEMA and the National Response Framework (NRF) partners, alerts and activates response teams, and notifies and reports known facts to FEMA and DHS staff about all-hazards events.

NAWAS is used to disseminate warning information on natural and technological disasters to nearly 2,200 warning points throughout the United States and the Virgin Islands. Warnings include acts of terrorism, aircraft incidents and accidents, earthquakes, floods, hurricanes, nuclear incidents and accidents, severe weather conditions, tsunamis, and winter storms/blizzards. NAWAS is able to issue warnings to one, several, or all stations as dictated by the situation.

National Weather Service (NWS)

The NWS Warning System transmits weather information 24 hours a day to the public. These weather-warning messages are issued by the

NWS Weather Forecast Offices and NWS weather centers, such as the National Severe Storms Forecast Center, and the National Hurricane Center. The NWS disseminates weather forecasts, watches, and warnings via the NOAA Weather Wire Service, which is a satellite communications system that broadcasts to specialized receiver terminals. A range of weather messages, from severe weather watches and warnings to amber alerts, is provided. The NOAA All-Hazards Network also sends life-saving information nationally, regionally, or locally.

State Government and Local Notification

A State Operations Center (SOC) may also issue warning messages from the governor or other key state officials to local governments within a state, such as hurricane evacuation routes along coastal areas for large-scale evacuations or travel alerts for a region. For example, the State of Texas has the *Texas Warning System (TEWAS)*, a state-level extension of NAWAS. TEWAS has a dedicated telephone warning system linking the SOC State Warning Point with the Department of Public Safety (DPS) and NWS Aviation Weather Centers (AWCs) located across Texas.

Notification of Local Officials

When local officials receive warning messages, the messages are disseminated to the affected population and notify those who are able to respond quickly to the scene and take the appropriate action. Local protocols dictate the communication methods used such as sirens for tornado alerts, reverse 9-1-1, and radio and television broadcasting.

Organizations and Individuals

Warnings of emergencies can come from organizations or individuals who have learned of an incident, such as an active shooter in the workplace or hostage situation in a bank. These calls usually are reported through 9-1-1 or an organization's in-house emergency telephone number or system. The number of calls of this type is increasing for a broad range of activities, given the use of cell phones with cameras and the national campaign "If you see something, say something" promoted by the U.S. Department of Homeland Security Secretary Janet Napolitano. This campaign was launched in February 2011 in partnership with the National Basketball Association (NBA).

Individuals have a general responsibility or duty to notify emergency resources when they become aware of an emergency that could pose a threat to public health, safety, or property. Individuals are expected to follow instructions given by government officials to protect people or property. It is important to confirm information reported by individuals before issuing a public warning regarding a situation. For example, at a college or university, it is not uncommon for a student to report that he or she has been infected with bacterial meningitis and therefore all should be notified. In speaking with the public health authority, you learn that the student may have been ill but had the flu instead. If you had elected to act based on the report from the student without confirmation from a trusted source before sending a warning message to the student population as suggested by the sick student, the likelihood that the false warning message would have led to unnecessary panic in the community and negative image for the organization was high. The situation would have required crisis management and the follow-up of subsequent communiqués to correct the false information.

DISSEMINATION OF WARNINGS TO THE PUBLIC

In the initial stages of an emergency, the Local Warning Points (LWP) will, within the limits of the authority delegated to it, determine if a warning should be issued, formulate the warning message, and disseminate it. When an Emergency Operations Center (EOC) has been activated, the EOC will normally determine who needs to be warned and how, and will normally formulate warning messages and public instructions. An LWP or EOC usually initiates delivery of warnings through activation of the appropriate warning systems available to them. Systems commonly found within most organizations are

- E-mail
- Company website
- Fire alarm system
- Switchboard recording
- Voicemail
- Paper postings/signage
- Text messaging
- Word of mouth

Systems that are often used within a community for disseminating messages given the probability and frequency of hazards within the area and budget are

- E-mail
- Community website
- Fire alarm system
- Radio/television broadcasting
- Switchboard recording
- Sirens/horns
- Word of mouth
- Print media
- Reverse 9-1-1 to registered telephone numbers
- Paper postings/signage
- Text messaging
- NOAA Alerts
- Highway Video Signage
- Public Address Systems (equipped on police and fire vehicles)

WHEN TO ACTIVATE

Activation of emergency communiqués should be used prudently. Emergency messaging should comply with rules and regulations established within an entity, community, state/providence, or nation. Overuse or misuse of emergency messaging can result in problems. Whether to activate or send emergency communications and the appropriate method to use should take into consideration the severity of the situation, how quickly the message needs to be received by affected persons, and the alternatives available for sending the message. The severity of the situation asks:

- Will the communiqué aid in reducing loss of life or substantial loss of property?
- How quickly does the message need to be received by the affected persons to avoid an adverse impact?

The alternative is a review of the methods available for disseminating information to determine whether they are adequate and can ensure rapid delivery.

It takes time to develop best practices. Notification is the last step in a series of tasks to assess the situation, and determine who should be notified, how quickly, and what the message should be. Failure to complete the prenotification activities can lead to overuse and misuse of emergency communications.

CASE STUDY: Overuse and Misuse of Emergency Communications

People complain that their voicemail box is full of dozens of messages from Emergency Alert Systems on what some felt are nonemergency events such as how hot it is outside. Many wish they could select the type of emergency messages that requires their attention affects them. It is not that they want to downplay the messages related to the dangers of extreme heat, but the information that a local cooling center is open is not something of interest to them. Understanding the importance of this type of communications they do not want to unsubscribe and miss important alerts in the future.

It is important to complete as much of the prenotification activities as possible before an event occurs. The process begins with identification of the threats that exist in a given area, the target audience you want to communicate with, and then development of prescribed messages where possible. The authority to release emergency warning messages for broadcast is restricted. Restrictions are based on roles rather than the individuals who fulfill the roles unless the individual assigned is unable to fulfill his or her role. The process should include how to handle the absence-authorized personnel, ideally having three individuals identified for each role and how to handle the delegation of authority.

The process may have multiple subprocesses for disseminating messages internal to an entity. Concurrently, there may be multiple subprocesses to transmit emergency messages to the media or an outside agency broadcast system. Technologies commonly used are telephone, e-mail, and fax. Upon receipt, the media may broadcast the message as given, truncate it, or relay a summary of it. Anything less than broadcasting the message as given can lead to its misinterpretation, which could result in the target audience taking inappropriate action. For example, a university sends a message to all media outlets stating:

> "North States University-Chicago campus is closed. The Bellaire campus opens tomorrow at 8:00 a.m. for normal operations. The Broadway campus is permanently closed."

The local media station has truncated the message and broadcasts the following message:

> "North States University is closed."

Truncating the message resulted in the target audience not going to any of the campuses thinking that all campuses were closed. This meant the Bellaire campus most likely experienced a fewer students and employees attending when the campus reopened than would have come. Also, the confusion most likely would generate calls from frustrated students, parents and employees unsure on what action to take.

Warning messages can start at the top and work their way down and throughout a community. Warning messages can also start at the ground level and work their way up through the community (see Figure 6.2 Initiation of warning communications). Earthquakes and hurricanes are among the natural disasters that are typically communicated down from the national or regional organizations responsible for tracking this activity. Conversely, a refinery fire, shooting, or nuclear reactor meltdown is communicated up and out to the community. Notification to local government agencies and officials becomes necessary. These agencies then become responsible for issuing time-sensitive warnings to inform affected populations and alert other government agencies such as regional, state, federal, or tribal.

Collection and dissemination of information should occur as soon as possible. Seconds can be vital, such as with a major earthquake or tornado. Activation of emergency communications can come from a varied list of resources, authorization usually based on the nature of the event. Appropriate personnel authorized to release messages commonly used within a community is the public information officer (PIO) or a qualified public information staff member. There are situations where an organization is legally bound to notify within a prescribed period of time such as

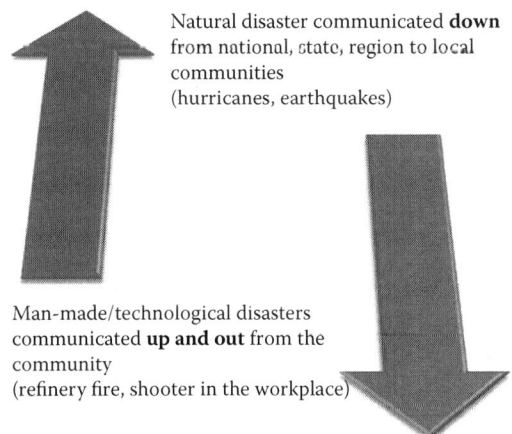

FIGURE 6.2 Initiation of warning communications.

the The Jeanne Clery Disclosure of Campus Security Policy and Campus Crime Statistics Act (Clery Act). The Clery Act is a U.S. federal statute requiring colleges and universities that participate in federal financial aid programs to give timely warnings of crimes that represent a threat to the safety of students or employees, and keep and disclose information about crime on and near their respective campuses. Compliance is monitored by the U.S. Department of Education, which can impose civil penalties up to $27,500 per violation and suspend participation in federal student financial aid programs.

Activations for People with Disabilities and Others with Functional and Access Needs

Emergency communication systems should have the capacity to reach every person within an affected area. While the designation "access and functional needs population" is used broadly to describe populations that are hard to reach, vulnerable, or otherwise grouped for purposes of description, emergency professionals recognize that no one term can accurately define these multiple groups of individuals.

> Access and functional needs populations are "... groups whose needs are not fully addressed by traditional service providers or who feel they cannot comfortably or safely access and use the standard resources offered in disaster preparedness, relief, and recovery. They include ... those who are physically or mentally disabled (blind, deaf, hard of hearing, cognitive disorders, mobility limitations), limited or non-English-speaking, geographically or culturally isolated, medically or chemically dependent, homeless, frail/elderly, and children."[2]

As a part of the prenotification process, the sender must know what subgroups make up their population, where the people in the groups live and work, and how they best receive information. On the surface, this may be thought of as obvious; research indicates that many subgroups are overlooked. Many communities have not comprehensively defined or located their subgroups. For those who have, it remains a work in progress. Many approaches can be used.

A method used for those who are geographically or culturally isolated is *route alerting* and door-to-door warnings. Vehicles equipped with sirens and public address systems are used for route alerting. Response personnel going door to door may also deliver warnings. Both of these methods are effective in delivering warnings. These methods are also labor intensive, time consuming, and may not be feasible for large areas.

To alert those who are visually impaired, good communication methods that are commonly available are voicemail, public address systems, radio, the audio of television and cable broadcasts, route alerting, and door-to-door notification. For those with assistive technologies, their computer interaction can include e-mail, text messaging, and Internet with audio readers or screen magnifiers.

For the hearing-impaired, e-mail, the Internet, video displays, text messaging, captioned EAS messages, television and cable broadcast, route alerting and door-to-door notification are effective communications methods. Assistive technologies, such as telephone calls using specially equipment telephones, are an effective approach for the hearing impaired.

Individuals who have limited English proficiency or limited proficiency in any language that is native to the community is a special population that requires consideration in an emergency. A message delivered in English when an impacted subgroup speaks German has little meaning without interpretation. Use of universal symbols or pictures is an effective approach when translation services are not available. This approach is also effective for those who have cognitive impairment and for children.

Warnings to Outside Agencies

The LWP is responsible for warning their target audiences. Target audiences in some instances may include adjacent or nearby communities that could be affected by an emergency originating within the LWP's community. For example, an earthquake in one country could trigger tsunami warnings in other communities far away, or an explosion at a nuclear power facility could also have far-reaching effects.

> **CASE STUDY: MAGNITUDE 9.0 EARTHQUAKE TRIGGERS TSUNAMIS AROUND THE PACIFIC OCEAN AND THE MELTDOWN OF A NUCLEAR POWER FACILITY**
>
> Thursday, March 10, 2011 at 11:46:23 PM (CST)—an 8.9 magnitude earthquake struck 24.4 km (15.2 miles) deep near the east coast of Honshu, Japan. The earthquake was preceded by a series of large foreshocks over the previous two days, starting with an M 7.2 and continued with three earthquakes greater than M 6 on the same day. The earthquake spawned numerous tsunami warnings across the Pacific and a ferocious tsunami that caused massive destruction—flattening whole cities, starting raging fires, and killing hundreds. Tsunami warnings were issued for the Pacific coastal regions outside California, Oregon, Washington, British Columbia, and Alaska. The tsunami, earthquake, and aftershocks have led to multiple catastrophes, including the meltdown of a nuclear reactor.[3]

Whether internal, external, or both, developing and maintaining a list of communication resources and key contacts before an event occurs is a critical step in the communication process.

PHASES OF EMERGENCY MANAGEMENT

The capabilities needed in a crisis are built upon the appropriate combination of people, skills, processes, and assets. The communications systems an entity has during a crisis, whether a minor incident or a catastrophic disaster, must be operable and satisfy daily internal communications and emergency communication requirements. Ensuring availability of these systems is a part of the preparedness stage. Once operable, systems should have communications interoperability, enabling internal resources to communicate with external resources within a given area in real time and as needed. An entity must continuously assess and improve their emergency communications capabilities by incorporating design, staffing, and resources to have and maintaining a rapidly deployable, responsive, interoperable, and reliable emergency communications capability.

To build the capabilities needed in a crisis means to consider the activities needed during each phase of emergency management—the disaster life cycle. The five phases of emergency management are prepare, response, recovery, protect, and prevent. The process is cyclical in that it is an ongoing process, as depicted in Figure 6.3 (Emergency management phases).

FIGURE 6.3 Emergency management phases.

Preparedness

Preparedness ensures that if an emergency occurs, people and organizations are ready to get through the event safely, and respond effectively. Preparedness includes planning on defining what should be done if essential services are disrupted, contingency planning, and practicing the plan. Everyone and every organization is responsible for safeguarding and being prepared. This phase within the crisis communications process is focused on developing and testing processes for getting information out to the target audiences. Key activities are

- Review emergency notification list of key officials and department heads.
- Develop and document communications procedures.
- Implement communication procedures using communications operating instructions.
- Acquire, test, and maintain communications equipment.
- Ensure there are spare parts on hand for mission-critical communications systems and prearrangements are in place for rapid resupply in the event of an emergency.
- Train personnel on appropriate equipment and communication procedures.
- Conduct periodic communications drills at least semi-annually, quarterly recommended.
- Regularly test warning systems at least quarterly, monthly recommended.

- Train emergency personnel, including public information staff, on transmitting emergency messages.
- Conduct public education on warning systems and the actions that should be taken for various types of warnings, including developing and distributing educational materials.
- Develop prescripted emergency information for release during emergencies.

Response

Response begins once an emergency is detected or threatens. *Response* communicates that an event is or will be occurring and the appropriate action to take:

- Get people to take the appropriate measures to get out of the way of danger or at least minimize its effects
- Mobilize and position emergency equipment and supplies
- Bring back online any damaged services or systems
- Provide needed food, water, shelter, and medical services

For many incidents, the first 72 hours after the occurrence of an incident are the most critical for individuals since services may be delayed. Local responders, local and state government agencies, and private organizations take action. When the scale of destruction is beyond local and state capabilities, federal assistance is available through government agencies such as FEMA in the United States or Danish Emergency Management Agency in Denmark. Key activities to accomplish in this phase of the disaster life cycle are

- When an EOC is activated, the Emergency Management Coordinator determines the communication personnel requirements given the incident.
- Initiate warning procedures identified.
- Ensure that all incident management entities use a common language during emergency communications.
- Activate the local warning system to alert the affected population of the emergency and provide instructions.
- Develop and publish public information.
- Monitor media sources to determine the need to clarify issues and distribute updated instructions.
- Discontinue warnings when they are no longer required.
- Conduct media monitoring to determine the need to clarify issues, control rumors, and distribute updated public instructions.

Recovery

Recovery consists of rebuilding after a disaster. Recovery activities can take months, even years. Services, infrastructure, and the lives and livelihoods of many may be affected, and the costs may be staggering for private and public entities and individuals. In many countries, including the United States, financial assistance from the federal level requires a president, prime minister, or other government leader to make an emergency declaration. In the United States, the president will make a Presidential Disaster or Emergency Declaration that will activate a host of services that are led by FEMA. Key activities carried out during the recovery phase are

- Activities in the emergency phase continue until emergency communications are no longer required.
- Advice to the target population when the emergency has been terminated, and issue of instructions for the safe reoccupation of damaged areas.
- Providing public information relating to the recovery process and programs.
- Gathering record of events.
- Evaluating the effectiveness of public information and any education programs used in preparation and prevention.

Mitigation

A hazardous material spill can be prevented, but an earthquake cannot. *Mitigation* of what can be prevented and that which we can be better prepared for is what this phase of emergency management addresses. Mitigation is achieved through partnerships at every level within a community and at the state and federal levels. It is ongoing and focused on risk reduction—lessening the impact disasters have on people and property through prevention and insurance. Mitigation efforts may involve moving computer equipment to higher floors within a building that is in an area prone to flooding, engineering buildings and infrastructures to withstand earthquakes, or updating or enforcing effective building codes to protect property. The key activities for this phase of emergency management as it relates to the crisis communications process are

- Assess vulnerabilities in the communications system.
- Establish plans to lessen the identified vulnerabilities.

- Identify emergency power requirements and alternatives if is not available.
- Develop contingency plans for interruptions in communications that include the use of no technology.
- Conduct hazard awareness programs to help emergency response and management team members on what they should do given their role and responsibilities.
- Develop systems and processes to enhance information dissemination during emergencies.
- Establish an effective public warning system that includes redundancy and incorporates the needs of special populations.
- Assess and update communication systems so that they remain current technology.
- Adopt new methods of warning that increase the ability to reach populations within the community that are underserved by current systems.
- Conduct public education designed to prevent individuals from taking unnecessary risks during an emergency, such as going to restricted areas or working around power lines.

Prevention

The focus of the emergency management phase, *prevention*, is to prevent harm to people, property, and the community, including the environment, from actual or threatened emergency events. This phase covers development, training, outreach, compliance, inspection, and enforcement. The key activities of prevention as it relates to crisis communications are

- Gather and apply intelligence and other information to activities including countermeasures as deterrence operations.
- Conduct comprehensive outreach regarding personal preparedness, and ensure public awareness of their responsibilities.
- Develop and implement emergency communications technology for, and collect comprehensive data on the community to validate how best to service their communication needs.
- Investigate major gaps in communications to identify and understand causes and potential mechanisms to prevent a recurrence.
- Analyze collected data to determine whether program changes are needed to prevent reoccurrences of any known gaps.
- Participate with other entities in information exchanges, training, and exercises.
- Develop a community-planning strategy to address high-risk areas.

- Ensure response readiness to mitigate threats.
- Enhance the coordination process with the media, hams, and translation service providers and any other external providers needed when activating warning messages and other crisis communications.
- Maintain training and any special teams capabilities.
- Assist other local agencies in emergency communication plan development.
- Provide program information and support using the office of emergency management (OEM) website, e-mail notices, and conferences/training to assist the community in preparedness and planning efforts.
- Bring high-risk and obsolete communication systems and processes into compliance.
- Establish criteria to identify areas of high risk in the communication process and systems, and develop methods for bringing these resources into compliance following inspections.
- Perform regular testing and inspections of systems and processes.
- Improve data collection and quality to ensure that the analysis of hazard and risk data is appropriately collected, analyzed, and disseminated to stakeholders.

Ensuring that needed capabilities are available requires effective planning and coordination. This is in conjunction with training and exercises in which the capabilities are realistically tested and problems identified. The outcomes are subsequently addressed in partnership with other federal, state, and local stakeholders. Knowing where to go to start and continue the activities defined for each phase of emergency management is to know the assignment of responsibilities within an entity.

ORGANIZATION AND RESPONSIBILITIES

The emergency communications system of an entity should be organized and coordinated by the LWP. Resources available from external entities that could become an extension of the internal communications systems are local OEMs, local media outlets (radio, television, cable, Internet), individual amateur radio operators or hams, social media sites such as Facebook and Twitter, and others with liaisons within other jurisdictions.

Ideally, the crisis communications program should have an executive sponsor within the organization who is able to assign resources that can establish general policies for crisis communications. A sound policy incorporates the following:

- Framework for the dissemination of information through a designated PIO
- Creation of information centers to execute predetermined processes and procedures for communication during and after a crisis
- Development of processes to verify, coordinate, and disseminate information during an incident
- Foundation for the management and coordination of publication information with community partners, the media, and the community
- Completion of all applicable training and oversight of the program

Assignment of Responsibilities

Department Heads

Department heads who operate communications system vital to the safety of the community should have representatives within their department to coordinate communications systems, including license and maintenance. The primary resource should have an alternate identified who is cross-trained and able to assume this responsibility in the absence of the primary resource.

Office of Emergency Management (OEM)

When the OEM is between activations, it is responsible for developing and maintaining a communications resource inventory that can be easily accessed upon activation. The OEM ensures that communications capabilities are operational among internal and external resources such as NOAA, law enforcement, fire, and emergency medical services (EMS). The OEM can assist other departments to ensure that they maintain a current recall roster for essential personnel, including radio and telephone operators. The OEM should ensure that hard copy forms are available for use in case no technology is available.

The Emergency Management Coordinator (EMC) or Director

The EMC or designated leader of the emergency communications center will staff and operate the LWP, assist in the development and maintenance of operations procedures, and coordinate efforts with other internal departments and external entities. The EMC ensures that the facilities used as the communications center can serve as the LWP. The EMC will verify and acknowledge warnings coming into the center related to an emergency, and then notify local officials of the event or conditions that could cause an event. The EMC will request activation of the warning system and determine if route alerting or door-to-door warning is needed. In addition, the EMC is the role responsible for the development and maintenance of hazard-specific warning procedures that cover receipt, validation, and dissemination of a warning.

Within some agencies, the Security or Risk Management Office performs the activities defined earlier in collaboration with the EMC. The activities specifically defined for an EMC are

- Develop an adequate warning system.
- Ensure that maintenance and periodic testing of warning system equipment are performed.
- Develop operating procedures for the warning system, coordinating as necessary with other departments and agencies. These procedures include the release of coordinated emergency public information.
- Assist in the development of warning messages and advisories when the Emergency Operation Center (EOC) is activated.
- Work with the PIOs and Incident Commander (IC) to provide senior officials with the information needed to conduct media briefings.
- Ensure that public information and educational programs are developed, conducted, and maintained.

External Entities

Outside agencies and organizations will report emergencies that merit warning the public as well as assist in disseminating information within their respective communities. For example, educational institutions can reach their students, parents, and employees. Hospitals, nursing homes, and even jails can reach their populations with communications tailored to their respective populations. External entities include media companies that are expected to disseminate warning messages and advisories as provided by the EOC or OEM to the public as rapidly as possible. Institutions, businesses, and places of public assembly are expected to monitor radio, television, and NOAA Weather Radio receivers for warnings and take appropriate actions to protect their populations. Depending on the nature of the emergency, the entity may elect to activate its PIO position in a *Joint Information Center* (JIC) and conduct media briefings.

Joint Information Center (JIC)

A JIC is made up of a pool of PIOs from internal departments and external agencies at any level of government or the private sector when needed—a multiagency approach to the emergency communications process. The JIC responsibilities include many of those performed by a single PIO but on a larger scale, such as directing public information efforts, serving as the primary source for disseminating official public information materials to the community, providing news releases for the media, coordinating communications efforts with senior officials, and maintaining a record of events. The JIC is considered a trusted source for information on an emergency and therefore will authenticate sources of information and verify its accuracy before releasing any communications. The JIC is also a venue to ensure the security of information related to the emergency and will handle unscheduled inquiries from the public or media.

DIRECTION AND CONTROL

Once an EOC or JIC has been activated, the overall authority for either operation (or both operations) is the senior official responsible for these operations. People who work within the EOC or JIC, while under the control of their own office, are responsible for knowing and following the procedures defined for each emergency operation. During EOC operations, the various code systems used by the different agencies represented, such as fire, police, or medical, for brevity should be discontinued, and normal speech should be used to ensure comprehension. During the transmission of emergency communications, local time is commonly used. A Universal Time (UTC) can be used later in documentation.

For specific time-sensitive emergencies, the LWP should be given the authority to determine if a warning should be issued, formulate a warning if necessary, and disseminate it. For other situations, the LWP should coordinate with one of the designated senior officials, who will determine if a warning should be issued and approve the general content of any warning message that will be disseminated.

When the EOC has been activated, the EOC staff will normally determine who needs to be warned and how. The PIO and other members of the staff will formulate warning messages and public instructions.

The line of succession for each department and agency represented in the EOC is according to the standard operating procedures (SOP) of each department and agency. At a minimum, the line of succession should be three levels deep.

READINESS LEVELS

Although northeast Japan was ravaged by a 9.0 earthquake in March 2011, the country is identified as being one of the most prepared for an earthquake. The country has strict building codes and regular training

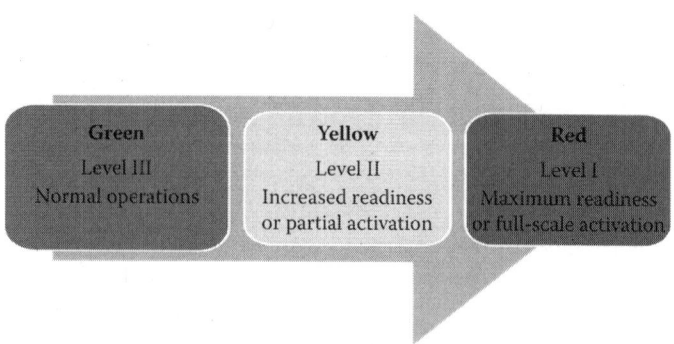

FIGURE 6.4   EOC activation levels.

and drills for the entire population. Schools regularly practice evacuation routes, classrooms and many businesses keep helmets in stock for everyone, and regular reminders of the safest place to be when an earthquake occurs (under tables or in doorways) are constantly reiterated. Simulation vehicles are also routinely used to educate the public on how a major earthquake would feel. Equally important are civil preparedness and security. This explains the general readiness for an emergency in Japan.

Readiness within an EOC and OEM is just as important, and in some respects is more important. An EOC or OEM must be prepared to act immediately upon notification for any type of emergency, as everyone must be prepared to act immediately when an earthquake occurs. Earthquakes offer little warning beyond that which can be scientifically calculated. Once an earthquake is detected, state-of-the-art technology can provide notice within seconds of its start. On the other hand, there are emergencies that allow for time to prepare and warn, such as an approaching hurricane. In an earthquake, readiness can quickly go from normal operations to the highest readiness level of activation within an EOC or JIC. There are many different naming conventions used for each level an entity elects to use. Most entities use three to five levels of activation, each level being color-coded. One of the easiest methods used is a color-coded system using green, yellow, and red to signal the level of activation as depicted in Figure 6.4 (EOC activation levels).

Green (Level III)—Normal Operations

The readiness level denoted by green signifies normal operations. Although no emergencies are active nor identified that require immediate action, this is a time to continuously monitor and assess a specific threat, unusual event or situation, and concurrently focus on prevention.

Prevention activities include the training of emergency response team members, public education and awareness, and expansion of supportive technologies, among other key activities. This is a time to test, upgrade, update, and refine technologies; process and train employees; and promote public awareness of emergency communications.

Yellow (Level II)—Increased Readiness or Partial Activation

The readiness level coded yellow is used to signal partial or limited agency activation. OEM staff and essential agencies with a role in the incident response are activated and required to report to the EOC. Dissemination of warnings is performed if necessary. Key activities germane to this readiness level for emergency communications are to monitor the situation closely, inspect warning systems to ensure they are operational, review communications procedures with personnel, and validate contact lists. The media is alerted to the increased threat so that they are aware of the situation and are prepared to disseminate warnings and public instructions if necessary.

This is a time to review any prescripted messages, determine requirements for additional pre-emergency public information and instructions, and produce and disseminate those materials for the impending threat. This level of activation is also used to identify requirements for warning communication methods, including route alerting and door-to-door warnings.

Red (Level I)—Maximum Readiness or Full-Scale Activation

In a full-scale activation, the EOC is activated on a 24-hour schedule because an imminent threat or a disaster has occurred. Warning messages and public information materials are updated as necessary, periodic communication checks are conducted, and information is provided to the media on local readiness activities. The EOC increases its monitoring of the situation, planning, and resource management, increases staffing when necessary, and staffs the public information positions. Each department and entity is responsible for ensuring adequate resources, including equipment that is available and operational during emergencies.

Should local resources prove to be inadequate during an emergency, requests are usually made for assistance from other entities and agencies in other jurisdictions, industry in accordance with existing mutual-aid agreements and contracts, or state or federal agencies. Although the government will allocate, use, and manage much of the resources for an emergency, private industry owns more than 85% of the resources needed.

EMERGENCY COMMUNICATIONS AND RECORD KEEPING

Prior to activation of an emergency communication system, a mapping of the available technology, the audience it serves, and who can activate a particular communication system should be developed and maintained (see Appendix B, Sample Mass Notification System Activation and Criteria Guidance Sheet). This matrix gives a quick visual of information that is critical in an emergency. Using a matrix can be a resource for developing a technology profile and ensuring that PIOs know what is available to them at their fingertips (see Table 6.1 Sample Warning System Matrix for a Private Business).

All data, hard copy or electronic, generated during an emergency should be gathered and filed in an orderly manner so that a record of events is preserved for use in determining response costs, and updating emergency plans and procedures. The records should include activity logs that were used to record the warnings received, key personnel notified and the actions they were directed to take. These records should also include the warnings disseminated to the community and how they were disseminated. The LWP, the Incident Command Post (ICP), and the EOC are each responsible for maintaining activity logs.

All vital records should be protected from the effects of disaster to the maximum extent feasible. Should records be damaged during an emergency, professional assistance in preserving and restoring those records should be acquired once it becomes possible.

In addition to the activity logs, the PIO should maintain a media contact roster, and compile and maintain copies of newspaper articles, video tapes of emergency operations and news broadcasts relating to an emergency, and other media materials that may be distributed for use in postincident analysis and future training activities.

TRAINING AND EDUCATION

The OEM serves as a leader in conducting disaster training and education to increase preparedness within the community and training on emergency communications equipment and procedures used within the EOC. The OEM typically obtains materials for disaster-related public education. Each department or external entity assigning personnel to the EOC for communications purposes is responsible for making certain they are familiar with their department's or entity's operating procedures.

TABLE 6.1 Sample Warning System Matrix for a Private Business

Mode of Communications	Employees		Community		Responsible Party (Implementation)
	First Responders	All Others	On Site	Off Site	
Common area and door-to-door notification and Emergency Vehicle Public Announcement System (paper postings or word of mouth)	Yes	Yes	Yes	Yes	Site Security, Maintenance, External Affairs, Office of Emergency Management, Joint Information Center
Company e-mail	Yes	Yes	Yes	Yes	Emergency Manager, IT Department
Company website/portal update	Yes	Yes	Yes	Yes	External Affairs, Web Administrator
Desktop PC notification—networked only	Yes	Yes	Yes	No	IT Department
Fire alarm annunciation	Yes	No	Yes	No	Maintenance, local fire department
Internal TV	Yes	Yes	Yes	Yes	External Affairs
Local/area media	Yes	Yes	Yes	Yes	Emergency Manager, External Affairs
Local/area media—ISC forms	Yes	Yes	Yes	Yes	Emergency Manager, Office of Emergency Management, Incident Commander
Main phone greeting update	Yes	Yes	Yes	Yes	Emergency Manager, IT Department
Marquees	Yes	Yes	Yes	Yes	External Affairs, IT Department
Rapid notify—text messaging	Yes	Yes	Yes	Yes	Emergency Manager, Site Security

continued

TABLE 6.1 (continued) Sample Warning System Matrix for a Private Business

	Employees		Community		
Mode of Communications	First Responders	All Others	On Site	Off Site	Responsible Party (Implementation)
Social Media Sites—Twitter, Facebook, etc.	Yes	Yes	Yes	Yes	Office of Emergency Management, Joint Information Center, External Affairs
Speakers/PA System	Yes	No	Yes	No	Site Security, President's Office
Two-way radio	No	No	Yes	No	Site Security, Maintenance, Emergency Manager
Video displays	Yes	Yes	Yes	No	External Affairs, IT Department
VoIP broadcast	Yes	Yes	Yes	No	IT Department

The public information staff should attend public information training that is offered in the community, by the state, or FEMA. The public information staff should also participate in drills and exercises at least annually.

It is desirable that preparation of warning messages, public instructions, and the activation of warning systems be included in emergency exercise activities (see Appendix C, Sample Messages). Ensure that these tasks are appropriate for the scenario being rehearsed in order to ensure that components of the system and operational procedures are adequate. If warning systems are activated at other than normal times for exercises, it is essential to give due notice to the affected community that such activations will or may occur.

SECURITY, SYSTEM MAINTENANCE, AND CAPACITY BUILDING

Due to the vital role of communications during emergency operations, a personal background check should be conducted for any personnel assigned to the EOC. This background check, if not conducted by the agency responsible for the EOC, should confirm that all volunteers have been vetted. As a minimum requirement the background check should determine that volunteers have not been recently convicted of a felony offense or a violent crime.

All warning systems should be maintained in accordance with the manufacturer's instructions for those systems. Often, warning systems are stressed during an emergency and could malfunction or fail to operate at optimal levels. Just as changing the oil in a car on a regular basis is important for good gas mileage, testing and maintaining technology by applying the appropriate security and patch management system helps ensure the health of technology when it is needed most. All components of the warning system should be tested on a regular basis, the minimum being every six months.

If the requirements exceed the capability of the local OEM communications resources, support from external entities should be requested. If personnel requirements exceed the capability of the EOC, additional representatives should be requested from external entities or from OEMs of another jurisdiction, state or federal agencies. This would initiate the formation of a JIC and activation of a Joint Information System.

SUMMARY

Emergencies will happen. Any organization or community can expect to experience emergencies that could threaten the health and safety of persons and property, necessitating the implementation of protective actions. Effective communications is a critical step in the management of an emergency within each phase of emergency management. The emergency communications framework begins with the tools you have and creating a plan. The plan includes assignment of resources, designation of an LWP and spokesperson, identification of communication needs, preparing the first announcements (prescripted messages), timing and broadcasting warning messages, monitoring information flow and public response, education and training, and administration of the plan throughout the disaster life cycle. Emergency communication systems, as a best practice, should be reserved for emergency preparedness, advisories, warnings and alerts, evacuations, sheltering, and testing of the system. These same systems should not be used for non-emergency purposes such as marketing, building contact lists, or sales.

ENDNOTES

1. Federal Communications Commission. (2011, January 25). *FCC takes action to advance nationwide broadband communications for America's first responders.* Washington, DC: FCC.
2. *Pennsylvania Department of Health: Special populations emergency preparedness planning.* http://www.dsf.health.state.pa.us/health/cwp/view.asp?a=171&Q=233957 (accessed March 1, 2011).
3. West Coast and Alaska Tsunami Warning Center. (2011). *Updated tsunami warnings Japan earthquake.* http://wcatwc.arh.noaa.gov/ (accessed March 11, 2011).

CHAPTER 7

A Crisis Occurs ... Now What?

> "Each crisis has its poster child"—from financial meltdowns and Ponzi schemes to Middle East unrest and EF5 tornados.
>
> Stephen Roach
> *Financial Times, May 18, 2010*

When a crisis occurs, many organizations attempt to protect their reputation over protecting stakeholders from harm. The first priority in any crisis, however, is to protect stakeholders from harm—ensure their personal safety. If the organization is perceived to be responsible for the crisis, or a contributor to it, any injuries or deaths may result in financial and reputation loss. The loss of reputation will have a financial impact on the organization. This loss is greater when there is delayed communications and response. Often times this delay is due to a fear or liability concerns.

Public safety, financial loss, and damage to the reputation are interrelated threats a crisis can create. The threat to public safety is focused on injuries or loss of life. The threat of financial loss could be caused from the disruption of operations, spawning litigation related to the crisis, or loss of market share. The third threat is damage to the reputation of the organization as perceived by the stakeholders of the organization. How the events related to the incident unfold, what information is made available, how resources are coordinated, how and when decisions are made, and what updates are provided can determine how much any of the three primary threats will impact an organization.

When a crisis occurs, crisis management can provide an overarching command and control process for managing its impact. The need to issue time-sensitive communications at the onset, during, and after a crisis is the operational foundation of crisis management. It is important to communicate decisions and resulting actions to those who need to know in a timely manner, or it will have the same effect as not having made a decision at all. Timely communications means

- Providing time-sensitive information to ensure the safety and well-being of staff
- Keeping Incident Management Teams (IMTs) and first responders focused
- Providing all IMTs with the information necessary to make strategic and tactical decisions
- Providing action or no-action instructions to targeted stakeholders (IMTs)
- Managing rumors, speculation, perception, and the tendency to treat assumptions as facts
- Mitigating real-time operational risks
- Demonstrating that proactive corporate due diligence was applied throughout an incident.

Exercises often reveal that during a crisis, humanitarian actors not local media prioritize information delivery. Generally, the energies of international agencies are first concentrated on service delivery and then on getting the message out. Conversely, local journalists are focused on the getting the "top story" of the day. Broader and more effective collaboration could lead to a more informed audience and a better response. As an organization, it is important to understand this and work through the challenges this presents by adhering to the simple process called The Ws of Effective Communications.

THE Ws OF EFFECTIVE COMMUNICATIONS

There are many approaches to crisis communications planning. The primary need for effective crisis communication planning is the need to provide timely and accurate information in a manner that is transparent to the receiver. The foundation of any crisis communication plan should address the Ws for defining a situation:

Why	Who	What
When	Where	How

Each is interlinked with the other to provide a more complete picture of the situation and for the development of an appropriate message.

Why

It is time to exercise the organization's crisis communication plan. A key concept of an effective crisis communication plan begins with stating *why*, that is, the reasons for the plan. The reasons for a crisis communication plan for any situation should address the organization's internal and external audiences. There may be instances where only the internal audiences require a response and no response is needed for the external audiences. An effective plan takes both types of audiences into consideration before eliminating either of them. This section of the crisis communication plan should

- Address concerns regarding the physical safety of all those who potentially impacted
- Communicate the known basic facts of a situation
- Direct the audience to where accurate information is available (a trusted source) such as the company website, social media sites, and local news media outlets
- Address concerns regarding the safety of material assets
- Provide information on the current operational status and plans of resuming normal operations.

The Scenario

Setting the Ws into motion, the following scenario will be used to review each W:

> People are having problems with skin rashes after visiting the county hospital. An Emergency Room (ER) nurse to the attending physician in the emergency room reported the first incident of the rash at 6:00 a.m. Two women called, asking questions related to an unexplained rash around 10:30 a.m. Three men, who were leaving after a daylong visit, stated that they had a rash that looked unusual and were concerned whether their sick relative was contagious, at 6:30 p.m. The ER team discussed alerting hospital administrators and began surveillance activities, including contacting each victim for additional information and status. Between 9:00 p.m. and midnight, July 18, 2009, seven people had come to the ER with the same symptoms. The physician on duty alerted the chief of staff and contacted the state public health department and Centers for Disease Control and Prevention (CDC) for guidance. Key decision makers from each department were summoned to the board room for activation of the hospital's response and recovery plans. By midnight, the hospital had a crisis on its hands. This hospital serviced a small rural area in Colorado. Rumors had started, and the public would start to panic as their day started the next morning. The crisis communications plan was activated.

The initial public statement was given at a press conference held in the parking lot of the hospital:

> *As of 6:00 a.m. today, July 19, 2009, ABC Hospital has confirmed several mild cases of the AC2 Skin Rash from individuals who have recently visited our hospital. All reported cases involved adults; no children. These rashes appear as small red spots on the arm and hands with no discomfort reported. The rash clears within 48 hours with no intervention. There have been no deaths or hospitalizations due to this skin rash. It is spread upon direct contact with the contaminated surfaces we found in two areas near the side entrances. These side entrances have been temporarily closed and extra precautionary measures have been taken. Further spread of the rash is not expected. You cannot get the rash from an individual with the rash. Ongoing investigations are in progress. We must limit release of more detailed information to avoid jeopardizing the investigation at this time. We recommend that you monitor our website for the most up-to-date information at www.abchospitalxxx.com and the state public health authorities for additional information.*

This communiqué addresses physical safety, basic information, where to locate current and accurate information, and at a high level when the hospital plans to return to normal operations. Since the symptoms of the victims are mild and there are limited intensifying factors after one to two days, the response requirements are low.

Who

The next section of the crisis communication plan covers *who* are the target audiences, both internal and external. Understanding all the target audiences that should receive a message or messages determines those have an immediate need to know, those requiring continuous updates, those with a general need to know, and those only needing a one-time update related to the crisis. Using the foregoing scenario, Table 7.1 (Target Audiences) outlines the potential target audiences that should be included.

Each audience has its unique requirements. For example, it is important to maintain the confidentiality of survivors, whether internal or external, and any others involved in the incident. Health Insurance Portability and Accountability Act of 1996 (HIPAA) requires that permission be obtained from the survivor before any personally identifiable or other sensitive information is released. Adhering to HIPAA is required by law and important for maintaining the privacy of the survivor. When making

TABLE 7.1 Target Audiences

Audience	
Internal	**External**
• Survivors • Employees (staff, contractors, physicians, consultants, etc.) • On-premises occupants and customers (patients, volunteers, contractors, vendors, visitors)	• Victims • Off-premises customers • Regulators • Public health agencies • Insurance carriers • Media • General public • Vendors

a public statement, you may group clusters of individuals using generic terms such as *patients* or *visitors* and not unique identifiers such as "Ms. Donna Mills, the patient," or "David Jones, her 12-year-old son."

What

The next area to address in the crisis communication plan is the *what* about the incident and any related communications. A viral spread associated with a hospital will require both internal and external communication. This addresses the communications and the incident. The plan works through a list of possibilities from what is the known situation through worst-case scenario order. Possibilities to consider using the skin rash are included in Table 7.2 (List of Possible Scenarios).

Line 1 defines the current situation. As you progress through the remaining lines of Table 7.2, each line increases in severity because as the skin rash directly affects more people, the complexity of the illness

TABLE 7.2 List of Possible Scenarios—AC2 Skin Rash (Ranked in Order of Severity—[First: Current Situation; Last: Worst Case])

Visitor	Others	Employee	Skin Rash	Fatal	Severity	Other Ailments	Hospitalization
Yes	No	No	X	No	Mild	No	No
Yes	Yes	No	X	No	Mild	No	No
Yes	Yes	Yes	X	No	Mild	No	No
Yes	No	No	X	No	Severe	Yes	Yes
Yes	Yes	No	X	No	Severe	Yes	Yes
Yes	Yes	Yes	X	Yes	Severe	Yes	Yes

increases, or one or more fatalities have been reported. The current scenario identifies the situation as (Line 1):

- Visitors = Yes   ...*individuals who have recently visited...*
  - Others = No
  - Employees = No
- Skin Rash = Yes   ...*AC2 Skin Rash...*
  - Severity = Mild   ... *confirmed several mild cases...*
  - Other Ailments = No
- Fatalities = No   ...*no deaths or...*
- Hospitalization = No ...*no...hospitalizations...*

Situational Crisis Communication Theory and Attribution Theory

Situational Crisis Communication Theory (SCCT) contends that information about past crises is a significant factor that can affect perceptions of a more recent crisis, and the reputational threat presented by the current crisis. Collectively, the information and potential threat should guide the optimal communication responses for protecting organizational reputation.

What we have learned is that to focus more on the reputation of the organization over the physical safety of persons can lead to greater loss of reputation and increased financial losses for the organization in the long term. SCCT includes instructing stakeholders on what, if anything, to do to protect themselves physically from a crisis and what the organization is doing to prevent a repeat of the crisis.

Attribution theory is a guide to linking the crises to crisis response strategies. Attribution theory hypothesizes that people search for causes of an event (make attributions) both positive and negative, particularly the negative and unexpected. These attributions of responsibility can motivate an organization to move from silence and inaction to action and communications. The associated behavior responses are as follows:

Positive: Organization is judged not responsible; sympathy is evoked.
Negative: Organization is judged responsible; anger is evoked.[1]

Management has a stake in minimizing any negative outcomes. Attribution theory-based research has found that what should be communicated increases with the level of responsibility the organization has in the crisis and whether there are any intensifying factors. At a minimum, what should be communicated is denoted in Table 7.3 (Crisis Communication Best Practices Based on SCCT and Attribution Theory-Based Research).

TABLE 7.3 Crisis Communication Best Practices Based on SCCT and Attribution Theory-Based Research[9]

Practice	Minimal (Crisis Responsibility and No Intensifying Factors)	Low (Crisis Responsibility and Intensifying Factors)	Strong (Crisis Responsibility)
Instructing information including recall information and trauma counseling (care response)—Most Important	X	X	X
Providing an expression of sympathy, corrective actions, and care response	X	X	X
Reminder and ingratiation strategies to supplement a response	X	X	X
Denying and attacking the accuser strategies used for rumor and challenge crises (as needed)	X	X	X
Adding excuse/justification strategies to the instructing information and care response		X	
Adding compensation/apology strategies to instructing information and care response		X	X
Having a compensation strategy for any time victims suffer serious harm			X

Note: Ingratiation is a strategic attempt to gain acceptance for you through persuasive and sometimes subtle methods in order to obtain compliance with a request. Strategies include flattery, conforming to the beliefs or values of the target audience, or presenting yourself in a manner that would be liked by the target audience.

What to Avoid Saying

There is always the temptation to evade responsibility in negative situations. With avoidance, there is an increased likelihood that a message is delivered based on a short-term perspective and may include statements that are dead giveaways that more serious issues may lie beneath the surface of the current crisis. Common red flags are

- "It's just an isolated incident."
- "We need more time."
- "Let's not overreact."
- "If we say something, people will find out."
- "It's too soon to act."
- "It's only competitor criticism."
- "The standards are unreasonable or unachievable."
- "We obey the law."
- "We can't take responsibility; we'll be sued."
- "It will trigger copycats."

Using this approach can leave executives with few remaining options to address the root causes of the crisis, particularly if the organization is eventually found to be at least partially at fault. This approach can also cause a delay in responding to a crisis and making sound decisions.

When

Section four of the crisis communication plan begins with documenting *when* the incident occurred and when to begin communicating with those affected. The *when* is the date and time of the first incident (6:00 a.m., MT, Thursday, July 18, 2009). This is the first of many *whens*. To answer "when," the plan must take into consideration scenarios before, during, and after (short term and long term) the incident. Communicating with the affected parties in most crises is not a one-time event. Using the skin rash scenario:

- Incident
  - *Before:* July 18, 2009, 6:00 a.m., MT
  - *During:* July 18–19 (Last report—Midnight July 19)
  - *After:*
    - *Short term/immediately:* July 20, 2009
    - *Long term:* July 25, 2009
- Communication
  - *Before:* July 18, 2009, 6:00 a.m., MT

- [Internal communications—immediate staff]
  - *During:*　　　　　　　　July 18–19 (Last report—Midnight July 19)
- [All internal and external audiences]
  - *After:*
    - *Short term/immediately*: July 20, 2009 [All internal and external audiences]
    - *Long term:*　　　　　　July 25, 2009 [Internal audiences—employees and affected contractors and volunteers; External audiences—survivors, regulators, public health, insurance carriers, media (limited)].

When and *who* are closely linked. Knowing when to communicate helps to get the situation under control quickly.

Where

Section five of the crisis communication plan covers *where* the incident occurred and when to begin communicating with those affected. Following the rule "know all, tell all," it is important to create the core postincident statement with information on where the skin rash was first reported and any other information you have regarding "where." For example, what is known but not communicated is:

- Where:
  - Incident occurred:
    - *... on the arm and hands*
    - *...at the hospital*
    - *...contaminated surfaces we found in two areas near the side entrances*
    - *...visitors to use the main entry; side entrances closed*
  - Communicating the message
    - *...initial response given live in the hospital parking lot*
    - *...additional information available on the web*

Where and *how* are closely linked. An organization should use as many methods of communications as they have available to push the message out to their target audiences. Technology is always the first place most organizations consider for communications. Word of mouth and pen and paper are also means of communicating; these methods classified as "no tech." Places and suggested types of communications tools needed following the public health incident:

- Organization's website
- Social media sites
- Hospital television and video displays
- Receptionist desks (verbal)
- Shared computer displays (screensavers)
- Building entrances (signs)
- Parking lot (prepared statement to media).
- Employee e-mail (statement from management)

How

People want to know *how* a situation occurred. There are many *whens* to address in the crisis communications plan, and there are also multiple *hows*:

- How the incident occurred and progressed
- How the incident is being managed
- How to respond appropriately
- How communications will be handled at each phase of the incident

Using the skin rash scenario senior management, regulators, public health authorities, insurance carriers, and the employees involved in the surveillance, care, and treatment will need to know all the details to identify the root cause of the incident, take appropriate corrective action, and work to prevent it from happening again to the victims and hospital.

In most cases, organizations are better served by assigning an incident commander (IC) working with their team to serve as the sole source of information about the incident itself. The incident commander could be the chief of staff or a designated representative from the state public health authority, depending on the severity of the incident. Using the skin rash scenario and its limited scope, the chief of staff will serve as the incident commander. This is how the incident will be managed. The incident commander has a public information officer (PIO) who will be responsible for coordinating communications. This accomplishes how the situation will be managed and how communications will be handled. Adherence to this process ensures that released information to internal and external audiences does not jeopardize any investigation; and, it shifts the ongoing responsibility of communicating with the media to the PIO so that the incident commander and other team members can focus on response and recovery efforts.

If the skin rash had been elevated to the worst-case scenario involving injury or death, then it would have been important to provide more facts surrounding the situation while maintaining the confidentiality of any victims. For example, upon the death of an adult, you may need to state the gender and age, and with the permission of the next of kin, more information may be provided. Extra measures are needed to protect the identity of any minors. It is also important to recognize the family of any victims; they need reassurance and sympathy.

The second element of *how* are the methods used to communicate with the target audiences. *How* is closely linked with *when* and *why*, the time and reasons associated with the situation. As an example, we will use the list from Table 7.1 of internal and external audiences, to review *how* each is addressed.

- **Employees**
  - *Before*
    - Ongoing awareness, training, and preparedness on infectious disease control
      - Hospital e-mail, posting on intranet site, letter or memo from senior administrator, employee meetings, orientation, exercises, training (face to face or computer based)
  - *During*
    - Share information and instructions with employees throughout the event
      - Employee meetings, teleconferences, e-mail, postings on intranet, letters/memos from senior administrators working with the PIO
  - *After*
    - Speak directly with affected employees to provide physical and emotional comfort and give them instructions about what is expected of them next
      - Work with the human resources department, to offer resources for counseling or other services that may be needed
      - Post on the intranet, letters/memo from senior administrators, newsletters, employee meetings to communicate appropriately based on the incident
      - Continue ongoing awareness, training, exercises, and preparation communications
- **Survivors**
  - *Before*
    - Ongoing communications of hospital services and any awareness campaigns regarding visits to the hospital
      - Statement stuffers, information shared with employees that is repeated in the community, hospital

literature offered at registration, information regarding patients' rights, such as HIPAA and Privacy Statements upon admissions
- *During*
  - Contact during surveillance process for additional information, provide resources or list of resources for victims' physical and emotional needs; provide messages offering instructions and promoting the need to remain calm
- *After*
  - Use signage, and be aware of what you are saying
  - Create a campaign and take appropriate actions to assure victims and the public that it is safe to return to the hospital after an incident has occurred

THE MESSAGE AND THE MESSENGER

In a crisis, it is important for leaders to watch what they say. The words of an executive expressed in every medium, from meetings, memos, or annual reports to face-to-face conversations in the elevator, define the organization in terms of mission and method, set an ethical tone, establish and defend the rationale for the leader's regime, and secure the consent or compliance of those managed.

Company English and jargon is becoming less effective as an everyday medium for doing what an organization needs it to do—inform target audiences, motivate the audience to take the appropriate action, and explain the situation. *Jargon* consists of acronyms and technical terms that are mutually understood within a group. Jargon is a time saver when combined with shorthand and used in its original context. When used in other contexts, it becomes incomprehensible. This is a time when communication must be clear, concrete, and concise rather than wordy, abstract, and vague. Words matter! In communications within an organization, among organizations, and with the outside world, it is the words used that carry the most weight.

Every community develops a specialized language for its circumstances. This language defines and reinforces the culture, conveys technical information unique to the organization, fosters loyalties to teams, expresses and reinforces power relationships, can also obscure realities, and silence disagreements that cannot be addressed otherwise.

In the complaint process, the use of jargon and company English hampers clarity. Over time, many buzzwords, or jargon, hide the facts by obfuscating, confusing, or avoiding an issue. This can become a way to avoid communicating.

When talk regarding an incident is communicated in terms known to only those in the industry, company English, it may be effective to those who understand, which is usually a small subset of those who need the information to actually perform the desired action, even within an organization. Language that is vague, wordy, or abstract is vulnerable to the interpretation of the receiver, and may cause unpredictable results. Company English and jargon when used outside the company can lead to misunderstanding and can make people feel excluded or stupid. This may be the desired approach when looking to protect an organization's intellectual property, but is ineffective for emergency communications. Words shape thought—if the words are vague, the thought is likely to be muddled.[9]

A message from the President of ABC Nuclear Products Company to its employees over the company's public address system:

> *"Today one of our employees experienced a meltdown. Our EH&S Director advise everyone to either go and stay in your area or wait while we assess the situation. We will update you again once the meltdown has been addressed."*

Interpretator (Receiver)	Message as Interpreted by the Receiver
You (Manager)	"I got the message and it sort of makes sense. So, I guess I will lock up my desk, and tell the others to sit tight, then grab a cup of coffee and wait."
Your subordinates (Secretary and staff)	"Well, that says a lot. Typical. I am going to lock up and go home. Who knows how long it will be before they tell us anything."
Your neighboring businesses and customers	"I am not sure if we should evacuate. I am going to call 9-1-1 to let them know what is going on. Maybe they can tell us what to do. Since we are next door, I want to know what is going on. The line is busy when I call over there."
The first media report	**BREAKING NEWS...** "ABC Nuclear Power Company had an employee that suffered a mental breakdown and began shooting—killing one and wounding another. This started around 8:00 a.m. at the beginning of the morning shift. First responders are scrambling to get to the site. We will monitor this situation and provide more updates."
ABC Nuclear Products Switchboard	Switchboard overwhelmed with calls from employees on what they should do and from concerned citizens. Family members want to know if their loved ones are OK.

The message sent from the president in the sample above is vague and only made sense to the sender, not to any of the receivers. As you moved further away from the point of origin of the message, the interpretation of the message changed. What was wrong with this message?

- Instructions were vague.
- Company English was used to describe the situation.
- The message was delivered using one method.
- Misinterpretation.

CORRECTED:
A message from the Safety Manager of ABC Nuclear Products Company to its employees over the company's public address system, e-mail, and text messaging:

"There is an active shooter situation reported by the Environmental Health and Safety Director at ABC Nuclear Products Company. Leave immediate area if safe to do so or shelter in place. Remain calm. Stand by for further instructions."

Interpretator (Receiver)	Message as Interpreted by the Receiver
You (Manager)	"I got the message, and it makes sense. So, I guess I will stay in my office and wait until they tell us more."
Your subordinates (Secretary and staff)	"I am going to lock up and close my door. I hope that this is over quickly."
Your neighboring businesses and customers	"We will stay inside for now."
The first media report	EVENING NEWS (on a slow day)... "ABC Nuclear Products Company had an employee threatening other employees in the cafeteria that he would shoot. The company issued an active shooter alert for a short period.
ABC Nuclear Products Switchboard	The switchboard had a few more calls than normal from concerned citizens and the media requesting an update.

By comparison, the message above provides clear instructions and uses common English, leaving little room for misinterpretation.

Peter Drucker's thoughts on communications are that communications is in the mind of the receiver and not the sender. That is, effective communications is when the receiver can interpret the message as it was meant by the sender. For an organization, company English works less well when attempting to communicate an organization's explanation of an incident to the media, shareholders, customers, and the community. With leaders having less time to think about what to say, much less how to say it, it is important to have those trained in crisis communications speaking with the target audience. As more information becomes available about the situation, and the president or messenger has been appropriately briefed, communications should be restricted to emergency personnel trained in crisis communications.

The use of texting, tweeting, and other electronic communications formats has drastically increased the use of company English and jargon. More sources are creating their own jargon, particularly media sources such as cable TV and the Internet as well as traditional sources. The life of these terms is also much shorter.

Once an organization's leader is in front of an audience, his or her tone of voice, choice of stories or jokes, and body language can convey a meaning that contradicts the spoken word. Remembering that eventually it will all come out, considering the moral and ethical aspects of effective crisis communications is important.

MORAL AND ETHICAL REFLECTION

There is a moral and ethical element to crisis management. Organizations are expected to have a conscience and to act in ways that reinforce this expectation. This concept is important since whenever there is a victim, someone will be held accountable. Victims and the community will make a moral and ethical assessment of an organization; therefore, the organization should consider a self-assessment of its moral and ethical position.

When a crisis involves integrity, moral, or ethical dilemmas, moral reasoning and questioning should begin quickly. This is a time when difficult and direct questions arise regarding the integrity, morality, and ethical position of an organization. Acting appropriately and demonstrating the belief system through behavior will counter the negative impact of the situation. Questions an organization can use to assess its moral compass are

1. What did you know and when did you know it?
2. What are the relevant facts of the situation?
    a. What decisions were made?
    b. Who was involved or affected?
    c. What was sacrificed to benefit the victims?
3. Was there a firsthand attempt to uncover the truth?
4. What alternative actions are available?
    a. Who would be affected?

5. What ethical principles or standards of conduct are involved or at stake?
   a. How would each alternative action considered affect the principles or standards?
6. Is it really the problem of the organization (is the organization responsible)?
7. What is the duty of the organization to update and inform?
8. Who should be advised or consulted?
9. What was the fundamental cause: omission, commission, negligence, neglect, accident, arrogance, or other?
10. How could this situation have been avoided?
11. Are all the crucial ethical questions being asked and answered?
12. Are the actions open, honest, and truthful?
13. What affirmative action is under way or planned to remedy the situation?
14. Is there an institutional "code of silence" when morally questionable decisions or actions become known?
15. How will future unethical behavior be disclosed? To whom? How fast?
16. What lessons can the organization learn, as an event moves from short-term to long-term recovery or resolution?
17. Is the organization prepared to remediate the behaviors that led to ethical compromises?
18. How many of the "typical behaviors" listed as follows are known that can potentially cause issues or risks?
    a. Insufficient controls and oversight
    b. Lack of appropriate and validated compliance
    c. Underreporting of infractions
    d. Leadership that allows supervisors to overlook bad behavior
    e. Allowing employees to experiment with "unapproved methods"
    f. Encouraging a "do whatever it takes" mentality
    g. Structuring incentives in such a way that they compromise safety, public health, or product integrity
    h. Overlooking shortcuts
    i. Avoiding confrontation with managers
    j. Operating "on the edge"
    k. Ignoring signs of rogue behavior
    l. Tolerating inappropriate behavior of individuals identified as "critical to the organization"
    m. Belittling or humiliating those who suggest or seek ethical standards or the credibility of whistleblowers
    n. Dismissing employees or otherwise retaliating against employees who report bad or outright wrong behavior.[2]

Ethical responsibilities are moral rules accepted as good practice within a society. Legal responsibilities are responsibilities we are obligated to adhere to by law. Ethical responsibilities include respect and honesty between individuals, and to care for and assist others. It is the glue that holds society together—giving back to society through charity and goodwill. Businesses have an ethical responsibility to deal honestly with their partners, offer products and services as promised, and disclose any necessary information. Ethical responsibility for the environment is a moral duty businesses and individuals have to keep the environment clean and to be environmentally responsible.

PREPARATION IS KEY

The crisis communication plan should incorporate concrete steps to prepare the organization for the possibility of a crisis. In the case of a health care facility, there is the need for infectious disease control. The crisis communication plan should have documented the designated primary and backup spokespeople in the event of an emergency. The designee is usually the highest-ranking official of the organization that is available. This person should be equipped with prepared statements that can be delivered using technology they have on their person or at their location such as smartphone, fax machine, laptop, or hardcopy (on paper). Building relationships with the media and first responders in the community before an incident occurs is always helpful. This aids in developing trust, a key factor in working with the media during a crisis. This relationship building should go beyond the CEO. Relationship building should also extend to those who will serve as the public information officer and the leaders of the departments that are most likely to communicate directly with the public.

With the development of a comprehensive crisis communications plan, it is possible to react quickly and effectively to a crisis. Advanced preparation will yield dividends through the avoidance of business disruptions and the maintenance of good relationships with stakeholders, including the communities the organization services. Leveraging all available technology can help balance the information an organization can provide its target audiences: Posts on social media sites and word of mouth statements can get the word out to many quickly.

From a technology perspective, the move is toward digital and online content. This has been accelerated by online aggregators such as Google and Tencent, and computer hardware and mobile devices. While the use of digital and online content is increasing, traditional broadcasts on TV and radio are on the decline. Computer hardware and mobile devices and online aggregators have increased access to video. Consumer demand for computer hardware and mobile devices has increased the demand for high-quality computer technology for the home and on the go using Internet delivery. In times of a crisis, access to real-time information is available to end-users from almost anywhere they live, work, or play. Any crisis communication plan should consider this. An overreliance on traditional forms of communications prevents the organization from reaching out to much of its target audience, particularly vulnerable populations that need the information early to prepare and respond.

According to Oliver Wyman, an international management consultant firm, they found that adults 34 years of age and younger increasingly use online and mobile services for viewing and communicating in preference to traditional television and radio broadcasting. Mobile video is a communication tools that is growing fast. In the United States, the time spent viewing using mobile phones grew by 54%, and even higher for iPhone owners. Mobile video is expected to grow exponentially over the next few years as iPhones and other full-screen smartphones proliferate. While this service remains in its infancy, its use will continue to increase as traditional broadcasting platforms decline in use.[3]

The increasing role of the Internet in content distribution has affected crisis communications. As broadband mass-market penetration levels increase, a growing proportion of video consumption will occur when a crisis happens, and more services will be delivered across the open Internet rather than traditional television and radio platforms.

LESSONS LEARNED AND CONTINUOUS IMPROVEMENT

Most crises cannot be avoided. Institutional memories are short as people prefer to forget negative experiences, employees leave the organization or move to other areas, and technology shifts occur. It is ideal for an organization to learn as it executes its crisis response and remedial actions. The lessons learned approach teaches the organization how to forecast, mitigate, or significantly reduce the likelihood of a similar situation occurring or reoccurring.

The lessons learned approach is a critical component of any crisis response process and continuous improvement plan. The public expects organizations to talk about and describe the lessons they learned from mistakes, errors, accidents, or negligent acts as a part of the long-term crisis communication process. Speaking publicly about lessons learned is a major corporate step toward transparency and obtaining public and employee forgiveness. The lessons learned briefing should cover the following:

- Ethics, compliance, and standards of conduct
- Event timeline

- Lessons learned
- Open questions
- Operational issues
- Recovery issues
- Relevant patterns from similar previous events
- Response timeline
- Special actions
- Strategy gaps and failures
- Surprises: negative and positive
- Unintended consequences
- Visibility timeline
- Variations from approved procedures[4]

If nothing else, lessons learned from those who have been faced with "now what" as stated by Lori Musser, Director of Public Relations for the Tampa Port Authority, are as follows.

Crisis Communications Tips[5]

- Know all; tell all.
- Use the CEO or Incident Commander
- Get the situation under control quickly
- Own up to mistakes
- Situations:
  - *Layoffs:* How it helps and how you are mitigating
  - *Harmful releases:* How you are mitigating the situation and thanks for standing by us
  - *Accidental death:* Importance of safety and how you are helping the family
  - *Violence:* Express concern and explain full cooperation with agencies
  - *Protests:* Don't ignore them; show how you are working with the public
  - *Strikes:* Tell of effects; how you are resolving
  - *Terrorism/catastrophes:* Express thanks for those alive; expect preparedness but calmness

As you go through the interview process—media briefings and follow-up question and answer sessions—consider the following interview tips Ms. Musser suggests:

Interview Tips[6]

- Thank the media
- Ask who else the reporter is interviewing, what they said, and when it will air

- Discover the story slant; adjust your key message
- Do not use slang
- Talk in complete sentences
- Never speak for more than 15 seconds and 3 sentences
- Do not use "no comment" and "I don't know." Instead try "I will get back to you with an answer"
- Do not say "I'm sorry"; try "it's unfortunate that ..."
- Never go "off the record." If it was said, then it is on the record.
- Do not panic, remain calm
- Always be polite. Never lose your temper
- Sing from the same hymnal
- Be succinct
- Be helpful
- Refer to nonorganizational issues
- Think long term—there will sometimes be bad stories

Solving problems and "winning" in crises is a function of speed, decision making, action, reaction, collaboration, and swiftly applied common sense. Any approach should take into consideration not only what is legal, but also what is ethical and morally correct.

- *Speed*—act quickly
- *Decision making*—plan and preauthorized
  - Understand the difference between crisis communication management and crisis management
- *Action*—Effective responses are incremental in nature
- *Reaction*—Show compassion and transparency
  - Compassion
    - Offer the assistance as though the victims were family
    - Use humane words and deeds from the start, beginning with the end in mind
    - If you or your organization is in the wrong, apologize and help the victims no matter what
  - Transparency through statements that are important, brief, and worth being heard and repeated
- *Collaboration*—work as a team
- *Common sense*—
  - Everything comes out eventually
  - Bad news never improves with age
  - Fix it now[10,11]

SUMMARY

Solving problems and "winning" in crises involves planning and common sense. Any approach should take into consideration not only what is legal, but what is ethical and morally correct. Organizations need a comprehensive crisis communications strategy that addresses community needs and the expectations that a crisis may cause. The factors that are critical to the success of the strategy are

- Assist the survivors and those directly affected.
- Address the issues causing the crisis.
- Communicate with and solicit the support of employees.
- Inform those indirectly affected.
- Efficiently manage the media and other self-appointed outsiders.

The key challenge remains achieving the critical success factors of the strategy on as many levels as possible as quickly as possible. As operations return to normal, assess the lessons learned, and identify gaps and areas to target for continuous improvement. Operate ethically and maintain transparency as everything will come out.[7,8]

ENDNOTES

1. Weiner, B. (2006) *Social motivation, justice, and the moral emotions: An attributional approach*, Lawrence Erlbaum Associates, Mahwah, NJ.
2. Lukaszewski, James E. (1999), Seven dimensions of crisis communication management: A strategic analysis and planning model. *Ragan's Communications Journal*, January/February 1999, http://www.e911.com/monos/A001.html (accessed May 1, 2011).
3. Carpenter, Guy and Wyman, Oliver (2011). In search of new value growth, communications, media, and technology 2010. *State of the Industry*. New York: Marsh Mercer Kroll.
4. Adapted from a news story in *The Westchester County Journal News*, Sunday, October 25, 1998, "Hamburger meal nearly kills mom: *E. coli* bacteria in meat caused liver failure in Croton woman."
5. Musser, Lori A. (2005, May 2). Crisis communications—Controlled, consistent, caring, concerned. http://aapa.files.cms-plus.com/SeminarPresentations/05_ExecMan_Musser_Lori.pdf.
6. Musser, Lori A. (2005, May 2). Crisis communications—Controlled, consistent, caring, concerned. http://aapa.files.cms-plus.com/SeminarPresentations/05_ExecMan_Musser_Lori.pdf.

7. Lukaszewski, James E. (1999), Seven dimensions of crisis communication management: A strategic analysis and planning model. *Ragan's Communications Journal*, January/February 1999, http://www.e911.com/monos/A001.html (accessed May 1, 2011).
8. Adapted from a news story in *The Westchester County Journal News*, Sunday, October 25, 1998, "Hamburger meal nearly kills mom: *E. coli* bacteria in meat caused liver failure in Croton woman."
9. Florida Department of Health. (2009, June 5), Florida Vulnerable Populations Communications Work Group. Communication considerations for vulnerable populations: Before, during and after a disaster. Final report. http://www.doh.state.fl.us/demo/BPR/PDFs/CommunicationsResourceGuide2_8-11-09.pdf (accessed May 1, 2011).
10. Coombs, W. T. (2007). Crisis Management and Communications. Institute for Public Relations. http://www.instituteforpr.org/topics/crisis-management-and-communications/ (accessed May 2, 2011).
11. Coombs, W.T. (2006) The protective powers of crisis response strategies: Managing reputational assets during a crisis. *Journal of Promotion Management*, 12, 241–259.

CHAPTER 8

International, Federal, State, and Local Laws, Regulations, Systems, Plans, and Structures

> While food, water, and medical supplies were vital to meeting people's needs when disaster strikes, those familiar building blocks for mounting an effective humanitarian response were, however, missing one critical element: information. For crisis-affected populations, weather reports, health bulletins, and directions to emergency shelters, played an equally crucial role in helping save or rebuild lives.[1]
>
> United Nations

The September 11, 2001, terrorist attack on the United States highlighted the need for all levels of government, the private sector, and nongovernmental agencies to prepare for, protect against, respond to, and recover from events that could exceed the capabilities of any single entity. Events that require multiagency or multijurisdictional support need a unified and coordinated approach to planning and incident management. From borderless cyber-attacks to local tornadoes, the fundamental needs remain the same.

When a disaster strikes, the immediate needs of food, water, shelter, and medical supplies are obvious. Getting these necessities to survivors requires a largely invisible communication network that must be set up quickly. The network will enable first responders and other relief workers to save lives. The right information is crucial in making the right decisions. The United States has developed a national preparedness architecture encompassing a full spectrum of prevention, protection, response, and recovery efforts to prepare the nation for all hazards, whether man-made or natural incidents.

In 2003, President George W. Bush signed a series of Homeland Security Presidential Directives (HSPDs) intended to develop a common approach to preparedness and response: HSPD-5, Management of Domestic Incidents, and HSPD-8, National Preparedness. Both have local implication, each requiring the U.S. Department of Homeland Security (DHS) to coordinate its efforts with other entities to achieve its stated objectives. To date, there are 25 active HSPDs.

The Management of Domestic Incidents (HSPD-5) denotes key objectives needed to improve coordination in response to incidents. HSPD-5 requires the coordination of DHS activities with other federal departments and agencies and state, local, and tribal governments to establish a *National Response Plan* (NRP) and a *National Incident Management System* (NIMS). The NRP was replaced in 2008 with the *National Response Framework* (NRF) that defines the principles, roles, and structures to organize how as a nation, the United States responds.

National Preparedness, HSPD-8 describes the way federal departments and agencies will prepare. HSPD-8 required DHS to coordinate with other federal departments and agencies—and with state, local, and tribal governments to develop a National Preparedness Goal. HSPD-8 was replaced by Presidential Policy Directive (PPD)-8: National Preparedness. This updated directive is intended to strengthen the security and resilience of the United States using organized preparation for the threats posing the greatest risk to the security of the nation, including acts of terrorism, cyber-attacks, pandemics, and catastrophic natural disasters. PPD-8 also defines national preparedness as a shared responsibility for all levels of government, the private and nonprofit sectors, and individual citizens.

Collectively, the NRF (formerly the NRP), ICS, NIMS, and the National Preparedness Goal define the actions needed. These programs also define to what degree action is needed to prevent, protect, respond, and recover from a major event. The efforts identified align government entities, the private sector, and nongovernmental organizations (NGOs) to provide an effective and efficient national structure for preparedness, incident management, and emergency response.

NATIONAL RESPONSE FRAMEWORK

The NRF's purpose is to ensure that government, private-sector, and NGO leaders, and emergency management practitioners understand the domestic incident response roles, responsibilities, and relationships in order to respond more effectively to any type of incident. The NRF defines the principles, roles, and structures that organize how the United States responds as a nation. The NRF

- Describes how communities, tribes, states, the federal government, private sectors, and nongovernmental partners work together to coordinate national response whether limited to a single jurisdiction or national catastrophe.
- Makes clear specific authorities and best practices for managing incidents.
- Builds upon NIMS providing a consistent template for managing incidents.
- Identifies special circumstances where the federal government exercises a larger role, including incidents where federal interests are involved and catastrophic incidents where a state would require significant support.

The NRF is designed to have operational capabilities that are adaptable, flexible, and scalable through unified command. This means the response is adjusted to meet an incident changing requirements given changes in size, scope, and complexity. Through unified command and a readiness to act, roles and responsibilities of each participating organization are clearly understood.

NRF outlines the U.S. response doctrine, responsibilities, and structures. The NRF replaced the NRP that was an all-discipline, all-hazard plan for the management of domestic incidents. The NRF also supersedes the Federal Response Plan (FRP). The FRP established a process and structure for the systematic, coordinated, and effective delivery of federal assistance to address the consequences of major disasters or emergencies declared under the Robert T. Stafford Disaster Relief and Emergency Assistance Act (the Stafford Act). The FRP included the U.S. Government Interagency Domestic Terrorism Concept of Operations Plan (CONPLAN).

The purpose of the CONPLAN was to ensure the policy in Presidential Decision Directive (PDD) Number 39 and PDD-62 is implemented in a coordinated manner: PDD-39 defined the U.S. Policy on Counterterrorism as written in 1995, and PDD-62, which followed in 1998, created a new and more systematic approach to fighting the anticipated terrorist threat of the twenty-first century. The CONPLAN was designed to provide overall guidance to federal, state, and local agencies of the NRP concerning how the federal government would respond to a potential or actual terrorist threat or incident that could occur in the United States, specifically an incident involving weapons of mass destruction (WMD).

The CONPLAN was replaced by the Terrorism Incident Law Enforcement and Investigation Annex. Its purpose is to facilitate effective federal law enforcement and investigative response to threats or acts of terrorism within the United States. The response is based on whether threats are credible, or a threat or act of terrorism should escalate to an Incident of National Significance.

The NRP also included within its scope the Federal Radiological Emergency Response Plan (FRERP). The FRERP was replaced by the Nuclear/Radiological Incident Annex of the NRP. This annex provides guidelines for radiological incidents that are considered Incidents of National Significance and for those that fall below the threshold of an Incident of National Significance. *Incidents of National Significance* as defined in the NRP are high-impact events that require an extensive and well-coordinated multiagency response to save lives, minimize damage, and provide the basis for long-term community and economic recovery.

In summary, the NRF enables first responders, decision makers, and supporting entities to provide a unified national response. Through engaged partnerships, leaders at all levels can communicate and actively support the partnerships through shared goals and aligning capabilities so that no single resource is overwhelmed in times of crisis. The NRF promotes use of a tiered response. A *tiered response* means that incidents are managed at the lowest possible jurisdictional level, and additional capabilities offered when needed elsewhere within the jurisdiction or external to the jurisdiction. The NRF incorporates the National Incident Management System (NIMS), which establishes a systematic approach for managing incidents nationwide.

NATIONAL INCIDENT MANAGEMENT SYSTEM

NIMS introduces a consistent framework for incident management at all jurisdictional levels, regardless of the cause, size, or complexity of the incident. NIMS builds upon ICS, giving first responders and authorities the same foundation for incident management, whether it is a terrorist attack, natural disaster, or any other emergency first responders will face.

NIMS represents the areas of incident management clustered into the following groups:

- Command and management
- Multiagency coordination systems
- Public information systems

- Preparedness
- Resource management
- Communications and information management and sharing
- Technologies and technological systems
- Ongoing management and maintenance

All modules within NIMS are equally important and depend on each other. NIMS mandates the interoperability and compatibility of first responder communication systems that touch upon each NIMS module. *Command and management* is the application of incident command structures and organizational systems. For emergency communications, the NIMS command structure and management is used in designing the architecture for command, control, communications, computers, coordination, intelligence, and interoperability. Command and management is used for early situational awareness and to customize operational pictures based on common and shared data.

Multiagency coordination systems define the operating characteristics, interactive management components, and organizational structure of supporting incident management entities engaged at the federal, state, local, tribal, and regional levels through mutual-aid agreements and other assistance arrangements. Multiagency coordination of emergency communication systems requires interoperability among the communication systems used by public officials, first responders, field operations, and Emergency Operations Center (EOC), among others. Today, interoperability of emergency communication systems is at best a patchwork system where seamless integration of systems is a vision of the future rather than a reality for today.

Public information systems refer to processes, procedures, and systems for communicating timely and accurate information to the public before, during and after an emergency. Public information systems include technology (electronic solutions) and no technology (non-electronic solutions) for delivery of emergency alerts to the public using, where feasible, commercially available services including text messaging, e-mail, radio, television, and cable broadcasting, digital signage as well as posters, flyers, and route alerting.

Effective incident management begins with a host of preparedness activities conducted during normal operations, in advance of any potential incident. *Preparedness* involves an integrated combination of planning, training, exercises, personnel qualification and certification standards, equipment acquisition and certification standards, and publication management processes and activities. Preparedness activities include acquiring and building new capabilities to meet existing public alerting gaps in current public information systems such as reaching a more diverse population not well served by current systems.

NIMS defines *resource management* as standardized mechanisms and requirements for processes to describe, inventory, mobilize, dispatch, track, and recover resources over the life cycle of an incident. Many venues are needed to effectively alert and warn the public that incorporates both public and private emergency alert communication systems. The goal of resource management is to effective manage interoperable multi-agency enterprises, federal, regional, state, local, and public resources to delivery emergency notifications to as many people as possible, as quickly as possible, anytime, anywhere.

NIMS identifies the requirements for a standardized framework for *communications, information management* (collection, analysis, and dissemination), and *information sharing* at all levels of incident management. Emergency communications information management and sharing requires the surety of the data shared and managed: information security. *Information security* is protection of the availability, privacy, and integrity of data. Surety includes the security, authentication, and timely delivery of data.

Although NIMS requires that plans are to include solutions that do not require technology, technological systems and technology are considered critical to NIMS. Technology and technological systems provide supporting capabilities essential to implementing and refining NIMS. These include voice and data communications systems, information management systems (e.g., recordkeeping and resource tracking), and data display systems. Also included are specialized technologies that facilitate ongoing operations and incident management activities in situations that call for unique technology-based capabilities.

Ongoing management and maintenance of NIMS establishes an activity to provide strategic direction for and oversight of NIMS, supporting both routine review and the continuous improvement of the system and its components long term. The ongoing management and maintenance of emergency communication systems is to address the ongoing challenges of interoperable systems and crossing ownership boundaries, addressing the technical and policy aspects of the systems, including access control and security. Developing and refining prearranged groundwork for interoperability among diverse participants having different policy and security contexts. Ongoing management and maintenance of emergency communication systems also includes the incremental addition of services and participants. Emergency communication systems become resource multipliers as new stakeholders are added.

More and better information improves disaster response and recovery operations. Information is a vital form of aid. Timely exchange of information can save lives, livelihoods, and resources. For many, just having accurate information of what is going on around them and actionable information to help affected populations make choices are the only form of relief an affected population receives.

NIMS is the national approach to managing the disaster life cycle and meeting the challenges of timely emergency communications. It is also NIMS reinforced by the use of its defined Incident Command System (ICS) that should be institutionalized and used to manage all U.S. incidents.

INCIDENT COMMAND SYSTEM

ICS is a nationally recognized, standardized management system designed to enable management of incidents by integrating a combination of facilities, equipment, personnel, procedures, and communications of agencies operating within a common organizational structure. It is designed to be applicable across all emergency management disciplines, to help organize near-term and long-term field operations, and to be used for a broad spectrum of emergencies ranging in size and complexity. ICS should be used for all planned events, including field training exercises, public events, planned activities, political forums, and special events and unplanned incidents requiring a response. ICS provides a core mechanism for coordinated and collaborative incident management.

The ICS organizational structure is built around five major components: command staff, operations, planning, logistics, and finance/administration (see Figure 8.1 Incident Command System (ICS) structure).

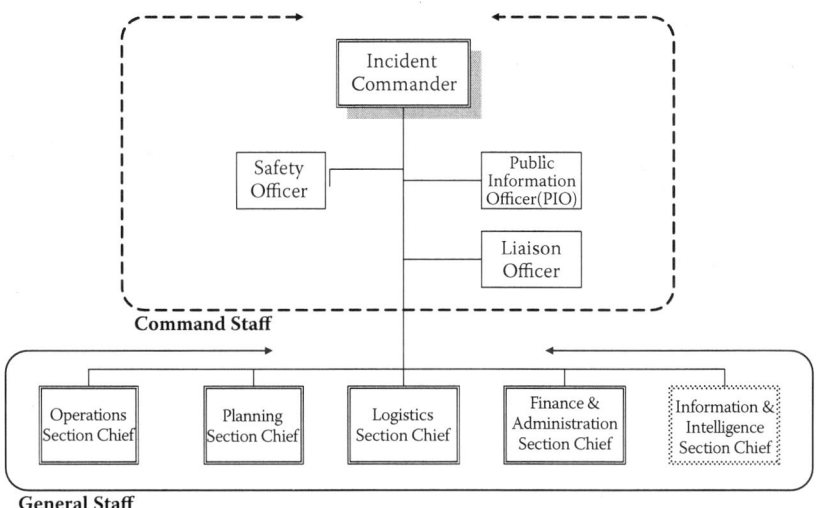

FIGURE 8.1 Incident Command System (ICS) structure.

For some events a sixth component is added called information and intelligence.

ICS defines the operating characteristics, interactive management components, and structure of incident management and emergency response organizations engaged throughout the life cycle of an incident. At the policy level, ICS is to be institutionalized by government officials as the official incident response system at the policy and operational levels. ICS directs incident managers and response organizations to train, exercise, and use ICS in their response operations and integrate ICS into their functional, systemwide emergency operations policies, plans, and procedures.

When an incident occurs within a single jurisdiction and there is no jurisdictional or functional agency overlap, a single Incident Commander (IC) assumes management responsibility and starts as the most senior first-responder to arrive at the scene. The responsible agency may assign a more highly qualified IC as the incident grows in size and complexity. If the IC determines that additional support from other agencies is required, he or she will request that support through the local Emergency Communications Center or EOC.

When multiple jurisdictions, a single jurisdiction with multiple disciplines, or multiple jurisdictions with multiple agencies are involved, then *Unified Command* is followed (see Figure 8.2 Unified Command). Unified Command permits agencies with jurisdictional authority having different legal, geographic, and functional responsibilities to operate efficiently together within a common organizational framework. Unified Command begins with using communications. Agency leaders having functional responsibility for any aspects of an incident and agency begin with a brief initial meeting.

A third type of command structure is called *Area Command*. Area Command is used to oversee the management of multiple incidents or a very large incident being handled by separate ICS organizations, incidents that are not site specific, and when there are a large number of the same types of incidents in the same area.

If incidents under the authority of area command are multijurisdictional, a *Unified Area Command* should be established (see Figure 8.3 Unified Area Command). One of the key responsibilities of an Area Command is to ensure effective communications. Other responsibilities include

- Setting overall agency incident-related priorities
- Allocating critical resources
- Ensuring incidents are managed properly
- Ensuring management objectives are met and do not conflict with each other
- Identifying critical resource needs and reporting them to the EOC

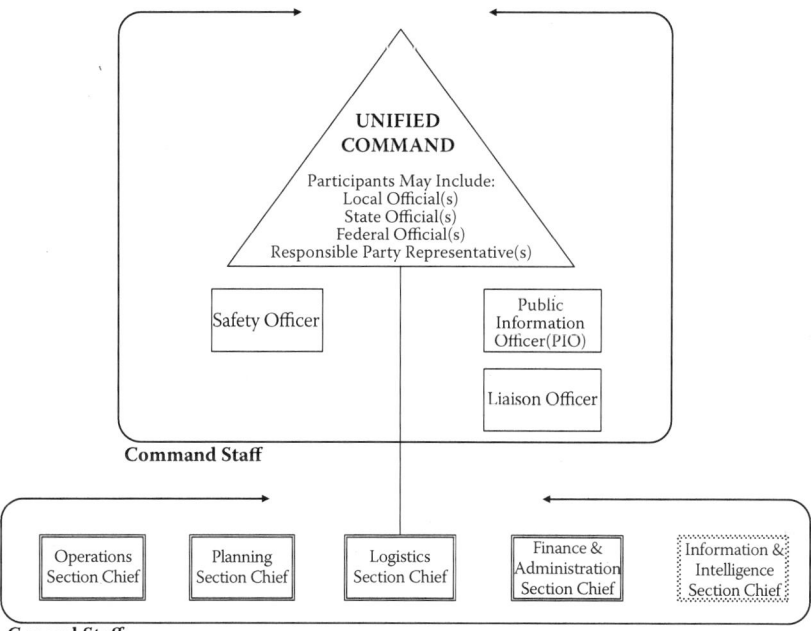

FIGURE 8.2   Unified Command.

When using the ICS incident management organization, there is the Command Staff and General Staff. The Command Staff consists of the Incident Commander (IC) and the special staff positions of Public Information Officer, Safety Officer, Liaison Officer, and other positions as required, who report directly to the IC. They may have an assistant or assistants, as needed.

The IC is the individual responsible for all incident activities, including the development of strategies and tactics and the ordering and the release of resources. The IC has overall authority and responsibility for conducting incident operations and is responsible for the management of all incident operations at the incident site. Working with the IC are the members of the *Command Staff*—the Safety Officer, Liaison Officer, and Public Information Officer—each reporting directly to the IC and support and advise the General Staff (see Figure 8.1).

The *Safety Officer* is the point of contact within the command staff that is responsible for monitoring and assessing safety hazards or unsafe situations and for developing measures for ensuring personnel safety. The Safety Officer advises the IC on issues regarding incident safety. If the Operations Section has been activated, the Safety Officer will work with Operations to ensure the safety of field personnel.

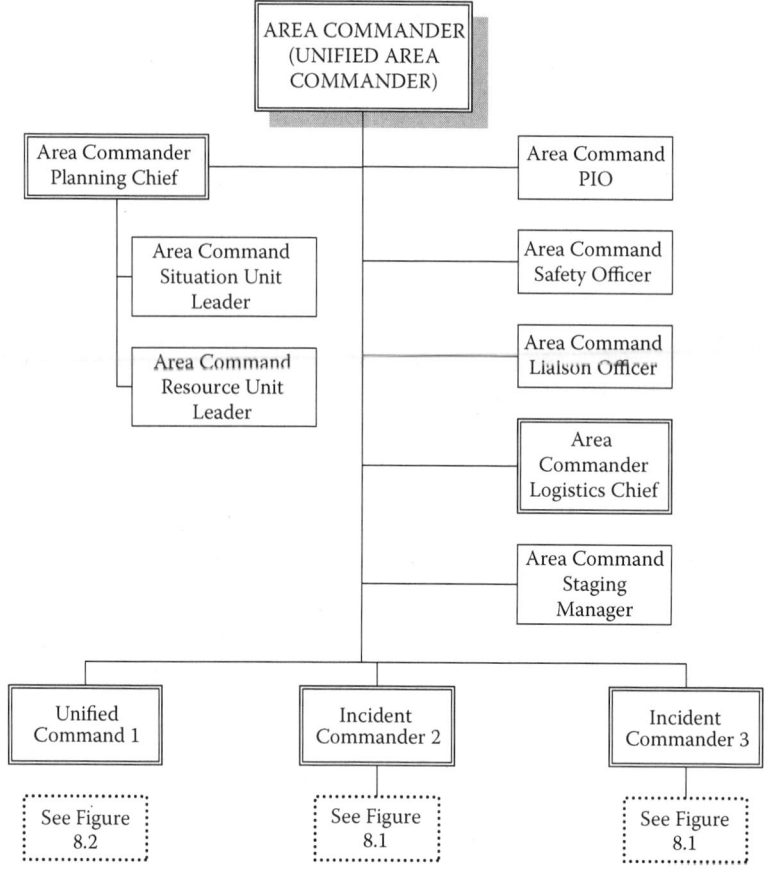

FIGURE 8.3 Unified Area Command.

The *Safety Officer* is responsible for monitoring and assessing hazardous or unsafe situations, and developing measures for assuring personnel safety. The Safety Officer has emergency authority to stop or prevent unsafe acts when immediate action is required or attempt to correct unsafe acts or conditions through the normal chain of command. The Safety Officer maintains awareness of actively developing situations and periodically briefs the Incident Commander.

The *Liaison Officer* is the point of contact for assisting agency representatives (e.g., local, state, and federal agencies; Social Services personnel, industry resources, etc.).

The *Public Information Officer* (PIO) is to prepare accurate and complete information and release the information about the incident to the news media and other appropriate agencies. The PIO should develop accurate and complete information regarding incident cause, size, current situation, resources committed, and other matters of general interest and the point of contact for the media and other entities desiring information directly from the incident. The PIO may elect to establish a single center for incident information, away from the Command Post and staging area at all incidents.

Large-scale incidents may require a *Joint Information Center* (JIC) for better dissemination of information to the public and press. The JIC requires workspace, materials, telephones, and staffing and will prepare an initial information summary as soon as possible. The IC will provide instruction of the release of approved information to the news media and posting, updating news releases, facilitating tours and photo opportunities in an area where the safety of the press can be guaranteed. The PIO is also responsible for maintaining an activity and unit logs as needed.

The next level of the ICS organizational structure is the general staff as developed by the IC upon arrival. The size and complexity of the organizational structure will be determined by the size and nature of the emergency. This group of incident management personnel is organized according to function. There are four general staff sections: Operations, Planning, Logistics, and Finance/Administrative, each having its own section chief. A fifth section, Information and Intelligence, is added to some incidents where the management and flow of information is beyond the capability of the Information Group under the Planning Section.

The Operations Section Chief is responsible for the management of all operations directly applicable to the primary mission, and the safety and welfare of the personnel working in the Operations Section. This chief activates and supervises operations, organizational elements, and staging areas' resources (personnel and equipment) as defined in the incident action plan (IAP). The Operations Section Chief, the 15 Emergency Support Functions (ESFs), and several recovery program groups are organized functionally under branches to provide a coordinated approach and seamless delivery of assistance to survivors and affected states (see Figure 8.4 Operations section organizational structure). A similar organizational structure is used at the state and local levels.

The Planning Section Chief is responsible for the collection and evaluation of information about the incident and the status of resources. He or she is to anticipate future needs for equipment and manpower, assemble information on current and alternative strategies, identify needs for special resources, provide periodic predictions on incident potential, ensure that normal information collection and reporting requirements are being met, prepare recommendations for release of resources, and compile and display incident status information.

The *Logistics Section Chief* is responsible for providing facilities (a place for people, equipment, and supplies), services (needed or used by responders and other field personnel, i.e., procurement, staging), and

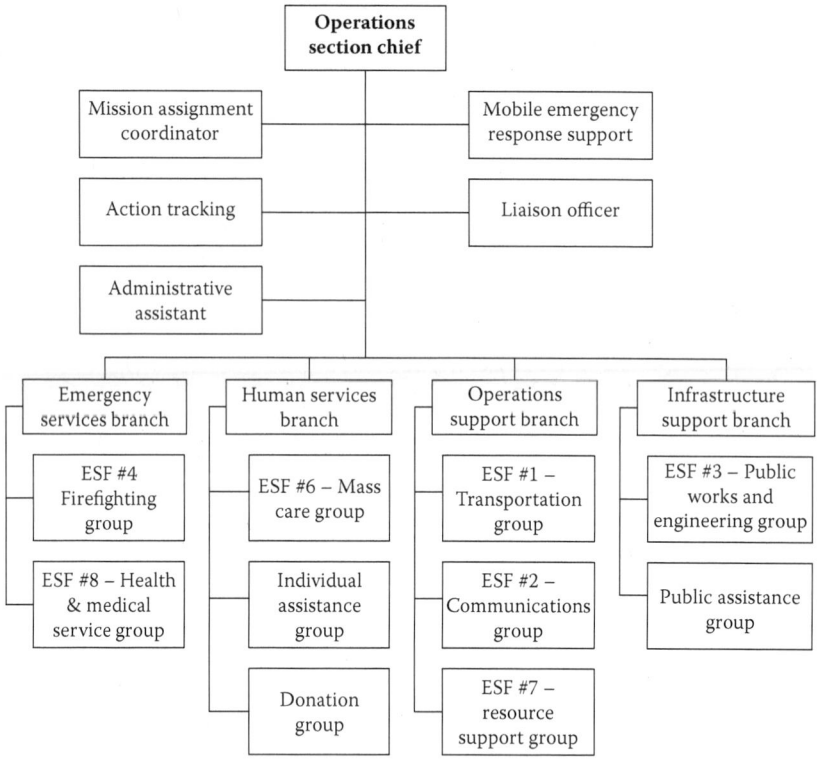

FIGURE 8.4 Operations section organizational structure.

materials in support of the incident. Although the responsibility can be neatly stated in a single sentence, this role has many complex, divergent, and dynamic parts. The areas of responsibility include sheltering and mass care, volunteer and donation management, and medical support for first responders and field personnel. Logistics also includes acquiring resources for field personnel, transportation, and other items needed during the incident.

The *Finance and Administrative Section Chief* is responsible for the management of all financial and cost analysis aspects of the incident. It is recommended that the person normally responsible for such activities under nonemergency conditions continue to serve in this function.

It is important to remember that most incidents do not require a general staff, since their scope is limited to a single jurisdiction, last less than 12 hours, affect only a few persons or entity, and have minimal resource requirements, such as a residential fire at a home or early school closure due to loss of power. The assignment of a general staff is dependent on the duration and complexity of the incident.

You can also take a specific element from the general staff to augment the command staff, such as staging operational communications, a deputy to handle operations for another location, or transportation support. Similar to the general staff, not all command staff positions need to be filled. The positions not filled are handled by the IC. This too is dependent on the incident.

Communications

Communications is the key to success in most any disaster. Communications among the section chiefs and command staff, the general staff with its staff, communications with persons and entities affected and responders define an inclusive communications continuum as depicted in Figure 8.5 (Communications continuum). Communications among members of the incident team are needed to ensure that information is available to resources. The data is used to estimate times of arrival, the intent of command, and communicating the dynamics of a situation. It is critical that all are informed of the situation on an ongoing basis, bringing each member into the situational awareness arena.

Once the IC has established command and briefs the general command staff, the IC is identified over the radio by using the location of the incident, followed by the term *Command*, for example, West Los Angeles Command. From this point forward, the dispatcher will relay all information concerning the incident through the command. For radio communications, the IC may designate additional operational talk groups for the incident. Use of "company speak" and jargon is replaced with clear text voice transmission to avoid confusion upon activation of the command. Communication over the airways is minimized to the details in orders, and the orders are repeated to ensure the message is

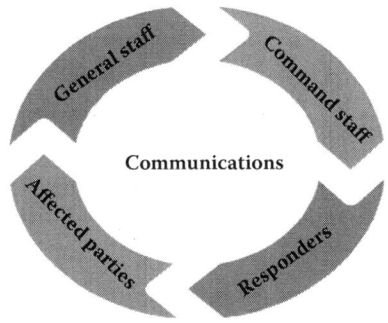

FIGURE 8.5 Communications continuum.

received. Face-to-face communications is preferred when possible over electronic or written communications.

Debriefing

Debriefings may be conducted on an ongoing basis during the incident for intelligence gathering and to enhance operational effectiveness. All participants should be debriefed at the conclusion of the incident and the information used to develop a complete and accurate understanding of the response activities for future planning purposes and training needs. Field personnel can be debriefed as they return to the staging area or team leaders who have previously debriefed their team members may be debriefed. For complex, multijurisdictional operations, a debriefing session may be held after the incident is concluded, led by the Incident Commander.

After action reports (AARs) are used for significant events and exercises. AARs are not required for preplanned events that have had a plan of operation prepared, approved, and implemented. A hostage incident, a plane crash where there is a loss of life, or a building fire are events that may require an AAR.

Emergency Support Functions (ESFs), annexes, and appendices provide concepts for operations, procedures, and structures for achieving response objectives. These are sections of the Basic Plan as illustrated in Figure 8.6 (Basic plan structure).

The ESF annexes provide the structure for coordinating federal interagency support for a federal response to an incident. Incident annexes establish the context and overarching strategy for implementing and coordinating an accelerated, proactive response to an incident by type, such as a hazardous material spill or aircraft-down situation.

ESFs are mechanisms for grouping functions most frequently used to provide federal support to states and federal-to-federal support, both for declared disasters and emergencies under the Stafford Act and for non-Stafford Act incidents as denoted in Table 8.1 (ESF Annexes Roles and Responsibilities). ICS provides for the flexibility to assign ESF and other resources according to their capabilities, tasking, and requirements to augment and support a Joint Field Office (JFO), Regional Response Coordination Center (RRCC), or National Response Coordination Center (NRCC) as needed for incident response in a collaborative manner. ESFs are usually assigned to a specific section for management purposes yet the resources may be used anywhere within a unified coordination structure. Collaboration is needed to ensure that the appropriate planning and execution of missions happen.

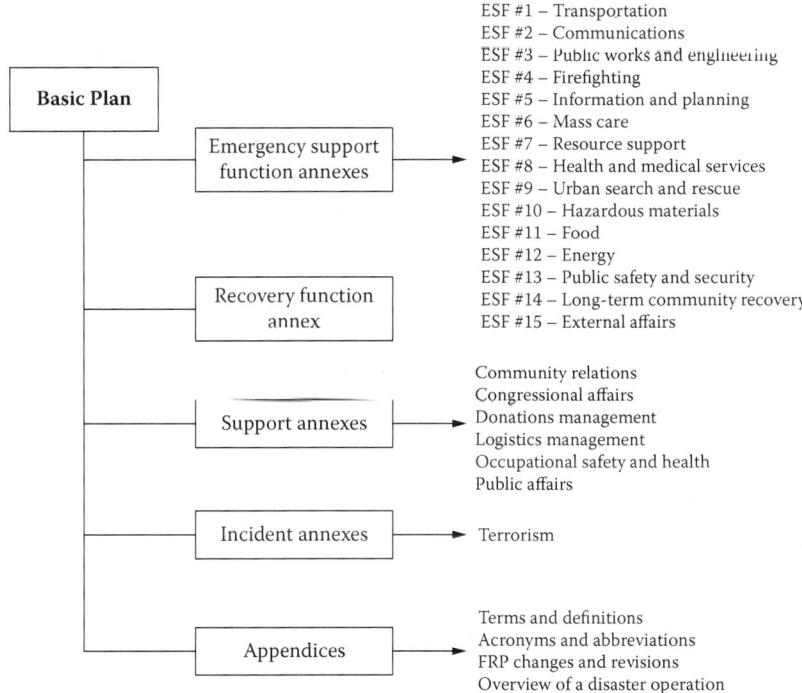

FIGURE 8.6 Basic plan structure.

Although all ESF functions are critical to emergency operations, we have learned that communications is critical to the success of any operations. Although all ESFs require attention to effective communications, ESF #2 *Communications* and ESF #15 *External Affairs* are specific to communications infrastructure and communication with all affected persons and entities. Emergency communications also encompasses the expectation that entities or individuals will request resources that may be lacking in preventing, preparing for, responding to, and mitigating an incident.

The *National Preparedness Guidelines* supersedes the *Interim National Preparedness Goal* that defines all hazard preparedness for the United States (see Figure 8.7 National Preparedness Guidelines). The guidelines also include lessons learned from past disasters, establish a readiness metric to measure progress, and guide investments in national preparedness. There are four critical elements of the guidelines: vision, scenarios, universal tasks, and target capabilities.

The *National Preparedness Vision* gives a concise statement of the core preparedness goal for the United States. The vision is a statement that

TABLE 8.1 ESF Annexes Roles and Responsibilities[7]

Emergency Support Function (ESF) Annexes	Scope
ESF #—Transportation	• Aviation/airspace management and control • Transportation safety (land, sea, air) • Restoration/recovery of transportation infrastructure • Movement restrictions • Damage and impact assessment
ESF #2—Communications	• Coordination with telecommunications and information technology industries • Restoration and repair of telecommunications infrastructure • Protection, restoration, and sustainment of national cyber and information technology resources • Oversight of communications within the federal incident management and response structures
ESF #3—Public Works and Engineering	• Infrastructure protection and emergency repair • Infrastructure restoration • Engineering services and construction management • Emergency contracting support for lifesaving and life-sustaining services
ESF #4—Firefighting	• Coordination of federal firefighting activities • Support to wildland, rural, and urban firefighting operations
ESF #5—Emergency Management	• Coordination of incident management and response efforts • Issuance of mission assignments • Resource and human capital • Incident action planning • Financial management
ESF #6—Mass Care, Emergency Assistance, Housing, and Human Services	• Mass care • Emergency assistance • Disaster housing • Human services
ESF #7—Logistics Management and Resource Support	• Comprehensive, national incident logistics planning, management, and sustainment capability • Resource support (facility space, office equipment and supplies, contracting services, etc.)
ESF #8—Public Health and Medical Services	• Public health • Medical • Mental health services • Mass fatality management

TABLE 8.1 (continued)   ESF Annexes Roles and Responsibilities[7]

Emergency Support Function (ESF) Annexes	Scope
ESF #9—Search and Rescue	• Lifesaving assistance • Search and rescue operations
ESF #10—Oil and Hazardous Materials Response	• Oil and hazardous materials (chemical, biological, radiological, etc.) response • Environmental short- and long-term cleanup
ESF #11—Agriculture and Natural Resources	• Nutrition assistance • Animal and plant disease and pest response • Food safety and security • Natural and cultural resources and historic properties protection and restoration • Safety and well-being of household pets
ESF #12—Energy	• Energy infrastructure assessment, repair, and restoration • Energy industry utilities coordination • Energy forecast
ESF #13—Public Safety and Security	• Facility and resource security • Security planning and technical resource assistance • Public safety and security support • Support to access, traffic, and crowd control
ESF #14—Long-Term Community Recovery	• Social and economic community impact assessment • Long-term community recovery assistance to states, local governments, and the private sector • Analysis and review of mitigation program implementation
ESF #15—External Affairs	• Emergency public information and protective action guidance • Media and community relations • Congressional and international affairs • Tribal and insular affairs

preparedness requires a coordinated national effort involving every level of government, the private sector, NGOs, and individuals. The vision for the guidelines addresses capabilities-based preparedness for a full range of homeland security missions, from prevention through recovery.

The *National Planning Scenarios* documents 15 diverse sets of high-consequence threat scenarios of potential terrorist attacks and natural disasters that focused on contingency planning for preparedness work throughout government and the private sector. The scenarios are used

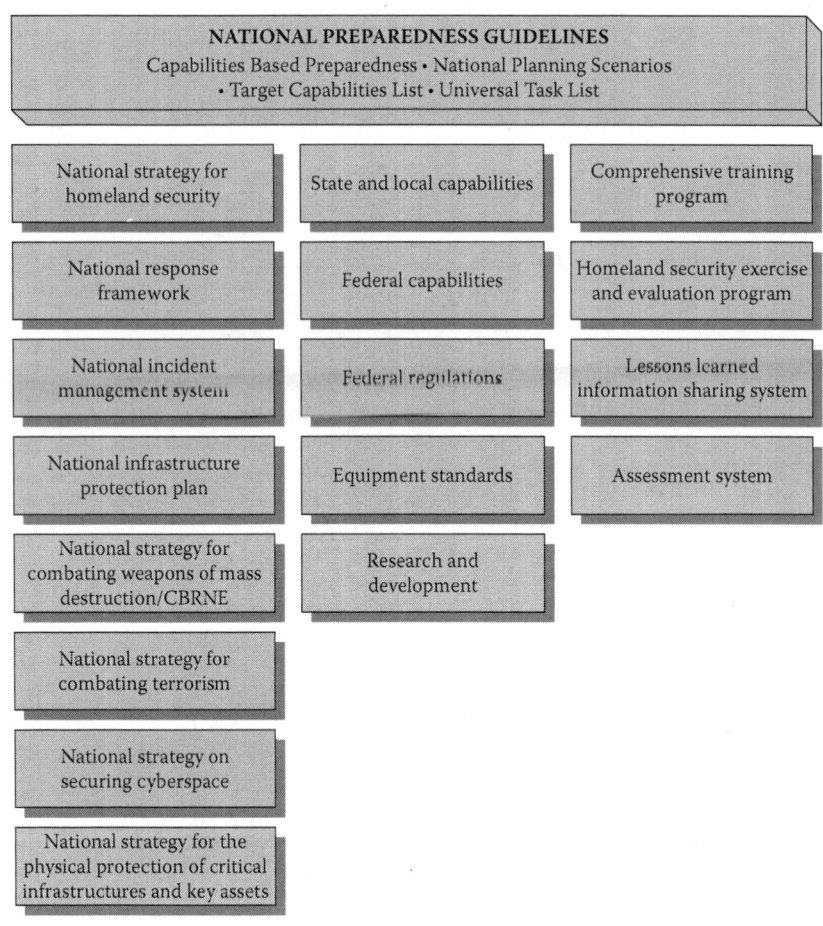

FIGURE 8.7 National Preparedness Guidelines.

for planning, training, exercises, and grant investments needed to prepare for emergencies of all types.

The *Universal Task List* (UTL) is a menu of more than 1,600 unique tasks that can facilitate efforts to prevent, protect against, respond to, and recover from major events based on the National Planning Scenarios. It includes a common vocabulary and identifies key tasks that support development of essential capabilities among organizations at all levels. Of course, no entity will perform every task. *Target Capabilities List* (TCL) defines 37 specific capabilities that communities, the private sector, and government should collectively possess in order to respond effectively to disasters. The guidelines also reinforce the reality that preparedness is a shared responsibility (visit www.dhs.gov for more information on any of the documents listed earlier).

State and local emergency operations plans are continuously developed, updated, and revised to reflect the principles and concepts outlined in the *National Response Framework*. Private sector organizations play a key role before, during, and after an incident. In many areas of incident response, the government works directly with private sector groups as partners in emergency management. The *NRF* describes the process of how the private sector interfaces with other response organizations during an incident and better articulates the private sector's relationships with other response entities.

Following a catastrophic event, segments of state, tribal, and local authorities as well as NGOs and the private sector may be severely impacted. The federal government will employ a proactive federal response to expedite resources to the impacted area. In rare instances resulting from a catastrophic incident, local and state jurisdictions may not be able to establish an effective incident command structure and lead the response.

There may also be serious gaps in continuity of government and public and private sector operations. In these situations, the federal government may temporarily assume certain roles typically performed by state, tribal, and local governments. As soon as state, tribal, or local authorities reestablish the incident command structure, the federal government will transition to its normal role supporting the incident.

For *Stafford Act* incidents (i.e., presidentially declared emergencies or major disasters) and upon the recommendation of the Federal Emergency Management Agency (FEMA) Administrator and the Secretary of Homeland Security, the President appoints a Field Coordinating Officer (FCO). The FCO is a senior FEMA official trained, certified, and well experienced in emergency management, and specifically appointed to coordinate federal support in the response to and recovery from emergencies and major disasters. The FCO executes Stafford Act authorities, including commitment of FEMA resources and the mission assignment of other federal departments or agencies.[2]

Use ICS to manage all incidents, including recurring or planned special events, such as professional sporting events or parades. ICS includes the integration of all response agencies and entities into a single, seamless system, from the Incident Command Post, through Emergency Operations Centers (EOCs). It is critical to develop and implement a public information system before an event occurs that is flexible and adaptable and that employs no-tech, low- and medium-tech, and where feasible, high-technology solutions. NRF enables the identification and typing of resources according to established standards.

It is important to provide training so that personnel can properly perform their job before, during, and after an event. Understanding that communications can define the success of a mission; and ensure that communications interoperability and redundancy are addressed. Remember that, depending on the cause, size, and complexity of an incident, the number of responders participating can vary widely. Regardless of the circumstances, it is always critical that information be shared and that those responsible for making information public "speak with one voice."[3]

NATIONAL EMERGENCY COMMUNICATIONS PLAN

The ability to communicate real time is critical to establishing command and control, operations, and maintaining event situational awareness for a broad range of incidents. Despite investments in technology, processes, and people, communications deficiencies still exist that affect the ability of responders to manage routine incidents and support responses to natural and human-made incidents. Communication deficiencies have repeatedly been cited as a major point of failure and challenge.

To address these shortfalls, the DHS Office of Emergency Communications (OEC) in 2008 developed the *National Emergency Communications Plan* (NECP) under Title XVIII of the Homeland Security Act of 2002 (6 United States Code 101 et seq.) The purpose of the NECP is to facilitate a national emergency communications interoperability plan. The NECP begins with three strategic goals for the country:

- *Goal 1:* By 2010, 90% of all high-risk urban areas designated within the Urban Areas Security Initiative (UASI) are able to demonstrate response-level emergency communications within one hour for routine events involving multiple jurisdictions and agencies.
- *Goal 2:* By 2011, 75% of non-UASI jurisdictions are able to demonstrate response-level emergency communications within one hour for routine events involving multiple jurisdictions and agencies.
- *Goal 3:* By 2013, 75% of all jurisdictions are able to demonstrate response-level emergency communications within three hours, in the event of a significant incident as outlined in national planning scenarios.

The 75 largest urban and metropolitan areas in the United States maintain policies for interoperable communications. All 56 states and U.S. territories have also developed *Statewide Communication Interoperability Plans* (SCIPs) that identify near- and long-term initiatives for improving communications interoperability within a state or territory. SCIPs are

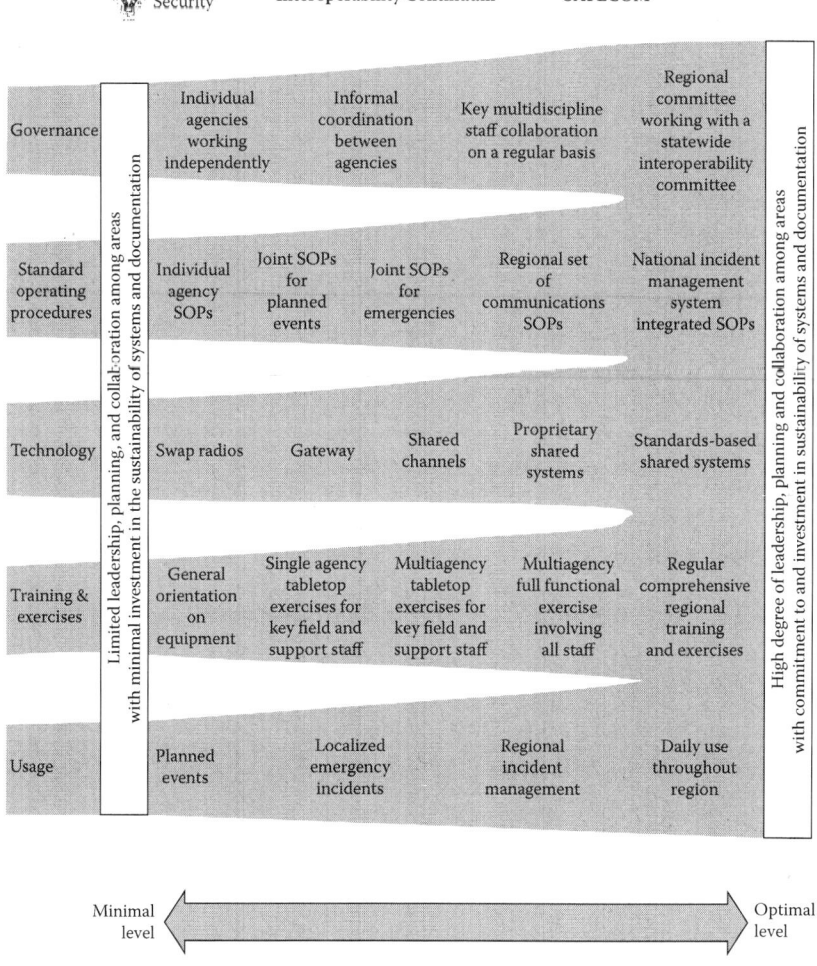

FIGURE 8.8 DHS SAFECOM Interoperability Continuum.

locally driven, multijurisdictional, and multidisciplinary statewide emergency communications plans. Each SCIP reveals the complexity of the state's or territory's interoperable communications environment as determined by the lanes of the *Interoperability Continuum*: governance, SOPs, technology, training and exercises, and usage as illustrated in Figure 8.8 (DHS SAFECOM Interoperability Continuum).

The DHS *SAFECOM Interoperability Continuum* is used by emergency response entities to tackle planning and implementation of solutions that work to improve multijurisdictional and intergovernmental communications interoperability. The degree of interoperability and success depends on improvement in all lanes.

Governance

Governance structures provide the framework used for collaboration and decision making among stakeholders. Governance begins with agencies working independently, and then moving through the continuum where informal coordination between agencies is adopted. Governance continues to transform toward regional committee working together to form a Statewide Communications Interoperability Plan Framework where multidisciplinary jurisdictions work together across a region promoting optimal interoperability.

Standard Operating Procedures

SOPs are used in developing and deploying interoperable communications solutions. Individual agencies have SOPs that are not shared, creating a situation of uncoordinated procedures or incompatible data systems. These silos hamper effective multiagency/multidiscipline response. As SOPs and governance matures, agencies work together to develop SOPs initially for planned events and emergencies, eventually moving toward development of SOPs for regional communications and interoperability. Ultimately, the regional SOPs conform to the elements of NIMS.

Technology

Technology is a tool that is critical for improving interoperability. Once governance and SOPs are developed, the best solution can be identified to meet the need. Many think in terms of selecting a tool, and then building the business around it rather than the tool fitting the business. Interoperability requires the successful implementation of data and voice communications technology among participating agencies and jurisdictions. Technologies must take into consideration the needs of responders in the field, regional requirements, existing infrastructure, cost versus benefit, and sustainability. Technology requires a review of data and voice elements.

Data Elements

Data elements encompass swapping physical and electronic data that is at rest in stand-alone and networked systems where security, integrity, and the confidentiality of the data and applications are critical. The synchronization and age of the data are also important. Common applications and use of open architecture are often preferred over proprietary applications that require agencies to purchase, coordinate operations, and perform ongoing maintenance. Custom-interfaced applications allow multiple agencies to link disparate proprietary applications using single, custom "one-off" links or a proprietary middleware application. This increases access to information, improves user functionality, and permits real-time information sharing among agencies.

One-way standards-based sharing enables applications to "broadcast/push" or "receive/pull" information from disparate applications and data sources. This system enhances the real-time common operating picture and is established without direct access to the source data; this system can also support one-to-many relationships through standards-based middleware. Two-way standards-based sharing approach permits applications to share information from disparate applications and data sources and to process the information seamlessly. This approach can increase access to information, improve user functionality, and permit real-time collaborative information sharing between agencies. This form of sharing allows participating agencies to choose their own applications.

Voice Elements

Swapping radios, or maintaining a cache of standby radios, is a basic solution that is time-consuming, management-intensive, and likely to provide limited results due to channel availability. Ideally, agencies can migrate toward gateways that can retransmit across multiple frequency bands, providing an interim interoperability solution as agencies move toward shared systems. Shared channels is when agencies share a common frequency band or air interface (analog or digital) and are able to agree on common channels. Eventually, Proprietary Shared Systems and Standards-Based Shared Systems used across a region are the optimal solution for interoperability. Interoperability becomes a by-product of system design creating an optimal technology solution.

Training and Exercise

*Training and exercise*s to practice communications interoperability is vital to verifying that technology will work and responders are able to effectively communicate during emergencies. The first step is to provide equipment and application training for users. Training is important for consistency. The next stage is to conduct a tabletop exercise (TTX) for those who will use the solution. The TTX can be used to reinforce earlier training and identify response gaps. Step three is the use of multiagency full functional exercises that includes the users, management, and supervision. At this level, all who will use or deploy the solution should receive training and participate in the exercises. As agencies and regions strive toward optimal interoperability, training and exercise on solutions at time of hire or in an academic setting is considered a best practice.

Technology has risks: extended loss of power and nonworking communication towers, for example, can negate the best of planning and preparations. Within ICS, Communications Unit Leader is a critical function that requires the ability to respond to unexpected events, and it should prepare them to manage the communications component of larger interoperability incidents by applying the available technical solutions to the specific operational environment of the event.

Usage

Usage is how often interoperable communications technologies are used. Ongoing progress and interaction among the other elements of the Interoperability Continuum defines success. Opportunities to use interoperable communications technologies are through planned events such as concerts and parades, and daily use of interoperability systems for managing routine activities in a region. Interoperable communications technologies are also used for local emergency incidents that involve multiple intrajurisdictional responding agencies such as a high rise fire, or road collapse on a highway; or regional incident management such as routine coordination of responses across a region such as response to flooding in a low-lying area.

FEMA, under the *Homeland Security Act of 2002*, established *Regional Emergency Communications Coordination (RECC)* working groups within each FEMA region to coordinate multistate efforts and measure progress in the areas of survivability, sustainability, and interoperability of communications within each region. Group members include both federal and nonfederal agencies. Federal agencies include FEMA, FCC, and other federal departments and agencies with responsibility for coordinating interoperable emergency communications. Nonfederal agencies include law enforcement agencies at all levels, state officials, local fire departments, public safety answering points (9-1-1 services), government emergency managers and homeland security directors, and other emergency response providers as appropriate.

The RECC scope is preparedness and response activities related to emergency communications. Preparedness begins with planning by the ECPC of coordination and information sharing that is used in organizing, training, and equipping first responders to respond. The preparedness life cycle includes conducting exercises, evaluating the outcomes, and using the information to improve plans and enhance the technical assistance offered by the OEC. Once the RECC moves from preparedness to response, the response life cycle begins once an incident occurs or upon knowledge that an incident has occurred with the deployment of federal resources. First responders come to the scene then notify local officials that will active the local EOC. Through the EOC, local officials will request mutual aid and state assistance.

The governor of the state will activate the State Operations Center (SOC), which will assess damage, request mutual aid from other resources within the state and, if needed, request a presidential declaration. The FEMA regional coordinator will work the SOC to assess the situation and make recommendations to the FEMA Administrator. The Administrator working through the DIIS Administrator will make a recommendation to the President to declare a federal emergency or major disaster. If the President gives the declaration, then through FEMA additional resources are deployed within the impacted area. A Joint Field Office is established to coordinate resources for the incident, including emergency communications. Emergency communications is highly dependent upon the radio spectrum, a global resource that must be managed.

NATIONAL TELECOMMUNICATIONS AND INFORMATION ADMINISTRATION

The radio spectrum is a limited natural resource that is accessible to all nations. The U.S. Department of Commerce (DOC) *National Telecommunication and Information Administration* (NTIA), The Office of Spectrum Management (OSM) is responsible for managing the federal government's use of the radio frequency spectrum. OSM receives assistance and advice from the *Interdepartment Radio Advisory Committee* (IRAC). OSM responsibilities include

- Establishing a policy on the allocations and regulations governing the federal spectrum use
- Developing plans for peacetime and wartime use of the spectrum
- Implementing the results of international radio conferences
- Assigning frequencies
- Maintaining spectrum use databases
- Reviewing federal agencies' new telecommunications systems and certify that spectrum will be available
- Providing the technical engineering expertise needed to perform specific spectrum resources assessments and automated computer capabilities needed to carry out these investigations

- Participating in all aspects of the federal government's communications-related emergency readiness activities
- Participating in the federal government telecommunications and automated information systems security activities.

Federal Spectrum Use

The OSM publishes a summary of the U.S. federal government radio frequency spectrum usage in the 30 MHz to 3,000 GHz frequency bands used for emergency communications. The spectrum summary includes spectrum use by the federal agencies such as the military agencies, Federal Aviation Administration (FAA), Department of Justice (DOJ), Department of Interior (DOI), and the National Science Foundation (NSF). The Federal Communications Commission (FCC) maintains information on nonfederal spectrum including the private sector.

Emergency Communications for Which an Immediate Danger Exists to Human Life or Property

Where immediate danger to human life or property exists, the government may grant an exception for an agency to operate temporarily on any regularly assigned frequency in a manner that was not previously agreed to. This alternate use of an assigned frequency can continue as long as is necessary to ensure that the danger to human life or property has passed. Once the situation has passed, normal operations can be reestablished. The DHS *National Interoperability Field Operations Guide* (NIFOG) guides interoperable communications among different levels of government and entities. The NIFOG is a collection of technical reference materials for radio technicians responsible for radios that will be used in disaster response applications.

National Security and War Emergency Communications

Once the President has made a proclamation of war, threat of war, state of public peril or disaster or other national emergency, or in order to preserve the neutrality of the United States, the President may exercise war emergency powers as defined in 47 U.S.C. § 606 under the President's war emergency powers. NTIA authorizes and assigns radio frequencies.

NTIA has developed the *Telecommunication Service Priorities for Radio* (TSP-R) for agencies to designate the appropriate TSP-R for their spectrum-dependent systems. TSP is a FCC mandated program that is used to identify and prioritize telecommunication services that support *National Security/Emergency Preparedness* (NS/EP) goals and objectives. TSP-R is a priority services program for restoring and allocating circuits required by entities with *National Security/Emergency Preparedness (NS/EP)* responsibilities and duties. All common carriers under the FCC's jurisdiction are required to offer TSP. The federal rules governing the NS/EP TSP System are found in Title 47 of the Code of Federal Regulations (CFR), Part 64.

NS/EP is based on a set of telecommunications policies and procedures established by the National Communications System (NCS) under Executive Order 12472, to ensure that critical government and industry needs are met when an actual or potential emergency threatens the security or socioeconomic capabilities of the United States. There are 14 basic functional requirements for NS/EP as defined in Table 8.2 (NS/EP 14 Basic Functional Requirements).

NS/EP users are provided with access to the *Government Emergency Telecommunication System* (GETS), a dependable and flexible switched voice and voice-band data communications service for use during an emergency or crisis. GETS uses existing features and services of the public switched telephone network (PSTN) with selected NS/EP augmentation and enhancements. GETS architecture provides emergency access and specialized processing in local and long-distance telephone networks using a simple dialing plan and Personal Identification Number (PIN). GETS traffic receives priority over normal traffic. The capability of NS/EP calls generated through GETS increases the likelihood that they will be completed in congested networks. GETS does not preempt public traffic. GETS calls are first come, first serve—there are no levels of precedence in GETS. Using a GETS card number will give your call a calling queue priority over regular calls, increasing the probability (90%) that your wireline call will get through the network even with congestion.

It is recommended that senior management and the Emergency Management Coordinator for the EOC or the Public Information Officer obtain a GETS card. There is no cost to applying for or obtaining a GETS card. A charge of 7 to 10 cents a minutes applies to all calls made through the GETS. Special arrangements can be made for testing. Your card will come with a universal GETS access number and a PIN.

In addition to GETS, there is also the *Wireless Priority Service (WPS)*, which is managed by DHS. WPS does provide priority call process on wireless networks to authorized persons approved by the *National Communications Service (NCS)*. The telecommunication industry solution for emergency wireless communications includes methods to allow increased capacity over the different wireless technologies available and minimizes capacity issues.

To get priority status over cellular communications networks, enrollment in both the WPS and GETS is needed. Your GETS and WPS can

TABLE 8.2 NS/EP 14 Basic Functional Requirements

Functional Requirement	Explanation
Affordability	The service must leverage network capability to minimize cost (e.g., use of existing infrastructure, commercial off the-shelf (COTS) technologies, and services)
Broadband service	Broadband service must be provided in support of NS/EP missions (e.g., voice, imaging, web access, multimedia)
Enhanced priority treatment	Voice and data services supporting NS/EP missions have priority over other traffic
International connectivity	Voice and data services must provide access to and egress from international carriers
Interoperability	Voice and data services must interconnect and interoperate with other government or private facilities, systems, networks which will be identified after contract award
Mobility	Voice and data infrastructure to support transportable, redeployable, or fully mobile voice and data communications, i.e., Personal Communications Service (PCS), cellular, satellite, and high-frequency (HF) radio
Nationwide coverage	Voice and data services must be readily available to support the national security leadership and inter- and intra-agency emergency operations, wherever they are located
Nontraceability	Selected users must be able to use NS/EP services without risk of usage being traced (i.e., without risk of user or location being identified)
Reliability/availability	Services must perform consistently and precisely according to their design requirements and specifications, and must be usable with high confidence
Restorability	Should a service disruption occur, voice and data services must be capable of being reprovisioned, repaired, or restored to required service levels on a priority basis
Scalable bandwidth	NS/EP users must be able to manage the capacity of the communications services to support variable bandwidth requirements
Secure networks	Networks must have protection against corruption of, or unauthorized access to, traffic and control, including expanded encryption techniques and user authentication, as appropriate

TABLE 8.2 (continued)   NS/EP 14 Basic Functional Requirements

Functional Requirement	Explanation
Survivability/ endurability	Voice and data services must be robust to support surviving users under a broad range of circumstances, from the widespread damage of a natural or human-made disaster up to, and including, nuclear war
Voice band service	Voice band service must be provided in support of presidential communications

be used in combination. Authorized users dial *272 on a WPS-enabled device to receive calling queue priority. WPS is available through most major cellular service providers, including AT&T, T-Mobile, Sprint, and Verizon Wireless. There is a one-time activation fee of $10, a monthly fee of $4.50, and charges of 75 cents per minute.

Enrollment for either a GETS or WPS account is available by contacting the National Communications System—GETS Operations and Administrative Support. Their information is as follows:

Telephone: 866-NCS-CALL (627-2255)
DC Metro Area (Local): 701-760-2255 or 703.607.4950 (24/7)
FCC (24/7): 202.418.1199

GETS
E-mail: gets@ncsc.gov or gwids@saic.com
Website: gets.ncs.gov

WPS
E-mail: wps@ncs.gov
Website: wps.ncs.gov

NS/EP includes requirements upon the industry to provide critical telecommunication services, including new services that are to be provisioned as early as possible regardless of costs. The industry must also be prepared to install the solution and restore existing solutions rapidly rather than using normal business procedures.

NC/EP also has within its scope the *SHAred REsources (SHARES) High Frequency Radio Program* in cooperation with the NCC in support of the federal emergency response community. SHARES offers a single, interagency, emergency message handling system for the transmission of NS/EP information. The SHARES program joins existing federal and federally affiliated entities' high-frequency radio resources when normal communications are destroyed or otherwise unavailable.

Emergency Use of Nonfederal Frequencies

In an emergency, a federal radio station can use any frequency authorized to a nonfederal radio station, under the *FCC Rules and Regulations, Title 47: Telecommunication, Part 90 – Private Land Mobile Radio Services* (49 FR 36376) when needed for an emergency in progress. Part 90 states that the licensee of any station authorized under this part may, during a period of emergency in which the normal communication facilities are disrupted as a result of a natural disaster, use the station for emergency communications. This emergency communication can be in a manner other than that specified in the station authorization or in the rules and regulations governing the operation of such stations. Restrictions do apply when emergency use of nonfederal frequencies occurs. Specifically, the restrictions are

1. Concurrence has been given by the licensee, whether verbal or written.
2. Adherence to FCC Rules and Regulations.
3. Use is limited to the area and stations of licensee.
4. Control of operations by the licensee is terminated during the time of use.
5. Operations are limited to a maximum of 60 days.
6. The agency must provide a written report of the event to the FCC as soon as is practicable.

Coordination and Use of Emergency Networks

After coordination with the *FEMA National Radio System* (FNARS) program manager, the federal high-frequency (HF) radio stations can communicate with stations operating on the *FEMA National Emergency Coordination Net* (NECN) has predesignated and ad hoc frequencies to support response efforts including tests and exercises. These frequencies are a virtual "meeting place" to coordinate activities, exchange operational information, and receive support (such as relay, phone patch, information lookup, and third-party message handling) from the FEMA radio operators or other stations on the net. A list of frequencies is provided in Table 8.3 (Frequencies for the Safety of Life and Property).

Interoperability between Federal Entities and Nonfederal Public Safety Licensees

A Memorandum of Understanding (MOU) is used to establish interoperability between federal entities and nonfederal public safety licensees. Achieving interoperability involving the use of a passive *cross patch switch* that is installed on the nonfederal public safety entity's transmitter is one possible method. A cross patch switch or any other device could

TABLE 8.3 Frequencies for the Safety of Life and Property

Entity Type and Purpose	Frequencies
Ships for radiotelephone distress and safety traffic	2182 kHz 4125 kHz 6215 kHz 8291 kHz 12290 kHz 16420 kHz 156.8 MHz (VHF FM Channel 16)
Ships using Digital Selective Calling (DSC) for distress and safety calls	2187.5 kHz 4207.5 kHz 6312.0 kHz 8414.5 kHz 12577.0 kHz 16804.5 kHz 156.525 MHz (VHF Channel 70)
Ships for Distress, Urgent and Safety traffic using radio-telex or narrow band direct printing (NBDP)	2174.5 kHz 4177.5 kHz 6268.0 kHz 8376.5 kHz 12520.0 kHz 16695.0 kHz
Ship, aircraft, and shore stations using radio-telephony during coordinated search and rescue operations	3023 kHz 5680 kHz 123.1 MHz
Aircraft stations in addition to their normal air/ground communications channels for distress, urgent, safety, and calling purposes	121.5 MHz 243.0 MHz
Mobile earth stations for distress, urgent, and safety communications	1626.5–1645.5 MHz 1645.5–1646.5 MHz

alter how the transmitter operates. The use of a passive cross patch switch is less likely to alter how the transmitter operations. When a passive cross patch switch is used, a special permit from the FCC is not required.[4]

Federal Aviation Administration (FAA)

The FAA regulates the *Aviation Radio Service* in cooperation with the FCC. The Aviation Radio Service is an internationally allocated radio services group designed to improve the safety and protection of life and property in air navigation. There are three key services provided by Aviation Radio Service: aeronautical mobile, radio-navigation, and fixed services as defined in Table 8.4 (Aviation Radio Service).[5]

TABLE 8.4 Aviation Radio Service

Aeronautical Services	Scope of Services
Mobile	Aeronautical advisory and en route stations, airport control stations, and automatic weather observation stations
Radio navigation	Stations used for navigation, obstruction warning, instrument landing, and measurement of altitude and range
Fixed	System of fixed stations using point-to-point radio communications for aviation safety, navigation, or preparation for flight

Under the Aviation Radio Services is the airport emergency communications system. The airport emergency communications system provides the primary, and sometimes an alternate, method for direct communications between the

1. Alerting authority, Airport Traffic Control Tower (ATCT), Flight Service Station (FSS), Airport Manager, fixed-base operator, or airline office and the Aircraft Rescue and Fire Fighting (ARFF) service.
2. ATCT or FSS and the ARFF responders' en route to an aircraft emergency and at the accident or incident site.
3. Dispatcher and ARFF vehicles at the accident/incident site.
4. ARFF Incident Command (ARFF IC) and appropriate local and mutual aid organizations located on or off the airport, including an alert procedure for all auxiliary personnel expected to participate
5. ARFF IC and the emergency aircraft.

The airport emergency communication system includes a *Discrete Emergency Frequency* (DEF). The DEF establishes a direct link between the emergency aircraft and the ARFF IC, enabling the exchange of critical information on the emergency aircraft status such as fuel on board, number of passengers, and condition of aircraft. The ARFF IC will relay information to the pilot about the external situation of the aircraft, whether or not evacuation is recommended, and other hazards that may not be readily apparent to the pilot. Information transmitted on this frequency, is to be limited to ATC, the pilot of the emergency aircraft, and the ARFF IC.

Emergency communication with aircraft also includes emergency hand signals that are used by ground crews in the event of radio communications disruption or failure of the DEF. The three primary hand signals used for emergency communications with the pilot are evacuate, stop, and emergency contained. The hand signals used during the day are the same used at night with a wand in each hand.

- *Evacuate.* An arm is extended from the body and held horizontal. The hand is raised at eye level. Execute beckoning arm motion angled backward. The nonbeckoning arm held against the body.
- *Stop.* The "Stop" hand signal is used to advise the pilot to halt aircraft movement or other activity in progress. The arms are crossed at the wrists and over the front of the head.
- *Emergency Contained* hand signal is given when there is no outside evidence of a dangerous condition or the situation has the "all clear." The arms are extended outward and down at a 45-degree angle. The arms are to move inward below the waistline simultaneously until the wrists are crossed, then extended outward to the starting position and then repeated until no longer necessary. This arm position is similar to that used by an umpire in baseball in giving the "safe" signal when a player reaches home plate.

The initial notification alarm from an alerting authority to the primary responders is from the aircraft to the ARFF station dispatch room at airports with an ATCT or via link from a non-ATC two-way radio and direct-line telephone to the ATCT, the FSS, or other ATC point. Once the IC communications network is established, a non-ATC emergency frequency network should be used for internal communications.

If communications is lost between the emergency aircraft or ARFF responders and the ATCT, then a universal ATCT light gun signal is given to the aircraft when it needs clearance to land and to the ARFF responders on the airport movement area to grant clearance to cross active runways and taxiways. Once the aircraft is on the ground and electronic communications is not available, then the standard emergency hand signals should be used. These hand signals are known and understood by all aircraft crews and ARFF firefighters.[6]

Federal Communications Commission

The FCC was established by the *Communications Act of 1934* and charged with regulating interstate and international communications by radio, television, wire, satellite, and cable for the United States, including U.S. possessions and the District of Columbia. One of its many responsibilities includes licensure of the radio spectrum. Distribution of these licenses is across the spectrum. There are more than two million active licenses in 2011 with more than half held by individuals (see Table 8.5 Active Licenses by Category and Entity Type).

TABLE 8.5 Active Licenses by Category and Entity Type as of April 2011

Active Licenses by Category		
Category	Number	Percentage (%)
Personal Use	791,738	39.1
Safety of Life	696,898	34.4
Land Mobile Radio	331,826	16.4
Fixed Wireless	86,496	4.3
Broadcast Support	31,903	1.6
Broadcast Support	30,972	1.5
Mobile/Fixed Broadband	19,257	1.0
Paging and Messaging	18,166	0.9
Satellite Earth Station	17,784	0.9
Other	1,712	0.1
Total	2,026,752	

Active Licenses by Entity Type		
Category	Number	Percentage (%)
Individual	1,308,498	64.6
Business	515,882	25.5
Government	197,307	9.7
Other	5,065	0.2
Total	2,026,752	

Public Safety and Homeland Security Bureau (PSHSB)

There are seven operating bureaus and ten staff offices, each having an impact on emergency communications. The bureau established under the FCC to address specifically emergencies is the Public Safety and Homeland Security Bureau (PSHSB). The PSHSB is responsible for developing, recommending, and administering the agency's policies pertaining to public safety communications issues. These policies include

1. 9-1-1 and Enhanced 9-1-1 (E911)
2. Operability and interoperability of public safety communications
3. Communications infrastructure protection and disaster response
4. Network security and reliability

PSHSB is the clearinghouse for public safety communications information and emergency response issues. This bureau is focused on supporting and advancing initiatives that further strengthen and enhance the security and reliability of the nation's communications infrastructure and public safety and emergency response capabilities. The goal of PSHSB is to better enable the FCC to assist the public, first responders, law enforcement, hospitals, the communications industry, and all levels of government in the event of a disaster.

The Common Air Interface

The FCC in 2011 adopted Long Term Evolution (LTE) as the common air interface for nationwide interoperable broadband network and roaming for public safety within the 700 MHz band. LTE sets the minimum technology and standards requirements used in developing the nationwide interoperable broadband network. The adoption of LTE also addressed a number of issues related to achieving nationwide interoperability. Significant issues include the architectural framework, the roadmap for the network and the capabilities the network will offer its users. Just as the functional principle underlying all homes remains the same though there are different styles of houses: single or multi-level, small or large houses, etc. Houses provide shelter. The architectural framework adheres to the same principle: the network can be designed in many different ways, yet its primary focus is to facilitate emergency communications.

9-1-1 and E9-1-1 in Tribal Lands

Within the scope of the FCC and a significant concern is the quality and availability of 911 and Enhanced 911 (E911) service in tribal lands. Many *public safety answering points* (PSAPs) serving tribal lands are incapable of receiving E911 service. For example, in the State of New Mexico, all PSAPs are capable of receiving Phase II E911 service except the PSAPs serving the Navajo Nation and the Jicarilla Apache Nation in 2011. Many of these tribal lands have residences without traditional street addresses. Wireless E911 capability, including Phase I caller identification information, is available at an even lower rate. Enhancing the technology presents funding challenges since fee collection for wireline and wireless telephone subscribers often uses one-off solutions such as the community paying for an exchange that is used within the community. Funding for ongoing support remains an outstanding issue.

Other FCC Emergency Communications Initiatives

Also under the reach of the FCC is the *Communications Systems Analysis Division* (CSAD). The CSAD works with the communications industry to develop and implement improvements focused on the reliability, redundancy, and security of the nation's communications infrastructure. This division manages and analyzes network outage reports submitted by communications providers to identify trends in network disruptions. Armed with this data, CSAD collaborates with communications providers to facilitate improvements to communications infrastructure reliability. CSAD emergency communications initiatives are

1. Commercial Mobile Alert System (CMAS)
2. Disaster Information Reporting System (DIRS)
3. Network Outage Reporting System (NORS)

Disaster Reporting System (DIRS) CSAD manages the FCC's *Disaster Information Reporting System* (DIRS). DIRS is a voluntary, web-based system used by communication service providers, including wireless, wireline, broadcast, and cable companies to report infrastructure status and situational awareness information during a crisis. CSAD performs detailed technical studies of public safety and commercial communications systems during times of disaster to identify and share best practices for emergency preparedness and response.

Commercial Mobile Telephone Alerts (CMAS) The FCC took a number of steps in facilitating the ability of consumers to receive emergency alerts through their wireless phones. In 2008, the Commission issued a series of orders adopting requirements for a Commercial Mobile Alert System (CMAS), a system for commercial mobile service (CMS) providers to voluntarily transmit emergency alerts to their subscribers.

The CMAS is an end-to-end system where an Alert Aggregator/Gateway would receive, authenticate, validate, and format federal, state, tribal, and local alerts and then forward them to the appropriate CMS Provider Gateway. The gateway and associated infrastructure processes the alerts and transmit them to their subscriber handsets. Three types of text-based alerts can be transmitted via CMAS: Presidential, Imminent Threat (e.g., tornado), and Amber Alerts. Subscribers with a CMAS-compatible handset will automatically receive these alerts since subscriber opt-in requirements are automated.

For people with disabilities, CMS providers must provide a unique audio attention signal and vibration cadence on CMAS-compatible handsets. CMS providers must transmit alerts to areas no larger than the targeted county, and yet may transmit to smaller areas if they choose to do so.

Subscribers receiving services pursuant to a roaming agreement will receive alert messages if

- The operator of the roamed-upon network is a participating CMS provider
- The subscriber's mobile device is enabled to receive alert messages from the roamed-upon network

CMAS messages will not preempt calls in progress.

The FCC's actions implements provisions of the Warning, Alert and Response Network Act ("WARN Act") that allow CMS providers to voluntarily transmit emergency alerts to their subscribers. The FCC's actions also implement one of its highest priorities: to ensure that all Americans have the capability to receive timely and accurate alerts, warnings, and critical information regarding disasters and other emergencies irrespective of what communications technologies they use.

The *Personal Localized Alerting Network* known as PLAN is a technology developed by FEMA allowing individuals using wireless mobile devices to turn them into personal alert systems as provided by their wireless carrier. The alerts are geographically targeted. People receive these alerts based on where they are when the emergency occurs rather than on where they live. The system will provide local, state, and federal emergency communications. This approach is more practical and useful since it enables getting emergency information to you when it matters most. An important feature is that these alerts will be able to get through to phones even if cell towers nearby are jammed. To encourage its use, these messages are free. People do not need to sign up. Some new phones have the new PLAN technology and will have the alerts already activated on their phones. All newer models will have the chip installed to enable this service. You can opt out of local alerts but not any presidential alerts. Implementation began in May 2011, ahead of schedule in New York and Washington, DC. Other metropolitan areas will follow throughout 2011.

On January 1, 2013, all public safety and business industrial land mobile radio systems operating in the 150–512 MHz radio bands must cease operating using 25 kHz efficiency technology. They must begin operating using at least 12.5 kHz efficiency technology. Migration to 12.5 kHz efficiency technology (once referred to as *Refarming*, but now referred to as *Narrowbanding*) will allow the creation of additional channel capacity within the same radio spectrum, and support more users.

After January 1, 2013, licensees not operating at 12.5 KHz efficiency are in violation of the Commission's rules and could be subject to FCC enforcement action, which may include admonishment, monetary fines, or loss of license. All government agencies are subject to this mandate. This includes schools, public utilities, transportation, mass transit, and community watches. In addition, the following radio types should be included in the Narrowbanding efforts:

- Low-power "Dot" radios (operating under Part 90)
- Cache Radios–Transportable Systems
- Command Post/Communications Vehicles
- Mutual Aid Gateways.

Accessibility Act

The *21st Century Communications and Video Accessibility Act* (Accessibility Act) was signed into law October 8, 2010 by President Obama. The Emergency Access Advisory Committee (EAAC) is focused on the most effective and efficient technologies and methods enabling access to NG911 for persons with disabilities. The Video Programming and Emergency Access Advisory Committee (VPEAAC) focuses on matters pertaining to the accessibility of video programming. The VPEAAC is responsible for

- Developing recommendations on closed captioning of Internet programming previously captioned on television
- The compatibility between video programming delivered using the Internet protocol and devices capable of receiving and displaying such programming to facilitate access to captioning, video description, and emergency information
- Video description and accessible emergency information on television programming delivered using the Internet protocol or digital broadcast television
- Accessible user interfaces on video programming devices
- Accessible programming guides and menus

Providers must make information available visually as well as aurally, either through closed captioning or some other method so the hearing impaired can understand the nature of the emergency. For the visually impaired, if the emergency information is provided in a screen crawl or through some other nonverbal manner, there needs to be alert tones broadcasted identifying that emergency information is being conveyed. Visually impaired viewers can then make arrangements to find out what the emergency is. Some broadcasters and public emergency management agencies are unaware of their legal responsibilities to modify their information procedures.

2-1-1

In 2000, the FCC assigned the phone number 2-1-1 for community information and referral nationwide. The phone number 2-1-1 provides callers with a way to access information about and referrals to human services for everyday needs and in times of crisis. The United Way is usually the manager of the service in the areas where it is activated. Currently, 2-1-1 systems are funded by state or local government, businesses, nonprofit organizations, and other agencies. Twenty percent of the population has access to 2 1 1 telephone service in 21 states in 2011.

STAFFORD ACT

The U.S. Congress concluded two key findings about disasters. First, Congress deduced that disasters often cause loss of life, human suffering, loss of income, and property loss and damage. It also discerned that disasters often disrupt the normal functioning of governments and communities, and adversely affect individuals and families with great severity. Congress declared that special measures, designed to assist the efforts of the affected states to expedite the rendering of aid, assistance, and emergency services, and the reconstruction and rehabilitation of devastated areas, are necessary. This led to the *Robert T. Stafford Disaster Relief and Emergency Assistance Act*, Public Law 93-288 (42 U.S.C. 5121-5207), and Related Authorities, also known as the *Stafford Act*, signed into law in 1988. It was last updated in June 2007. The Stafford Act replaced the *Disaster Relief Act* of 1974, PL 93-288.

The intent of the Stafford Act is to provide an orderly and continuing means of assistance from the federal government to state and local governments in carrying out their responsibilities to alleviate the suffering and damage that result from disasters. Six fundamental activities are within the scope of the Stafford Act:

- Revise and expand the scope of existing disaster relief programs
- Promote development of comprehensive disaster preparedness and assistance plans, programs, capabilities, and organizations by all levels of government
- Achieve increased coordination and responsiveness of disaster preparedness and relief programs
- Encourage individuals, states, and local governments to protect themselves by obtaining insurance coverage to supplement or replace governmental assistance
- Advocate hazard mitigation measures to reduce losses from disasters, including development of land use and construction regulations
- Provide federal assistance programs for both public and private losses sustained in disasters.

The Stafford Act also opened the doors for state and local governments to use GSA Federal Supply Schedules to purchase products and services needed in recovery. This applies only to recovery from a major disaster that has merited a presidential declaration. GSA is critical to providing disaster recovery products and services to federal agencies. These

FIGURE 8.9 Presidential disaster declarations 2000–2010.

schedules provide speed and savings for those who can take advantage of this purchase program.

There have been more than 1,982 presidential disaster declarations since 1953. That is an average of 34 a year. In the first five months of 2011, there were 32 declarations. 2010 was the most active year on record with 81 declarations. A map of declarations for 2000–2010 is offered in Figure 8.9 (Presidential disaster declarations 2000–2010).

OTHER LAWS AND REGULATIONS

There are a number of laws that impact what you can say in an emergency and what you must share. A review of a few of these laws and regulations will be covered. These are laws and regulations that are common across many industries.

Health Insurance Portability & Accountability Act (HIPAA)

HIPAA is a federal law with rules setting standards for the protection of protected health information (PHI). The purpose of HIPAA is to ensure the confidentiality, integrity, and availability of PHI and to detect and prevent anticipated errors and threats caused by malicious or criminal actions, system failure, natural disasters, and user error. The failure to comply can be costly:

- *Class 6 Felony:* $50,000, 1 year in prison, or both
- *Class 5 Felony:* False pretenses: up to $100,000, 5 years in prison, or both
- *Class 4 Felony:* Intent to sell, transfer, use information for commercial advantage, personal gain, or malicious harm: up to $250,000, 10 years in prison, or both
- *Class 3 Felony:* $100/violation, to $25,000 for all violations of an identical requirement or prohibition during a calendar year
- *Willful neglect:* $10,000–$50,000 for each violation, up to $1.5 million per calendar year for one "identical violation," if corrective action is not taken

The concern for first responders and PIOs is an event that may generate a lot of questions, such as a pandemic scare involving serious illness, hospitalization or quarantine, to mass casualty. People want to know all the details. To protect the privacy of PHI, it is best to only provide information on a need-to-know basis. For example, consider a student with meningitis who lives in a dorm. Depending on the situation, it is best to involve the public health authority to help you create the appropriate message or determine if a message is needed. They will take the lead in contacting personally any individual who may have had contact with the patient or who may be suspected of exposure. If a case is not confirmed, no action may be required. Confirmation first is most important. For a confirmed case, mass notification to the school population could be in the form of a letter or an update on the website indicating confirmation of a case on campus, and instructions on where to seek care and additional information. The personal information of the student with the case of meningitis, such as name, description, classes they take, etc., may not be disclosed without the expressed permission of the patient.

Gramm–Leach–Bliley Act (GLBA)

GLBA is also known as the *Financial Services Modernization Act*. This is a federal law with provisions to govern the collection and disclosure of customers' personal financial information by "financial institutions." This includes businesses that provide loans to clients such as pay day and title loans, schools providing student loans, as well as traditional financial institutions. The law requires that financial institutions handle and store personal financial information securely. The law requires that customers are advised when sharing their information and to give customers an "opt-out" of sharing personal financial information. The failure to comply can lead to civil penalties for institutions up to $100,000 per violation. Individuals found personally liable can incur penalties up to $10,000 per violation.

There may be an incident involving a security breach compromising customers' personal financial information. You shared the information with the CFO to help him or her determine the impact. You left this file on your desk, and by mistake it was picked up, placed in an envelope, and mailed to a financial services marketing firm that will use the information for advertisement. You learn a week later of this error as customers began receiving solicitations by mail and telephone. You need to disclose that this occurred by sending letters of notification to all customers whom you suspect were compromised. Was this accidental? If so, then possibly the firm may avoid penalties, but it will still incur the cost associated with the notification. This is a crisis situation for an organization.

Fair and Accurate Credit Transactions Act (FACTA) and the Red Flag Rule

FACTA is a federal law with rules to help consumers fight identity theft through accuracy, privacy, limits on information sharing, and consumer rights to disclosure. The Red Flag Rule consists of guidelines and requirements for credit and debit card issuers to assess the validity of a change of address request and procedures to reconcile different consumer addresses. The failure to comply and willful noncompliance, may require the organization to give consumers actual, statutory and punitive damages. Statutory damages are $100 to $1,000 for each occurrence, not per consumer, and punitive damages may include attorney's feeds, credit monitoring services, and others as agreed to or court ordered.

Family Educational Rights and Privacy Act (FERPA)

Specifically for educational institutions and those organizations with access to student information is the Buckley Amendment (20 U.S.C. § 1232g; 34 CFR Part 99), also known as the *Family Education Rights and Privacy Act* (FERPA). FERPA is a federal law protecting the privacy of student education records. Students have the right to

- Inspect, review, or amend records
- Consent to disclosure with few exceptions
- File violation complaints with the Department of Education

The failure to comply could lead to withholding of federal funds for the institution, including financial aid.

Many organizations have databases containing student contact information. The marketing or public relations department may want access to this information so they can send materials about upcoming events and activities. Be careful! The law clearly states the use of student information for notification of a health and safety matter is an exclusion within this law. What this means is that you can use the database information for emergency notifications, but you cannot use the same for promotional materials. Students can "opt out" of contact via their personally owned devices. You can require that any e-mail, voicemail, or phone number of your organization is mandatory and cannot be excluded using the "opt out" process. You can also provide students, parents, employees and others an opportunity to "opt in" and provide more information.

Clery Act

The *Jeanne Clery Disclosure of Campus Security Policy and Campus Crime Statistics Act*, as a part of the Higher Education Act of 1965, is a federal law requiring colleges and universities to disclose certain timely and annual information about campus crime and security policies. All public and private institutions of postsecondary education participating in federal student-aid programs are subject to it. The Act includes

- Publishing an annual report disclosing campus security policies and three years' worth of selected crime statistics.
- Giving timely warnings to the campus community about crimes that pose an ongoing threat.
- Keeping a public crime log.
- Upholding the basic rights of victims of sexual assault.
- Making accurate crime statistics available to the U.S. Department of Education, which centrally collects and disseminates campus crime statistics at the national level.

Schools that fail to comply face possible fines by the U.S. Department of Education.

> **CASE STUDY: VIRGINIA TECH FINED $55,000 BY DEPARTMENT OF EDUCATION FOR DELAYED NOTIFICATION TO STUDENTS IN 2007**
>
> In early 2011, the U.S. Department of Education fined Virginia Tech $55,000 for waiting too long to notify students during a 2007 shooting rampage. The amount was the maximum fine Virginia Tech faced for two violations of the federal Clery Act. This law requires timely reporting of crimes on campus. In announcing the fine, DoE officials said the

violation warranted a fine "far in excess" of $55,000. Violations and fines are rare, with only a few dozen over the past two decades. DoE found in December that Virginia Tech violated the law when officials waited two hours to notify the campus after a gunman shot two students in a dormitory in the early morning of April 16, 2007. He killed 30 others two hours later, just after the alert went out. The university plans to appeal.

SUMMARY

The number of disasters continues to increase, that led to more presidential declarations in 2010 than any other year. There are many laws that affect emergency communications. There are laws, rules, and regulations pushing for the use of existing technology regardless of platform, others streamlining the use of the radio spectrum, and those governing when we should relay critical information and what should be relayed. There are also laws, regulations, and rules enacted to expand coverage to people with disabilities and others with functional and access needs. To facilitate all the changes in the world of emergency communications are the presidential directives and the National Strategy for Homeland Security, which includes a comprehensive database of documents that you can use to develop your communication plans. Most all levels of government are involved. This is an exciting time as government, the private sector, NGOs, and the public come together to leverage technology so that all the stakeholders have more information.

ENDNOTES

1. United Nations. (2009). Press Release IHA/1273 "Left in the Dark: The Unmet Need for Communication in Humanitarian Response." Subject of High-level Panel Held at United Nations Headquarters. New York: United Nations Department of Public Information, News and Media Division. http://www.un.org/News/Press/docs/2009/iha1273.doc.htm (accessed April 7, 2011).
2. DHS. (2008). National Response Framework: Frequently Asked Questions. Washington, DC: Department of Homeland Security. http://www.fema.gov/pdf/emergency/nrf/NRF_FAQ.pdf (accessed April 7, 2011).
3. FEMA EMI Course 272: Warning Coordination; 2) "Talking about Disaster." www.fema.gov/pte/talkdiz (accessed April 7, 2011).

4. National Telecommunication and Information Administration. (20 10). Chapter 7. Authorized Frequency Usage. http://www.ntia.doc.gov/osmhome/redbook/ed200801r ev201009/7_9_10.pdf (accessed April 7, 2011).
5. National Telecommunication and Information Administration. (2010). Additional Resources. http://www. ntia.doc.gov/osmhome/redbook/redbook.html (accessed April 7, 2011).
6. U.S. Department of Transportation, Federal Aviation Administration. (2008). Aircraft Rescue and Fire Fighting Communications Advisory Circular. http://www.faa.gov/airports/resources/advisory_circulars/media/150-5210-7D/150_5210_7d.pdf (accessed April 7, 2011).
7. FEMA. (2008). National Response Framework—Emergency Support Functions Introduction. http://www.fema.gov/pdf/emergency/nrf/nrf-esf-intro.pdf (accessed April 7, 2011).

CHAPTER 9

Ripple Effect of Social Media and Social Networking

> We are still in the process of picking ourselves off the floor after witnessing firsthand the fact that a 16-year-old YouTuber can deliver us three times the traffic in a couple of days that some excellent traditional media coverage has over five months.
>
> Michael Fox
> *Founder of Shoes of Prey, 2010*

INTRODUCTION TO SOCIAL MEDIA

Social media is the second generation of websites providing users with their own mini-websites and web pages. There is a high degree of participation and interaction between users. The interaction among users has led to the majority of the content being user generated. Social media has changed the rules for interacting with end users. These sites easily integrate with other sites and have the ability and tools to create and publish data such as Facebook and Twitter pages, for example. Facebook and Twitter have become mainstream and are used as a business tool for communications. Web 2.0, often referred to as the second generation of the World Wide Web uses applications such as WordPress that facilitate active participation through information sharing with a user-centric design. Social networking sites, wikis, and blogs are examples of Web 2.0 applications.

Organizations first used the web to draw people to them online. Having an online presence was hip and meant you were an early adopter of technology in the 1990s. Throughout the decade, organizations next sought to turn the traffic to their sites into cash opportunities by selling products and advertising space, harvesting sales leads, and provide ecommerce solutions for payment, ordering, and tracking. The first decade of 2000 has ushered in the use of social media to form relationships with the community. Individuals have taken the lead in this space and brought it into the workplace. Many organizations are struggling to get their arms around social networking as they race to adopt and stay abreast of the digital world. Their challenge is to balance among the many privacy and confidentiality laws and regulations, intellectual property rights and copyrights, and the overall security and control of the increasing mountains of data.

Using social media for mass notification is a challenge for organizations that individuals have managed to overcome. Historically, technology used at work or school was brought home and adopted by individuals. Microsoft and Apple have become mega-organizations based on this early model. Apple has learned with its introduction of new technologies to first approach the individual early adopters who take the product home to learn and then apply it at work. Going forward with the challenge is to aggregate and analyze the social activity against known information about an event within a system. (See Figure 9.1 Social networking ripple effect.) Social media monitoring tools, such as TweetDeck, still require some human capital to conduct analytics in closing the gap between what is fact and what is fiction. Social media has changed the rules for interacting with end users.

Information regarding an incident can come quickly ... an organization can quickly become deluged by the thousands on Facebook, Twitter, blogs, and other sites. Don't miss the opportunity to engage with the

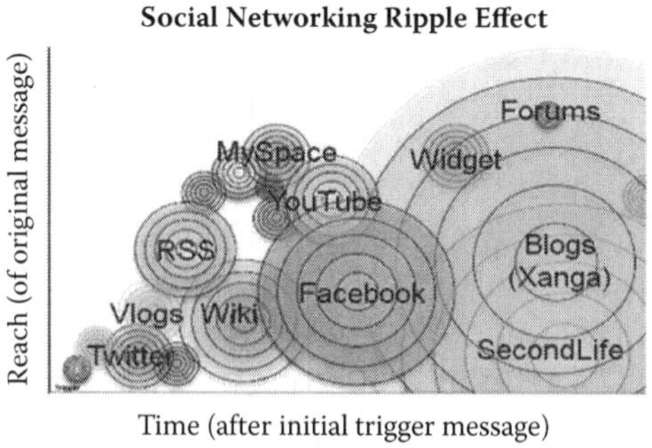

FIGURE 9.1 Social networking ripple effect.

> *FUN FACT OF 2011:*
> - Facebook has more than 500 million users; half log on to the site daily and, on average, Facebook users are connected to 130 friends—4.3 degrees of separation.
> - LinkedIn has more than 80 million members with a member joining every second.
> - Twitter has 106 million registered users who send an average of 55 million tweets daily.

online community. Social media by its very nature encourages users to find something to talk about—anything. Bloggers will espouse information, whether factual or not, speculate, and offer opinions and ideas. Just because a topic generates a lot of interest through social media does not mean it is real or is as presented. You have to weigh the data emerging from social media sites. This can be tricky. There are social media monitoring tools that can identify and capture online conversations that meet user-defined criteria. These tools remain in their infancy.

There is technology that can combine and integrate social media data (what people are saying online), the locations they visit using your demographic (e.g., personal e-mail addresses, telephone numbers, etc.), and business intelligence data to help draw a more complete picture on your area of research. There remains a void—a need for significant human intervention, such as to cut and paste the many messages captured for later analysis.

Organizations that plan have an advantage in appealing to increasingly tech-savvy end users through social network and mobile applications. This is a great opportunity to connect. The amount of data an organization can gather from social media chatter can overwhelm the organization. There is a lot of sharing going on.

Most social media sites like Facebook, Twitter, and LinkedIn, use a consumer-to-consumer(s) model, that is, people talk to each other rather than picking up the telephone to talk to an organization's representative—a consumer-to-business model.

Tools for scanning social media include Visible Technologies and Lithium. These tools today are unable to restructure the data to take advantage of analytics applications to decipher it. Today, this requires significant human capital.

Many organizations are either not ready nor do they want to collect, dissect, and absorb social media into their IT systems. Such organizations will instead monitor online conversation to focus on getting a head start on problems, such as rumor control during an emergency. For a single resource, many organizations follow just Twitter, a site where millions of posts appear daily.

Some organizations continue to ban the use of social media in the workplace due to concerns regarding data breaches, privacy and confidentiality, and compromised system. Other organizations have embraced social media and are building online communities—extending the organization into the online community, and promoting the sharing of information. Devising a social media strategy requires stakeholders in organizations to come together—IT, Marketing, External Affairs, Emergency Management, and senior leadership. Marketing and External Affairs can vet the tools, Emergency Management tests and use the tools, and IT installs and supports them. Senior leadership provides sponsorship and support, and the authorization required to make the tools effective within their organization.

Establishing a social media presence is easy; maintaining its currency and relevancy requires a team and is an ongoing effort. Students and interns are great ambassadors to assist. The ambassadors can share information about the organizations and links to relevant materials to generate additional traffic to the site. It is a way of creating interaction where you do not have to come to one site for everything, yet having a link-through is important. These links can create awareness of the site and the organization's brand.

There are many collaborative tools to share information such as Sharepoint but there are significant costs associated with many of them. YAMMER is a social media site for business to share information. Access to your information is limited to those who share a common e-mail address with you and who are invited. Using social media tools in an emergency means virtual collaboration is a reality. In 2011, the average American spent 25% of their time on the Internet on social networking sites for emergency information. For the Deepwater Horizon oil spill, the United States Coast Guard (USCG) reported the following regarding verifiable and actionable information provided during this event from people using social media:

- Ideas
  - 123,000 ideas submitted
  - 400 were real
  - 100 were field tested
  - 30 were massed produced
- Posting
  - *Flickr:* 1,200 videos/photos
  - *YouTube:* 300 postings
  - *Turtle Talk:* live chats

- Of all the information received, the top three categories were:
  - people wanting a job helping in the cleanup/recovery
  - people wanting to vent
  - people with questions for subject matter experts, e.g., "is it safe to swim," "can I drink the water," etc.

The media have embraced social media as a tool to learn where the stories are and where to go for more information. Social media has also introduced a new area for emergency management—online information and intelligence gathering. Historically this need was addressed under the Planning Section Chief and the Public Information Officer (PIO). In many large-scale events, Information and Intelligence has become the fifth section, along with Operations, Planning, Logistics, and Finance/Administration.

Social media as an emergency management resource requires that information is verified. Verification means looking for consistency in the information found online and having some familiarity with the person(s) who are posting the information. A person can be verified using interconnected sites such as sending Tweets, having a Facebook page, posting on Flickr, or having a LinkedIn account. The verification process continues with reading the profiles of the individuals and reviewing the groups they are attached to. Use of avatars also gives a high level of adoption.

In defining success as an emergency manager using social media, you and your team need to tap into the information provided through social media, mine the data, and then determine what is actionable and what is junk. Every organization should have a written social media strategy (see Appendix E: Sample Social Media Strategy). Select the tools that are complimentary to your operation. It is also important to remember that social media supports the transfer of knowledge that is in your head—being grammatically correct and correctly spelled with legible penmanship are less important than the message. In this ever changing world we must adapt and learn the new tools that are available to us. A brief discussion of commonly used tools within the social media space continues.

MORE ABOUT SOCIAL MEDIA

The use of social media is growing. Amazon and eBay were the starters for online sales and among the early social media systems; however, sites such as craigslist and backpage offer new and preowned items for sale using classified type ads. Classified ads used to only be available using paid subscription newspapers and magazines.

Social media growth will continue to explode; the mind-set of today says if it is not free, then why bother to use it? Meanwhile, print media that fails to modify its approach to the changing market influenced by free social media will continue to die. Unlike magazines that offer ratings once a month or quarter, social media can provide ratings on products and services in real-time. Ratings come as an aggregate count of ratings rather than individual ratings that are not usually trusted unless it is a friend you are following. The power of aggregated numbers is used when there is not a friend who gives the "thumbs up" or likes a product as marked on their page. Social media also offers speed in which you deliver your message.

Instant messaging, text messaging, blogs, electronic bulletin boards (EBBs), and early social networking sites like MakeoutClub and later Friendster linked users together at the start of the twenty-first century. These early social networking tools were used heavily during the September 11, 2001 events to share where people were, how they were feeling, and their individual perspectives on the events of the day. Blogs and bulletin boards, from a disaster perspective, became an alternate media for communicating news information that people began to trust. The individual accounts had names associated with them unlike the many stories reported on mainstream media. Social media changed with the decade. Now, move forward to 2007 and the Virginia Tech massacre. Before, learning the status of a survivor took hours or even days to confirm due to the layers of privacy and access to technology. Cell phones, text messaging, Facebook, MySpace, and other social media sites led to confirmation within four hours of the Virginia Tech massacre victims, unlike any disasters before.

This time around it was the power of social media that made this possible rather than traditional means of verification. Traditional media sources are bound by privacy and confidentiality laws, and regulations. The need to confirm information before releasing it requires more time to get the message out. Social media has changed the dynamics of crisis communications.

Today, anyone with a cell phone can get the facts out quicker than authorities are often prepared to respond to. These resources have become the new reporter who is "inside the yellow tape" while the traditional reporters are "outside the yellow tape." Social media gives a user a medium where they can say what they want, whether it's true or not, and irrespective of the many confidentiality or privacy laws and regulations that exist.

A plane was forced to make an emergency landing on the Hudson River on January 15, 2009 around lunchtime. It quickly became known as the "Miracle on the Hudson." Janis Krums, known as JKrums on Twitter, sent a tweet with a picture that said:

> "http://twitpic.com/135xa—There's a plane in the Hudson. I'm on the ferry going to pick up the people. Crazy. 12:36 PM Jan 15th, 2009 via TwitPic."

Since Krums was the first to report on the event he became the expert news media outlets sought because he was an eyewitness with a picture and a story to tell—not because he knew what caused the plane to land or even if all on the plane were safe. The cause was later reported that the commercial jet had struck a flock of birds during takeoff at LaGuardia Airport a few minutes earlier that forced an emergency landing. On the U.S. Airways jet there were 155 passengers, all saved as the plane slowly sank. Government, political figures, and corporations have not been able to put the genie back in the bottle. Today anyone at the scene of a crisis with a phone having a camera is a reporter; they are the tentacles of social media.

At the start of 2011 few would have thought that activists in the Middle East would use social media platforms to revolutionize the way information is used and disseminated. Twitter and Facebook played a pivotal role in providing disenfranchised Arab citizens a place to pressure regimes to democratize power and increase transparency. The impact that social media had on the Arab Spring is undeniable.

Crowdmapping along with Twitter and Facebook was used to share user created videos, images, and reports. This meant that through geo-tagging and geo-plotting using online maps by trusted online users, the information was considered verified. Soon after, other sites began allowing citizens across the Arab world to submit information and footage directly from the streets to their websites. Hundreds of incidents of human rights abuses by different governments were captured in this manner.

Crowdmapping has been used for relief purposes—for identifying the location of survivors, supplies, and hot spots. This social media tool has now migrated to other users that now include documentation on reports of violence from both sides of a conflict.

In a disaster, social media is faster than many nonprofit groups in requesting assistance for others. When the Icelandic volcano erupted in 2010, "I need a place to stay," "Get me home" were stated on Facebook pages, and Twitter had tags. The response people received was faster than any government response that could be put together. Through social media, strangers opened their homes to victims and did what they could to ensure those trapped got home. The response was organic; social media had established itself as a lifesaving tool, an effective crisis communications tool, and one that governments could not control without negative public feedback.

Within a few hours of the 2010 Deepwater Horizon oil spill in the Gulf of Mexico, BP had more than 16,000 Facebook followers. Social media had clearly taken another turn in crisis communications with this catastrophic event. An individual set up a fake Twitter account called "BP Public Relations," designating its location as "Global." At the same time, BP had launched its official site for the event. The fake site was first and quickly had 106,000 followers. This time around, the corporate brand was at stake and BP quickly learned that it could not control the use of its brand unless it followed the rules of social media.

At the start of 2011 few would have thought that activists in the Middle East would use social media platforms to revolutionize the way information is used and disseminated. Twitter and Facebook played a pivotal role in providing disenfranchised Arab citizens a place to pressure regimes to democratize power and increase transparency. The impact that social media had on the Arab Spring is undeniable.

Crowdmapping along with Twitter and Facebook was used to share user created videos, images, and reports. This meant that through geo-tagging and geo-plotting using online maps by trusted online users, the information was considered verified. Soon after, other sites began allowing citizens across the Arab world to submit information and footage directly from the streets to their websites. Hundreds of incidents of human rights abuses by different governments were captured in this manner.

Crowdmapping has been used for relief purposes—for identifying the location of survivors, supplies, and hot spots. This social media tool has now migrated to other users that now include documentation on reports of violence from both sides of a conflict.

The rules of social media and crisis communications in general is that people want accurate information quickly from what is perceived to be a trusted source. If the company is not perceived to be the trusted source, another source will surface to expose it. Other popular social media tools include Digg, Buzz up!, ShareThis, and subscribing by RSS or e-mail. To facilitate two-way communications, people like to add comments to blogs and websites, and start discussions in chat rooms. They use social media to push the message using any of the sites mentioned and others, by simply e-mailing to themselves or others, or by tagging the message as a bookmark for later reference. Finally, seeing the printed word on paper is important for some. To facilitate this process, websites have come forward with many different ways to incorporate much of what their target audience is looking for. Mashups and internal resources are coming to the forefront in pushing and pulling information across the Internet. (See Figure 9.2, Sample organization's interactive web page using social media.)

FIGURE 9.2 Sample organization's interactive web page using social media.

COMMONLY USED APPLICATIONS

Twitter

Twitter is a real-time information network connecting users to the latest information through small bursts of information called "tweets." A tweet is a maximum of 140 characters similar to the limits of a single SMS message. Connected to a Tweet is a pane providing additional information—more context and embedded media. In 2011, there are 175 million registered users and 95 million tweets written daily.[1] On Twitter there were 5.2 billion relationships in 2010 with people separated on average by 4.6 individuals. A tweet can be thought of as a billboard or a headline to a story told through a details pane containing photos, videos, live streaming, and other media content.

Twitter has become an excellent resource, and for many, the preferred source for quickly delivering immediate and emergency information with its mobile social network. Tweeting 140 characters a message can enable a testimonial, distribute information, or start or continue a conversation. For example:

- *Testimonials:* (I am going to work) or (campus closed)
- *Distribution:* Linking URL (blog, wiki, http://www.ready.gov/)
- *Conversation:* (@janedoe I am safe)

Twitter functions as a real-time (synchronous) network offering fast distribution and the broadcasting of the latest news and information.

Retweeting a message can ripple out. The average tweet is valid or useful for approximately an hour according to Sysomos. Sysomos is a provider of social media monitoring and analytics tools. Sysomos conducted an analysis of 12 billion tweets in 2010 and found that 6% were retweeted by others and 92% of these retweets occurred within an hour of the original posting. Less than 1% of tweets are retweeted after 3 hours.[2]

Demonstrating how quickly information is shared via social media versus traditional media resources, we review the time difference between tweeting and a television broadcast. Information reported on Twitter and many other sites gave the news of Bin Laden's death before the President's live broadcast on Whitehouse.gov and before it appeared on television on May 1, 2011. The first tweets were reported around 7:30–8:00 p.m. EST. The first broadcast was at 10:45 p.m. EST. The tweet was captured as:

> @keithurbahn Keith Urbahn
> (Chief of Staff for former Defense Secretary Donald Rumsfeld)
> So I'm told by a reputable person they have killed Osama Bin Laden. Hot ?!?.
> 1 May via Twitter for BlackBerry®

(Source: http://twitter.com/#!/keithurbahn/status/64877790624886784)

News appeared on Twitter more than an hour before the announcement. The tweet above is time stamped 9:24 p.m., as the first credible announcement of Bin Laden's death. Both Urbahn and CBS news producer Jill Scott's confirmation of the rumor tweets were re-tweeted hundreds of times before the first broadcast. Twitter experienced traffic spikes of 4,000 tweets per second. By May 5, four days after President Obama made the announcement on Twitter, he had:

- 7,791,343 Followers
- 698,049 Following
- 138,897 Listed
- 1,341 Tweets

Twitter is increasingly becoming the tool to use for crisis communications. Information gets out faster and spreads like a virus, not experienced in traditional media. Journalists are scanning Twitter, Facebook, and other social media sites for breaking news. As this trend continues, from an emergency communications perspective, journalists using traditional format roles will shift from being the source for breaking news to the historian of events, and the one responsible for serving as the scribe and the professional analyst of data captured for events.

Twitter offers other advantages. First, it enables messages to be distributed based on geo-location, interest, or reach. Second, tweeting can be two-way. Tweeting can be used to "push" the message out and receive "pull" intelligence. A single tweet does not have much depth in content with its restriction of 140 characters per tweet but as the conversation builds or the message is distributed, depth can be acquired. Third, Twitter offers what is called a "half-gated policy," that is, one can openly broadcast content while selectively following others. Finally, Twitter has open APIs and hundreds of applications and services for individuals and businesses.

Facebook

Facebook is a social utility that people use to communicate with their friends, family, and others in the global world. Like Twitter, Facebook fosters the open exchange of information. In 2011, Facebook stated it had 50 million active users—with 70% outside the United States. Half of the active users log on to Facebook on any given day. The average Facebook user is connected to 130 friends—4.3 degrees of separation. Monthly, people spend more than 700 billion minutes per month on Facebook. There are over 900 million objects that people interact with (pages, groups, events, and community pages), and there are more than 30 billion pieces of content (web links, news stories, blog posts, notes, photo albums, etc.) shared each month. Half of the active Facebook users, more than 250 million, access Facebook through their mobile devices.[3] Whether day-to-day or in an emergency Twitter can be thought of as the billboards on your route, Facebook as your gated community or neighborhood, and your website is home.

Facebook is a great tool for distributing emergency and advisory information. Facebook can be used to connect to others in a time of crisis, to find news on others, or to share community rituals and event around a situation. Information can be restricted since members stand at the gate and allow other members to connect or not. In sharing information, Facebook has many viral "touchpoints" to pass along information. This gives Facebook the role of distributor rather than its being just content based. Facebook is free and available for any individual or organization. It is not a blog. Any organization can create a Facebook page and information for distribution in the social network and incorporate applications such as "Clipin" for donations and fund-raising.

Blogs

In emergencies of the past, blogs were the place to post and people could check them out. Today, how many people read or post to blogs? The numbers were declining in favor of instant messaging (IM) and e-mail at the start of 2000, however, the blogisphere has been rebounding as it remains an effective tool to push information out on the Internet with hooks into many different social media applications and organizational websites. A blog is a one-to-many solution for emergency communications. Blogs are an effective tool for posting advisories, upload photos, videos, podcasts, and chat, and retaining an archive of information. It is a place for a series of articles that are written in a diary format with comments and embedded social tools. Blogs can be used to get your message out and control the message. Blogs offers the advantage of not requiring the download of software, and they are easily accessible on the web page. The key to blogging is for the writer to be aware of the audience, topic, tone, and content for community conversations.

A sampling of the many free and open sources available for blogging are WordPress, Blogs.com, and Thoughts.com. WordPress has become one of the most-used blog tools in this category. To demonstrate the power of blogs, here is a one-day snapshot of activity on WordPress:

- 19,733,442 websites are powered by WordPress
- 337,164 bloggers are posted
  - 440,542 posts
  - 410,713 comments
  - 306 posts per minute
  - 285 comments per minute
  - 112,818,879 words written
- Approx three times larger than the entire Encyclopedia Britannica (about 40 million words![4] (*Source*: wordpress.com accessed May 5, 2011)

Just as Twitter was one of the first resources to report on bin Laden's death, blogs were used to provide more information on who was the source of the first tweet and the first follow-up response to the Tweet. It is in blogs that many received the confirmation they needed to know that he was dead. TechCrunch Europe, a blog covering Web 2.0 and international news, featured a man from Pakistan who was sending tweets as helicopters flew past the Osama bin Laden's compound and an explosion followed. He later learns that he had become the subject matter expert (SME) who had recorded some of the U.S. forces helicopter raid on Osama bin Laden's compound. Google was updating its maps to depict the location where bin Laden was killed. This all occurred before President's Obama official speech a couple of hours later.

The speed at which news travels and is confirmed through these new formats is reshaping emergency communications, the need for transparency, getting the message right the first time, and the need for speed. The longer it takes to get the message out equates to more rumors that need to be controlled.

Widgets

Widgets are HTML code that can be embedded in other sites as a feed or data from other pages. Widgets are usually free and let you link to other sites' information. They are a great source for advisories related to emergencies. Also called a *snippet* or *gadget* (Google) and a way of extending RSS content, a widget can be graphical and extend into games and information boxes such as a list of weather reports from U.S. cities as an RSS feed. Another example is the pinning of live updates of the weather to area maps, with a map widget that can be clicked or zoomed in or out for a more interactive expericence. Widgets break up the Internet into "little bits everywhere." They are highly customizable and personal sites. FEMA promotes the placement of a ready.gov widget on websites to reach as many audiences as possible. FEMA, as a part of its communications arsenal for promoting emergency readiness and response, is increasing its use of social media and Web 2.0 tools.

Mobile Phone Applications and Mobile Web Widgets

Many who carry cellular phones use smartphones. Smartphones are mobile phones with advanced connectivity and computing capabilities over standard mobile phones. The power and spread of smartphones has led to countless mobile phone applications and mobile web widgets. A *mobile web widget* is a web widget design for use via a mobile device. The iPhone, for example has sold over three billion devices and more than 350,000 applications.[5] There are a broad range of functions available on these smartphones such as motion sensitivy, GPS, and touchscreen systems without any buttons. An increasing number of smartphones also offer touchscreens with and without buttons. Open coding and develop coding is available to anyone. Major players include Apple, Google Android, BlackBerry, and Microsoft 7. Many social media applications are streamlined to run on mobile devices such as Facebook and Twitter, which strips down the content. There are voice-activated applications enabling hands-free use, such as apps that can read e-mails aloud, and applications that allow a smartphone to take on new and different uses, such as applications to use the cell phone as a flashlight, read QR labels, or used to swipe for payment. By 2020 it is expected that the primary means to access data is by using a mobile device—one you can wear, carry in your pocket, or in a purse. The primary drawback to applications on mobile devices is security and the physical constraints of the device itself.

Open data sources are those used for third party application developers to create applications, for example, some municipalities. The City of San Francisco uses opened geographic information systems (GIS) layers and crime data layers called the "DataSF App Showcase" (http://datasf.org/showcase/) to prepare data to be released, as well as for reporting nonemergency situations to the City's 3-1-1 operator using applications like SeeClickFix on the DataSF App Showcase website. DataSF is a clearinghouse of data sets available from the City and County of San Francisco. The primary goal of DataSF is to improve access to data. The use of open data sources generates a lot of opportunity for inexpensive basic application builders such as Appmakr and SwebApps for the iPhone, and similar applications for other platforms to build and host applications for individuals.

Anyone can use open code to develop an application, however, the security and privacy issues remain. BlackBerry has a long-established brand associated with security of messaging with PIN communications, and conducts extensive testing of user developed applications before it is available for broad distribution. Smartphones are also vulnerable to heat and water. If exposed to either, degradation of performance may affect the device or it may not work at all. Manufacturers are developing rigid and rugged tablets and smartphones that can withstand extreme temperature variations, sand and dirt, and water. A fourth issue is battery life. As smartphones add more features, the life of device's battery are reduced. A backup power source is critical.

Really Simple Syndication (RSS Feeds)

RSS feeds are a very powerful tool for pushing emergency and advisory messages—breaking news. RSS is a format used to deliver information on regularly changing web content to whoever wants it. Many organizations, news sites, weblogs, and other online publishers syndicate their content as an RSS Feed, a subscription. The advantage for subscribers is that they can access content when and where wanted. Most of the major news outlets make news headlines available via RSS feeds to facilitate reading their stories using news aggregators.

RSS uses a special web coding called XML developed by the global online community. RSS contains a summary and links on new content available on a website, about a new product, or can be used as a way to alert the audience of a crisis. Anyone with online access can pick up the RSS codes and display the information automatically. The concept is that the information is published once by an organization then *syndicated*, picked up and displayed by other organizations.

In Web 2.0, content is separate from form (receive information as text headlines or images, without having to go to the host page). Individuals can see when they have been updated without having to open e-mail or visit your site. You can create content on YouTube, Flickr, and other content social media sites, and RSS allows fans to subscribe (be notified) of new content. RSS breaks the Internet up so that people can receive your latest headlines on their Facebook page, My Yahoo!, an RSS reader, e-mail, in their browser, etc.

Wiki Sourcing

What is a *wiki*? A wiki is a website that enables a user to create and edit interlinked web pages via a web browser using a WYSIWYG (what you see is what you get) text editor. Wikis began as a simple online database that has since seen its use grow. Wikipedia, a wiki, is now considered one of the largest online, open-based, and free encyclopedias. Wikipedia has replaced the hardcopy traditional encyclopedias for many. Hardcopy encyclopedias find that their information is obsolete before it is sent to press although much effort is spent to validate the information. Conversely, Wikipedia, is considered a good reference source to get you started—but much of the information requires verification. Verification occurs as people confirm the information online. For example, when Michael Jackson died, many did not consider him dead until it had been confirmed by a number of people who posted their information on Wikipedia. On a grand scale, this is a form of wiki sourcing. Microsoft offers wikis to address emergency-related information on servers using its operating system. These wikis include information on how to shut down a server in a safe manner and then implement disaster recovery in the case of an emergency.

Wikis are great tools for broadcasting advisories. There is no software to download and it is accessible on a web page. Anyone can upload different types of files—pictures, videos, FAQs, or links to other websites or web pages. They are used collaboratively by multiple users such as a community website, corporate intranet, knowledge management systems, and even personal note taking. Access control is available for some systems that can restrict the rights of others to create, edit, or remove data. An article on a web page can be edited by a group of people, and as use grows it becomes a broadcast or collaborate knowledge management system. Another feature of wikis is that you can have both an editable article and a discussion tab for comments. It is an alternate solution to e-mail.

The now famous WikiLeaks, not affiliated with Wikipedia, is an international nonprofit organization that publishes submissions of private, secret, and classified media from anonymous news sources, news leaks, and whistleblowers. It claimed to have a database of more than 1.2 million documents in 2007 and considers itself to be "an uncensorable system for untraceable mass document leaking." While the founder, Julian Assange, and the website have received some praise; WikiLeaks has also experienced significant backlash and pressure from the international community and businesses. The site has released information on the prisoners held at Guantánamo, thousands of leaked United States embassy cables, and even plans of an attack on its own site.

Wikis are a valuable communications tool for emergencies. Examples of its use are

1. The County Welfare Directors Association of California has a publicly available wiki for sharing information on Temporary Assistance for Needy Families (TANF) Emergency Contingency Fund (ECF).
2. Homeowner associations use wikis to share information specific to the community being services, from block captain contact information to the status of boil water notices or the latest information related to a rash of burglaries.
3. Microsoft's wiki containing articles on its Windows Server Emergency Management Resources.

Geo-Location Systems

Geo-location is the identification of an object's geographical location real-time. Geo-location technology is used in radars, mobile phones, and even an Internet-connected computer terminal. Geo-location can include assessing the location, or going to the actual assessed location. Social geo-location systems are innovative for marketers who are using the data to target information on where people are. Google can provide pin-point data with geotagging information such as coordinates provided in many pictures taken using a cell phone. Applications such as Google Latitude and Facebook Places let you see where people are real-time and share where you are with others. You can add this as a widget, similar to Latitude on your Android home screen, or other mobile devices.

Geo-location systems can also let you check-in at places, letting others know where you are, including businesses that will unlock discount offers or for tracking first responders while deployed. Brand functionality includes mobile maps application, person-to-person mobile awareness, and location awareness. Most of these applications allow for opt-in functionality. In an emergency, geo-location applications are a great resource to aid search and rescue operations, debris management, damage assessment, spotter deployment (such as tornados), and field accountability.

Geo-Targeting

Geo-targeting is a method used to determine the geo-location of a website visitor. Geo targeting is then used to deliver different content to visitors based on their location or other criteria (IP address, ISP, zip code, state, city, etc.) Online advertising and Internet television (iPlayer or Hulu) heavily rely upon geo-targeting to restrict content based on location such as a country using *digital rights management* (DRM). DRM represents access control technologies used to limit use of digital content and devices by publishers, copyright holders, individuals, and hardware manufacturers. Adherence to DRM does present its challenges with accurate pinpointing of locations since many websites are delivered using proxy servers and virtual private networks.

A *proxy server* allows a client (computer) to connect or request resources from another server through an intermediary, a proxy server. Proxy servers may or may not disclose the location of the request originator, making it difficult to track or take advantage of geotagging. Geotagging may only provide the location of the proxy server that could be located anywhere in the world. A client may use multiple proxy servers to obtain information, making it even more difficult to trace the location of the client as illustrated in Figure 9.3 (Proxy server—open).

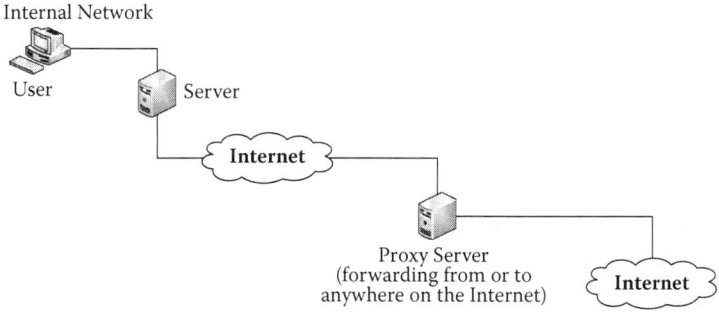

FIGURE 9.3   Proxy server—open.

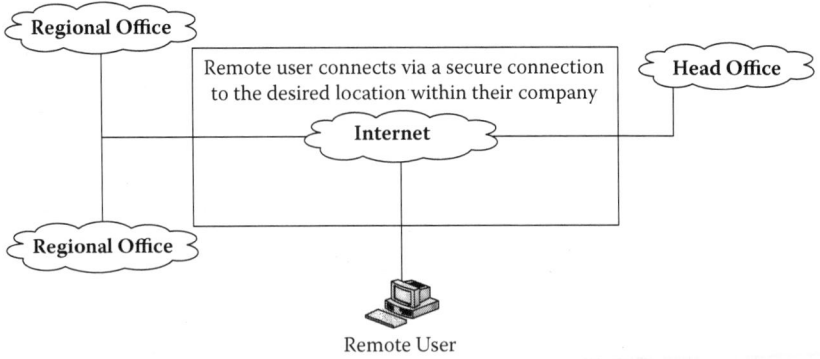

FIGURE 9.4 VPN connection.

A *virtual private network* or *VPN* is a method used to connect to a private local area network (LAN) at a remote location securely using the Internet or public network to transport network data packets privately. This secure transport is managed using encryption. VPN uses authentication to allow or deny access to the private network packets containing any kind of network traffic securely, including voice, video, or data. Remote workers and organizations with multiple locations can share private information electronically and network resources use VPNs. With the appropriate security, a user can bypass regional Internet restrictions such as firewalls and web filtering by "tunneling" the network connection to a different office or region. This capability also presents a challenge for correctly identifying the location of a client. In an emergency where sensitive information needs to be shared, use of a proxy server and VPN can restrict access to only authorized users. This capability is critical for military operations, incident command, and even to protect copyright and intelligence. VPN in its most simplistic form is shown in Figure 9.4 (VPN connection).

Geotagging

Geotagging is the process of adding geographical identification metadata (data about a record or data) such as photographs, video, websites, SMS messages, or RSS feeds. Metadata is used to provide information to data catalogs, clearinghouses, and brokerages to process and interpret data that has been received through a transfer from an external source. The role of metadata is the availability, access, fitness for use, and the transfer of data sets. Geotagging is a form of geospatial metadata giving the latitude and longitude coordinates, altitude, distance, bearing, accuracy data, and the names of place.

Geotagging is used in photos with GPS information of where the photo was taken or a photo is attached to a map. Geotagging can be used by schools to locate any student carrying a smartphone with a GPS chip and the feature enabled when a campus evacuation is issued related to a fire, a bomb, or a shooter. Geotagging could be used by smartphone owners to find location-based news, websites, or other resources as they move around, such as tracking severe weather warnings. Geotagging is now a requirement for U.S. wireless service providers to supply more precise location information for 911 calls by September 11, 2012. Some digital cameras have built-on or built-in GPS for automatic geotagging. For those without this feature, a stand-alone GPS with photo mapping software can write the location information to the image's header.

Geocoding

Geocoding is the process of finding latitude and longitude (geographic coordinates) associated with zip codes or street addresses. The geographic coordinates are mapped and entered into a GIS or embedded into media such as a digital photo. Geocoding can also be used in reverse, called *reverse geocoding*, that takes geographic coordinates to find the associated textual location (e.g., a street address). Examples of free geocoder systems include MapQuest, Yahoo PlaceFinder, Google Maps, USC Geocoder, or Bing Maps. There is *geohash* that uses a short URL for a location (latitude/longitude), and then appending the name to the geohash after a colon. The geocode for the U.S. Post Office at 401 Franklin Street, Houston, Texas, is:

geohash.org/9vk1mfs7spv5p
Top of Form

Address or coordinates:
Bottom of Form
29.7583272 -95.3704009

Characters can be removed from the end of a geohash; however, as you remove the end characters, you have a less precise coordinate. Geohashing can be used with wikis, such as OpenStreetMap.com, a free online editable world map.

Quick Response (QR) Codes

A *quick response* (QR) *code* is a high-capacity, black-and-white matrix barcode specific to an object or location that is readable by a camera phone or dedicated QR barcode reader. Conventional bar codes store up to 20 digits; QR codes can have hundreds times more information. A QR code can handle many data types, such as numeric and alphabetic characters, Kanji, Kana, Hiragana, symbols, binary, and control codes. Up to 7,089 characters can be encoded in one symbol. QR codes have

FIGURE 9.5 QR code for http://www.ready.gov.

been around since the 1990s and are seen around town and across the country on billboards, business cards, magazines, and website. Holding the camera up to the code, unlocks the QR code that contains information for identifying an object, providing a message, information on an event, customer reviews, location on where to buy, or whatever information the originator desires to communicate. They are easy to use with the free QR reader applications available for mobile devices. Many bar code applications will also read QR codes.

Creating a QR code is just as simple as scanning a code. Their ease of use means there really is no limit to what QR codes could be used for. For emergencies, QR codes could be used by logistical resources to manage inventory, materials, and supplies, and to request donations. The activities for sorting, tagging, tracking, and distributing donations can be simplified with QR codes. The infomation found in QR codes could take you to Facebook or Twitter for more information related to an event. QR codes could also be used to promote emergency preparedness by adding a link to www.ready.gov, for example. (See Figures 9.5 QR code for http://www.ready.gov and 9.6 QR code detail.)

Free resources for QR code generators include Kaywa, QRStuff, SPARQCode, or ZXing. There are numerous free QR code reader applications from the many different mobile device platform providers (i.e., iPhone, Android, BlackBerry).

Shared Content

Share content in times of an emergency can save time and resources while offering transparency. A Google Doc is one tool that is free. It enables users to access their documents from a mobile device, including viewing and editing documents, images, and more, and to convert photos. It

FIGURE 9.6 QR code stating "This is a test of QR Code readers that should take you to http://www.ready.gov if it works."

also enables the sharing and uploading of documents on the go. Google Docs works on most smartphone platforms. In emergency situations, first responders and Emergency Operations Center (EOC) operations have real-time collaborative editing, type-with-me, bulk e-mail senders hosted in the cloud, and e-readers. As the price comes down and more models that feature rich text become available, sales of the units will increase, diminishing the sales of printed books. Google Shared Spaces is seeking to become the Google Wave gadget. The shared space has a chat area (a wave gadget), and you can include an annotated shared map, a drawing board, or use a polling gadget. For emergency responders, this space could allow the EOC, for example, to pull up a map, annotate with location information about an event in progress, and have a white board for drawing simple images or write notes that field personnel can use real-time. The EOC could also insert video using a WaveTube, for instance, to watch the video or see others watching it.

Social Storage

With all the data coming at you quickly in an emergency, particularly when using mobile devices, storage on the devices or the storage card can quickly be consumed. Alternate storage is needed real-time when in the field. There are many social storage solutions online—backup, remote backup, data backup, and storage and file sharing solutions. MyOtherDrive.com, Carbonite.com, or a free service called Dropbox.com are effective solutions. Dropbox is a web-based file hosting service that uses cloud computing for storing and sharing data across the Internet with others using file synchronization.

Social Bookmarking

Social bookmarking is a method for Internet users to organization, store, manage, and search for bookmarks of resources online rather than the expanded resources. Social bookmarking can save time; it minimize the need to open a file to understand the contents within a document since a brief description or keywords are added to the bookmark in the form of metadata. When attempting to track information on a topic, such as "Osama Bin Laden Is Dead," social bookmarking along with search engine optimization can increase the visibility of websites containing the information you seek.

Internet Radio, Blogs, Talk Radio

Internet radio, talk radio, and blogging are audio services transmitted via the Internet; it is easily accessible and a tool for delivering a controlled message that may or may not be filtered. Free Internet Radio includes Pandora Radio, a personalized Internet radio that gives the user new music based on the user's favorites list. BlogTalkRadio.com lets users create and share audio on the web choosing from a broad range of categories—news, business, sports, life, politics, and comedy, among others. In an emergency, this can become another form of communications to broadcast your message reaching thousands. These tools can give you automatic pod casting and a controlled message, for example:

> Johnson County Radio Network promotes its
> "Johnson County—A Community Prepared!" at (http://www.
> blogtalkradio.com/jocoprepared)

The Internet radio show profiles first responder activities on its website, in addition to the many links for the county on Facebook, Twitter, e-mail, Share, RSS, and iTunes.

Maps

Maps are a powerful tool to communicate and share information. There are many free online resources such as ArcGIS Online Basemaps (free) or ESRI.COM/PUBLICSAFETY.ARCGIS.com has a gallery of maps and tools, and it can send maps to other sites such as Facebook and Twitter. Map types commonly used are:

- world imagery base map (using government sources and information provided by commercial providers)
- world street maps, including buildings
- topography base maps
- demographic maps
- USGS topography maps
- Microsoft Bing maps
- Open street maps

SMS Text Messaging and MMS

Short Message Service (SMS) is the text communication service component of a phone, the web, or other mobile communication systems, usually referred to as *texting*. Text messages are up to 160 characters, slightly longer than a Tweet, which is 140 characters. Depending upon the carrier, one can compose a message containing 960 characters; the carrier will split up the message into six messages, while other carriers will truncate the message, sending only the first 160 characters and dropping the remaining characters. In an emergency, text messaging has become a very popular tool for reaching the masses, particularly in a school or business setting.

SMS to cellular phones is an excellent resource for emergency communications since it is relatively inexpensive per message and, on average, messages are delivered within one to five minutes. SMS allows specific instructions for incidents in progress and can be sent from multiple platforms using any cell phone and e-mail address listed in a database. Within a college setting, more than 85% of students own cell phones and in a business setting, more than 90% of workers have one.

Along with its many advantages, it also has its negatives. SMS messaging is subject to the availability of cellular service that is prone to overload in major emergencies and has dead spots. This could cause a delay or even nonreceipt of the message by the intended parties. To receive messages on a cell phone, the end user is required to have a subscription. It is important to remember that not all cellular phones or service packages allow text messaging. Many cellular services charge to deliver text messages. Also, a single message is limited to 160 characters. If you are maintaining a database of users, upkeep can be difficult when it becomes a "opt in" or "opt out" system, and in adding, deleting, or modifying user profiles as they are hired or enrolled, separate from the organization or change numbers. Tailoring messages to the location of a person requires the use of geolocation technology associated with a subscription database. Also, text messaging only services the visually impaired when text-to-voice applications are turned-on to read the message aloud to the user.

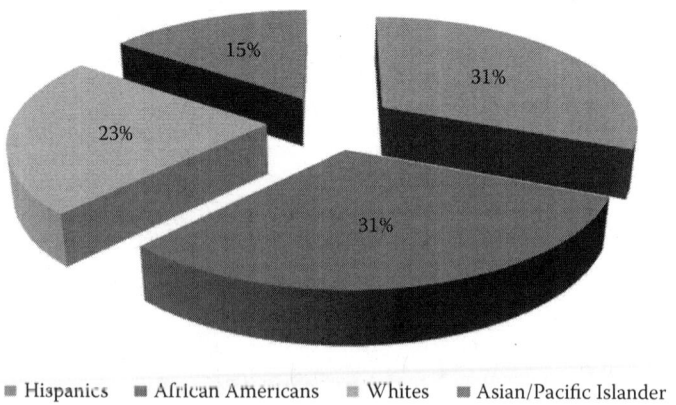

■ Hispanics ■ African Americans ■ Whites ■ Asian/Pacific Islander

FIGURE 9.7 Text messaging by ethnicity—U.S. 2010.

According to the Nielsen Company's yearlong study of mobile use, in 2010 they found that African Americans talked and texted more using their cell phones than other ethnic groups. African Americans average 1,300 voice minutes and send approximately 780 SMS messages a month. The breakout is denoted in Figure 9.7 (Text messaging by ethnicity).

One person in a classroom that receives a text message to evacuate can clear an area faster than many automated alarm systems. SMS messaging is valuable in relaying status information on survivors—how they are doing, where they are, and other important information to others. SMS messaging as a fund-raising tool recently experienced one of the greatest fund-raising campaigns conducted by the American Red Cross (ARC) for the Haiti Earthquake Relief Effort that was started on January 12, 2010. In addition to its traditional means of fund raising, including its website, the ARC was able to raise a total of $32 million, $5 million in the first 2 days through $10 donations submitted via text messaging. Donors were billed by their carrier. Record fund-raising was also experience for the Japan devastation. SMS can be received on most any mobile phone where the service is provided by the carrier.

Multimedia messaging service, called MMS, is used for sending pictures, audio, video, and rich text with a text message that can be longer than 160 characters. This feature is popular on camera-equipped phones. MMS may require mobile web to access it, meaning 3G or greater services.

Shared Video/Streaming Video

Video can be shared using many free tools available on the web including YouTube a video-sharing website where you can watch, upload, and share video or Skype, a software application for making calls, including video calls and chatting over the Internet. Both tools are free, although Skype has a paid component as well. Skype-to-Skype is free; for other services there may be a charge. Skype means you can talk from your phone, computer, or TV enabled with Skype.

There is Ustream.tv, a website for providing live video streaming of events online and live casting from a computer or mobile phone. This enables emergency management teams to take advantage of new media formats. It also means that the grainy video anyone "inside the yellow tape" usually gets can be pushed out faster and considered a trusted source by receivers who can view it than the video of traditional reporters who are "outside the yellow tape" with their high-end professional-grade cameras and video equipment. Ustream.tv and YouTube are the new media that enable anyone with a camera and an Internet connection to broadcast either real-time or on demand to a global audience. The service is free. Emergency managers, particularly public information officers and incident commanders, should be aware of anyone with a cell phone.

Forums

Forums are a good resource for communicating advisories. Forums are also called *Bulletin Boards* or *BBS*. They offer many-to-many communications, allowing anyone to start a conversation—unlike blogs. Topics are placed in relevant subforums that others can join. *Conversations* (called "threads") are not editable by members; however, each member can comment back in a linear fashion. Forums are a form of asynchronous communication—it is not real-time. This is the opposite of chat channels where conversations can occur real-time. An advantage that forums have is built-in social networking tools that the other social media tools do not offer (e.g., titles for commenters ["posters"]). It also gives power to the user; the user can ask a question and another can answer it. As a conversation develops and the number of subforums increases, it can become complex with multiple layers that are a challenge to follow.

Alternate Reality Games

Alternate reality sites, such as Second Life, are a venue to play games or narrate a story using real-world-like platforms and multimedia. They are used in experimentation with new models, public health and emergency management games, and training. These games are good for training but have limited appeal for crisis communications due to the work required for setup and training everyone on how to use it.

Using Social Media before It Is Needed in an Emergency

A huge advantage social media offers is that it is free. Developing a preparedness campaign to increase awareness of your preparedness efforts and help the community to be better prepared is inexpensive. Any program can quickly become viral by simply using a video camera.

Social Media Monitoring

Surveys are an effective tool to assess how the community in a disaster uses social media. The American Red Cross in 2010 conducted a survey and found that half of the respondents were signed up for e-mail alerts and want to use social media to say they were safe. Their survey also concluded that the public assumes that organizations are monitoring these sites and recommended that organizations monitor social media during an event.

Social media is growing in use and in the number of applications. The first generation of web use as a place to find static information remains; second-generation users with the advent of new applications are interactive with others real-time. The public assumes that agencies will monitor activity reported on the Net.

Social media monitor is critical, yet often overlooked. It is a way of tracking rumors and knowing what is being shared about an event that you may not know about. There are a number of great tools that are free to assist with monitoring social media. Monitter is a real time Twitter search tool for monitoring keywords on Twitter. Another is TweetDeck, which works well with Twitter and other social media applications including Facebook, LinkedIn, Foursquare, and Google Buzz for sending and receiving tweets and viewing profiles.

Software Integration

Social media means the integration of software is inevitable. There are a number of software tools for extracting and organizing the wealth of online data and media. OutWit is an all purpose web collection engine, including tools for collecting documents and images using a Firefox extension.

There is Google Calendar, another free online calendaring tool that can be used as a shared time management application where events can be tracked in a single location such as scheduling emergency crews. As a gadget, you can populate it. viewing schedules by the month or week.

There are bulk e-mail senders for sending e-mails and newsletters to hundreds every minute, some with a high-speed built-in SMTP Server for routing through multiple third-party SMTPs like Yahoo and Gmail. Doodle is another scheduling tool for meetings and other appointments that is free and easy to use.

Long URLs are hard to pass along, prone to typing errors, and can be difficult to remember. A number of government sites will reference long URLs such as

> http://www.whitehouse.gov/blog/2011/05/02/president-obama-presents-medal-honor-were-reminded-we-are-fortunate-have-americans-w

This needs to be shortened to something that is easier to remember. TinyURL.com and ReadthisURL.com are two free services that incorporate a one-click bookmarklet. Using any of the many free URL shortening services the above could be shortened to

> http://tinyurl.com/2wh99a

Mashup

A mashup is a web page or application that combines data and its presentation or functionality from two or more sources to create new services. A mashup can take data in its original form and use it as raw data to provide new services. It makes existing data more useful for the end user. An increasing number of web applications enable this functionality making it simple to be used by end users. Mashups contribute to a new vision of the web, where users are able to contribute.

One popular use for a mashup is interactive maps with geographic references to information that can be viewed in a standard web browser. MapBuilder (http://www.mapbilder.net) or ArcGIS (http://www.esri.com/software/arcgis/index.html) are two tools that provide a lot of information on creating mashups and are easy to use. Other resources include a live earthquake mashup offered by the O-Files (http://www.oe-files.de/gmaps/eqmashup.html), a mashup for monitoring health alerts from around the world such as droughts and disease outbreaks at Health Map (http://www.healthmap.org/en/), or the unofficial Google map mania website that provides street level maps to track flooding, construction, or other activities most anywhere in the world at http://googlemapsmania.blogspot.com/ (see Figure 9.8, Sample mashup).

The speed in creating, summarizing and disseminating information in a crisis with a mashup makes this an emerging tool for emergency management. Evacuation routes, current conditions, and other important information can be gathered and distributed to your target audiences without expensive software, in a short amount of time. Since the tool is interactive you can also have your stakeholders interact with the tool and provide additional information such as geo-location pictures of storm damage, and weather observations. Users can also send messages to loved ones about their safety.[6]

Fundamental Change

The fundamental change needed in emergency notification and crisis communications is that organizations must be proactive, find their unique voice and use it, plan for meeting growing public expectations, and be cost effective. Keeping the affected community informed is not about the technology you use; it is about the way you present information. Is it timely, accurate, and available from multiple sources?

There is a false perception that a huge amount of time is needed to get your message out. We must remember for every moment of delay there is high probability that someone else has already begun to fill the void. How to handle the information you have and getting established is an ongoing challenge. The time spent building your emergency communication systems before hand is the most time consuming. Hiring students and interns are a great way to contain costs during the setup process. They can design the system the way your customers will use the information. The emergency management coordinator should help to lead this setup effort. Once you have adopted social media as a tool, check twice daily for comments and engage your followers so they know who you are before an event occurs. This step is critical, because as they become familiar with you they will learn that you are a trusted source for online information regarding a situation. Guard this privilege by taking security of your ID and password seriously.

The lessons learned by the USCG for both the Haiti earthquake and BP oil spill is that social media took off on its own to the point that no single organization could manage the traffic, who was following who, and who could respond. If you are limited to your existing resources used for traditional communication resources, you have a challenge to overcome. Social media and the wealth of information it provides cannot be overlooked or taken for granted. This is where people go for information.

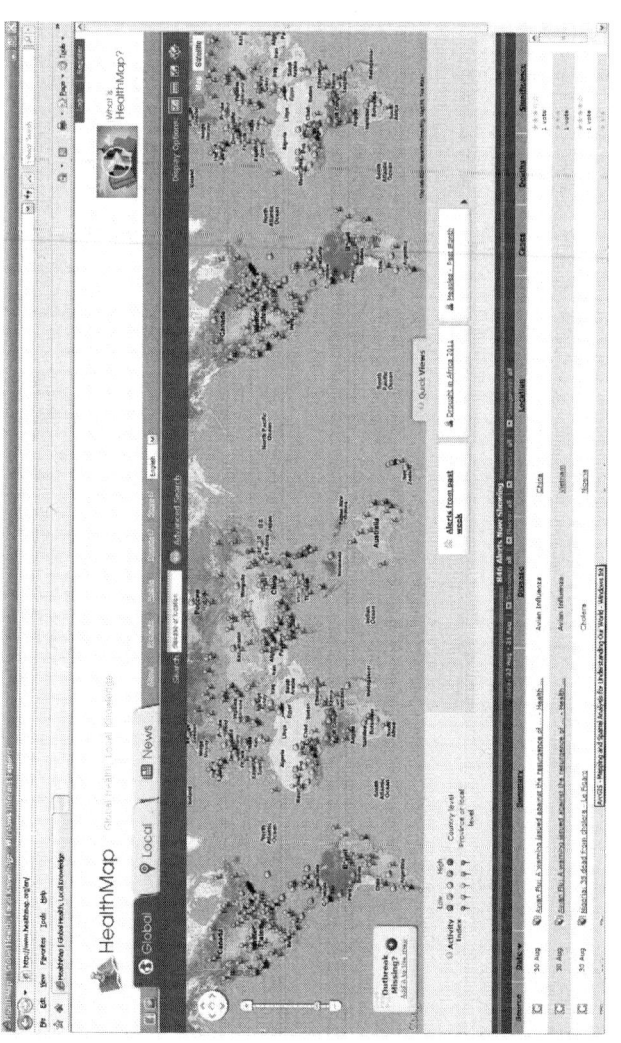

FIGURE 9.8 Sample mashup—health map—global.

If the PIO or Incident Commander (IC) elect to set up a Joint Information Center (JIC), at least one resource should be dedicated for social media. Although Incident Command System (ICS) does not include the role of social media, it is a role that is needed that can fall under Planning, the PIO, or a new section called Information and Intelligence. As the ICS model is updated, we anticipate that this section will be formally addressed.

Many organizations are concern about litigation and how to avoid it. You must be consistent in your use of social media tools and manage expectations. For example, you can state the hours of when you will monitor these tools except during emergencies or that you will use alternate system after hours, such as fire alarm systems, sirens, or e-mail.

Many PIOs are focused on traditional emergency communication methods and marketing the message. Adding a marketing spin to a message adds time to the delivery process. It is important to have the PIO engaged in the social media strategy, one of an organization's biggest challenges. Message mapping that incorporates prescribed messages that are preapproved can save time. Emergency notification requires speed and transparency. All communications should be tracked and records retained for auditing purposes or when there is a need to recreate what has occurred.

There are emergency management tools such as the PIER System that uses Application Programming Interfaces (APIs) to post to social media sites. Posting to different sites can be daunting, but the greater challenge is monitoring all the sites germane to your operation. The American Red Cross and the Canadian Red Cross use social media; many Offices of Emergency Management now use social media such as the Harris County Office of Homeland Security and Emergency Management. The real value of social media is to

- know the followers and their interaction to manage expectations
- engage them before a disaster regarding what to expect during a crisis
- be a part of the conversation before an event
- be prepared for defense against any claims of abuse

Social Media as a Technological Hazard

We have discussed at length the many benefits of social media. It's not all roses, though. Social media has its inherit risks and a potential for abuse that could affect a large number individuals. A major issue with social media is the accuracy of the information posted, both inadvertent or maliciously. Of the two, intentional misinformation presents the greatest issue.

The quest for speed, the need to be first, or learning you have the power to create a situation, is in the news all the time. An example of having the power to create a situation using social media is a paying customer at a movie theater, who sends a text message telling a freeloader when the theater door is unmanned, so the freeloader can enter and see the movie without paying.

The second area are those who tweet and text during meetings, relay confidential or sensitive information such as how an individual voted on an issue, or preempt a new product's release or new features that were to be unveiled by an organization. You have the events ongoing in the Middle East from the capture of a journalist, a man setting himself ablaze starting a new movement, or the capture and killing of Osama bin Laden.

A fourth area of concern are smartphones, particularly smartphones "inside the yellow tape" that can take pictures and quickly transmit them to others. These are only a few of the technological hazards social media presents.

We learned from the 9/11 Commission report that communication failures occurred due to our lack of imagination of what capabilities were available to us, how to use them, and the policy and management to ensure appropriate use. We also learned from this report that preparedness in both the public and private sector is required for every phase of emergency management, including having adequate communications capabilities. Interoperability has generated a huge gap in communications when first responders are unable to communicate with each other using their technology, and therefore must rely upon an intermediary and wait until face-to-face communications become feasible.

A growing phenomenon are antisocial behaviors organized via social media such as *flash mobs*. Flash mobs are spontaneous gatherings that are organized using online tools. These began as nonviolent events but over the last few years, some have escalated to violence. Recent examples include the Hammer Time Mob Dance on Sunset Blvd. at 3:31 p.m. (Los Angeles), or the Macy's fight involving 100 teens who stormed inside a Macy's store and start fighting. There were 14 arrested at this event. As a way to deal with flash mobs, some communities have implemented curfew on Friday and Saturday nights.

Geotags are of increasing concern as more pictures and video are uploaded from devices with this capability turned on. Geotagging is the process of adding geographical identification to photographs, video, websites, and SMS messages. The tags include the latitude and longitude coordinates of where the picture was taken. Using tools like Google Map or Google Earth you can pinpoint the physical address or obtain a satellite view of the physical address.

To test this theory, Adam Savage of the television show *MythBusters* in August 2010 took a picture of his vehicle using his personal smartphone, an iPhone, and posted the picture on his Twitter account. He was able to open the metadata associated with the picture to obtain the coordinates and plot them using Google Maps. The outcome matches the location. This information is great for forensics and for locating where you are when you want to be found; it is dangerous in the hands of the wrong person who may use it to burglarize or cause harm.

Social media is a great tool for soliciting donations and assistance. Likewise, it is also a tool that can lead to unsolicited or unorganized assistance, unneeded logistical support and requests, and the release of false or contradictory information. For example, someone hears that you will be opening a shelter and may need supplies. It is the dead of winter. Suddenly, you have at your doorstep a pile of coats and blankets, with no place to put them or means to distribute them, and a need to determine if they will even meet what is required.

The posting of false information is of concern. Someone could maliciously post an "All clear" after an event occurs, such as an active shooter or hazmat situation. People would return only to be reintroduced to a danger they otherwise had avoided. Today, these types of negative behaviors do occur but overwhelmingly in times of disaster people work together and do not act irrationally. Panic, looting, and other antisocial behaviors are not common, as witnessed in Japan while survivors stood patiently in long lines for necessities. In Japanese culture, it is a societal norm to be courteous in a group situation. This same level of courtesy is often found in the early days of a disaster.

As a growing number of emergencies affect more people or have a ripple effect, a shortage of first responders occurs and an emergent phenomenon begins—the surfacing of volunteers facilitated through social media. Volunteer categories that emerge are

- Supraorganization
- Quasi emergency
- Task emergence
- Mix of individuals
- People with latent knowledge
- Emergent citizen groups are the outcome of natural social processes
- Emergent citizen groups that are both functional and dysfunctional

The solutions to many of the challenges social media present involve

- Education/awareness
- Monitoring
- Active participation with use of these tools:
    - oneforty.com tool kit

- Twitter TXEM11
- Swiftriver
- Testing, using someone else's disaster to determine if you can "beat the media" and learn who are the trusted sources using this media

Social media does matter, whether you like it or not. Governments are using monitoring for these systems. The use of sites such as Twitter to organize demonstrations has made it easier for authoritarian regimes such as the Belarusian and Iranian governments to monitor the activities of anti-state protestors. Police forces in the United Kingdom use social media to conduct investigations into the activities of gangs. China has used social media to head off unrest.

What this means is that social networks and their methods of communications will continue to morph to meet the growing needs and desires of their members and users. So far, there has not been any landmark court cases in the United States where government agencies such as law enforcement have been successfully sued or charged with providing misinformation using social media. Until that day comes, social media should not be pushed aside as an effective communication tool, among the many tools available to you to relay crisis communications.

SUMMARY

Social media is the second generation of websites providing users their own mini-websites and web pages. There is a high degree of participation and interaction among users. Using social media for mass notification is a challenge for organizations; individuals have embraced it. Information regarding an incident can come quickly; an organization can soon become deluged by the thousands on Facebook, Twitter, blogs, and other sites. Establishing a social media presence is easy; maintaining its currency and relevancy requires a team and is an ongoing effort. Social media is not a problem for the media because they, too, are using the same tool to learn where the stories are and where to go for more information. Social media has also introduced a new area for emergency management—online information and intelligence gathering. The dominant rule of social media and crisis communications in general is that people want accurate information quickly from what is perceived to be is a trusted source.

Among the commonly used social media tools used today for crisis communications and mass notification are Twitter, Facebook, blogs, widgets, mobile phone apps, and geotagging. The sharing of content has also made these tools both attractive and security risks. The second area are those who tweet and text during meetings, relaying confidential or sensitive information such as how an individual voted on an issue, when a new product will be released, or what features they will unveil.

ENDNOTES

1. Twitter. (2011). What is Twitter. http://business.twitter.com/basics/what-is-twitter (accessed April 2, 2011).
2. Sysomos. (2011). Twitter statistics for 2010. http://www.sysomos.com/insidetwitter/twitter-stats-2010/ (Accessed April 2, 2011).
3. Facebook. (2011). Statistics. http://www.facebook.com/press/info.php?statistics (accessed April 1, 2011).
4. http://en.wordpress.com/stats/ (accessed May 5, 2011).
5. Apple. (2011). Apple Stats.
6. Robinson, Anthony (2010). Lesson 2: Hazards and disasters, emerging theme: Map mashups. The Pennsylvania State University. https://www.e-education.psu.edu/geog588/l2_p5.html (accessed August 28, 2011).

CHAPTER 10

Solutions—Some Solutions Are Better than None

> The advantage of modern means of communication is they enable you to worry about things in all of the world.
>
> Dr. Laurence Johnston Peter, 1986
> *Vancouver, British Columbia*

There are a number of solutions readily available to you today at home and at work—nontech solutions such as pen and paper, word of mouth, or a bullhorn. Most people have a phone, whether a landline, cellular, or VoIP service at home or at work, and a television. Many have access to a computer, enabling them to use e-mail and Internet access. As you move up the continuum, an increasing number of communications tools for day-to-day use are available on demand for crisis communications. Several of the solutions will be discussed in the following text.

The trends going forward are

1. 3D video use on mobile devices without the need for special glasses, and an explosion in HD live and on-demand video, giving cable and satellite providers serious competition. This will also lead to a major reduction in "bootleg" video due to the lack of consistent quality these products provide.
2. Corporate network traffic will be mostly web oriented rather than network centric.
3. Cyberattacks will be stealthier, more serious, and cause breach of government resources and national security from rogue gangs and other governments looking to spy and gather intelligence and to leverage the information for bargaining power.
4. Emergency communications for work or home will provide targeted messages so that you and the person sitting next to you

will receive a different set of instructions based on your profile for the same event.
5. Live and on-demand web video will take second place to social media in terms of traffic.
6. Mobility will continue to explode with more choices in mobile hardware (tablets, phones, e-readers) and millions of new mobile applications to use on them.
7. More cloud-friendly information infrastructures and the morphing of social media.
8. New data security standards as more government and critical infrastructure breaches attract public scrutiny.
9. Predictive analytics, real-time analytics for social media, and the digital universe that is expected to reach nearly two trillion gigabytes in 2011.
10. The FCC will begin to leverage social media and other technologies more to provide emergency communications to targeted audiences.

We have previously learned that emergency communications require documented processes that afford transparency, people who are trained in crafting and delivering the message at the right time, and the selection of the right tools for distribution of a message. Taking these trends into consideration, we have already discussed social media solutions in the marketplace today and, as with any new area with easy entry, new, stronger players will emerge. The same is true for applications and hardware used to provide mass notification of emergency communications and crisis communications. The trend in this area is to have one solution that can trigger events in multiple technologies with a few clicks. Subscription services will need to offer more than text messaging to a database of users that could become difficult to maintain and expensive if a catastrophic event occurs that requires the generation of multiple messages and the addition of new users. Cost is always a factor in what we choose. In this chapter we include solutions you already have and free apps available in the public domain to high-end solutions. We begin with what most organizations already have.

SYSTEMS YOU PROBABLY ALREADY HAVE

Cellular Phones and Smartphones

Cellular phones and smartphones in particular are excellent tools for almost any type of emergency communications, whether an immediate threat, general emergency, or an advisory message. There are many

> Here is how most people use their cell phone:
> - 76% Take a picture
> - 72% Send or receive text messages
> - 38% Access the Internet
> - 34% Play a game
> - 34% Send or receive e-mail
> - 33% Play music
> - 30% Send or receive instant messages
>
> (*Source*: Pew Research Center's Internet and American Life Project, 2011)

benefits to organizations and individuals to have a cell phone readily available. Within an organization or for a family, minutes can be pooled or shared to help in reducing costs. Cell phones can exchange text messages (SMS), video, and pictures and have its own contact list readily available. The next generation of cellular phones, the smartphone, considered a mini computer, can exchange data, audio, and video, and the user can browse the web and access countless numbers of mobile applications.

Cell phones in the United States are required to have carrier network access to make unlimited free 9-1-1 calls. The process is simple—ensure the unit has power, dial 9-1-1, and press "send" to connect immediately with your local 911 dispatch center. This enables emergency support for those unable to pay the monthly service costs associated with carrier network access. Cell phones are not without their drawbacks. Other than monthly service costs, battery life is limited, and therefore a backup battery or alternate power source must be available to the user to extend the usefulness of the device. Second, talk time, applications running in the background, display lighting, roaming services, and browsing the Internet are all functions that can quickly consume all available battery power faster than what is stated by the manufacturer. Third, cell phones are prone to outages due to dead zones in and around buildings, in remote areas that are sparsely populated, and due to network congestion during some emergencies.

It seems that every month a vendor provides a phone with more features, memory, and processing speed than the unit before it. People will stand in line overnight to be the first with the new device. The Apple iPhone is one product that captured the market in this fashion. The life of these units appears shorter than that of a desktop computer or tablet. Multiple generations of technologies have been introduced to increase bandwidth and speeds on cellular networks. The first generation, 1G

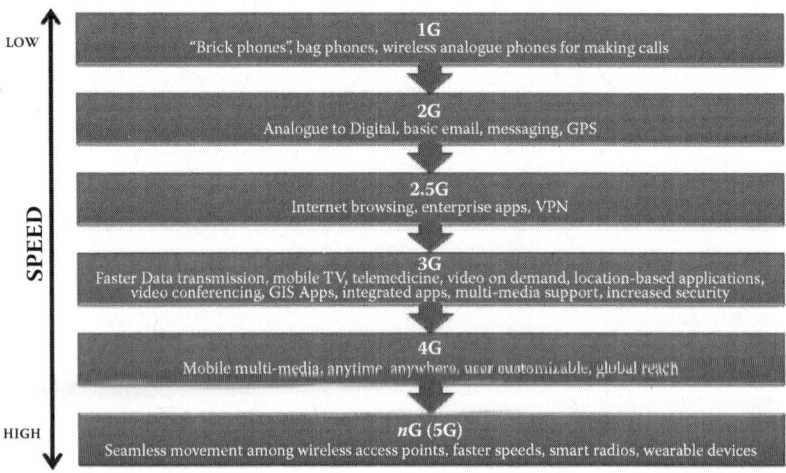

FIGURE 10.1 Bandwidth/speed cellular network.

wireless analogue phones, were used for making telephone calls wirelessly. The next generation was referred to as 2G, where voice was encoded to digital signals transmitted from radio transmission towers. The third generation network, called 3G, offers faster uploads and downloads, has GPS capabilities, and can handle streaming video. The devices with 3G capabilities operate on both the GSM and CDMA networks. The advances brought about by each generation are illustrated in Figure 10.1 (Bandwidth/speed cellular network).

Bandwidth and Speed

The increases in high-speed wireless technologies have enabled transformational changes in public safety and emergency communications. In 2010, the average user downloaded 7 GB of data, according to Sprint, using its WiMax 4G technology boasting speeds of 10 Mbps for downloads. T-Mobil uses HSPA+ technology, claiming speeds of up to 21 Mpbs for downloads, followed by Verizon LTE technology with downloads of 5 to 12 Mpbs in December 2010.

With 4G technology widely available, patrol cars are able to "connect" with mobile data computers in their cars, giving them access to video surveillance and wireless video cameras, auto license plate readers, National Crime Information Center (NCIC), Criminal Justice Information Services (CJIS)/Department of Motor Vehicles (DMV) databases, Information System (GIS), Global Positioning System (GPS), Remote Sensing (RS), and diagnostics. Public safety vehicles can also provide telemetry/GPS equipment to follow officers and quickly establish the car as a mobile hotspot for sending/receiving photos and to dispatch real-time streaming video.

Smartphones that are 3G and 4G capable have core applications installed on the base unit, addressing accessibility concerns for the hearing- and vision-impaired population. There are many models available in the market today, meaning, there is a cell phone available to satisfy the needs of almost anyone, including children, for which parental controls are available to restrict the use of the phone, phones with large button displays, phones with voice services, and units able to track and monitor the movement of children remotely.

For users of 4G technology, many different devices can be used to receive a message, including tablets pc, netbooks, and e-readers, in addition to smartphones, video displays, USB modems, and other network-attached devices. Communities in rural areas were often the last to adopt new technologies because they were the last to receive the infrastructure needed to support the newer technologies. It may be another 7 to 10 years before technologies introduced in large urban areas are available in smaller rural communities.

Communities are also looking for new ways to extend the use of systems already available to them. Taking this approach extends the life of a product and can help keep costs contained. Seguin, Texas, a suburban community east of San Antonio, had automated telephone notifications systems that were purchased in 2003. The tool has since been repurposed to become an emergency callout system as mentioned in the following case study.

Case Study: Situation—Emergency Manager Finds New Use for Cable and Cell Phones in Seguin, Texas

Fearing that residents along the Guadalupe River were not receiving flood warnings in a timely manner, Guadalupe County's emergency management coordinator Dan Kinsey developed and piloted an emergency callout system. "Here we could very easily have a situation such as a flash flood. If 20–30% of your population doesn't have the traditional home phone, you need to find a way to warn them," said Kinsey. He took advantage of an automated telephone notification system the county had purchased years earlier that was designed to place calls by zones. "We already had everything in place," Kinsey said. "It's a great tool with a lot of possibilities. It was just a matter of creating a database, collecting the information, and getting it into the system." Kinsey continued, "That database could not just rely on traditional landline telephone numbers, however. There are so many people using cable phones and cell phones nowadays. Your normal landline database just doesn't cover enough people."

SOLUTION

He drafted an Emergency Callout System Voluntary Registration form. Participants are required to list the location of their waterfront property, two phone numbers (designating whether they are land-lines, cell phones, or cable/Internet phones), and an e-mail address. Residents are asked to update their numbers in writing, or to notify the Office of Emergency Management if they move out of the flood hazard zone. However, being able to notify residents is only half of the system. The other half is being able to know when to notify them. Kinsey monitors water flows measured by the Guadalupe River Authority at its hydroelectric dams. Based on those numbers, he can predict when flooding is imminent. While not all emergencies can be avoided, Kinsey tries to prevent some and manage others in ways that minimize their impact.[1]

The federal government began sending text messages using the new emergency notification system starting in New York City in May 2011. NYC was the first metropolitan area to pilot the program with Washington, DC, following. Nationwide rollout occurs throughout 2011. The location of cell phone towers will serve as a way of isolating a message for a target area. Enabled mobile devices and those within the targeted cell phone tower range will receive emergency messages. Messages are up to 90 characters in length. Some phones and wireless devices already had the required chip. Cell phone companies are adding the chip to new phones and wireless devices as they are rolled out. Like other text messages, people who receive the message can forward to others or share by word of mouth the instructions provided, such as "Bomb threat in Times Square. Evacuate area now," or "Chemical release of nerve gas. Shelter in place. Turn off AC. Close all windows. Check local news for updates." Users can "opt out" of all emergency notifications except those from the President.

Cellular phone carriers are offering emergency to-go kits for first responders. These kits can include any number of phones that are pre-programmed for an area, have backup batteries for each unit, and can be quickly distributed. These emergency kits can contain a mix of cellular, satellite, and two-way radios, enabling a broad range of services, such as the Sprint Emergency Response Team Go-Kits used in support of catastrophic events like Hurricane Katrina, the earthquake in Haiti, and preparations for Hurricane Irene in August 2011. This offering can include thousands of charged mobile phones available for rent on-demand, ERT Go-Kits containing a cache of handsets, broadband devices and accessories, and Satellite Cell on Light Trucks (SatCOLTs) for restoring wireless services.

Fire and Gas Detector and Alarm Systems

Fire alarms are installed in almost all commercial and industrial buildings. They are easy to use and usually integrated with light bells, horns, hooters, sirens, and strobe lights. Collectively, these alarm systems provide a visual and audible warning to accommodate hearing- and visually impaired persons and the general population in the protected area. These systems are managed by an Alarm Control Panel that receives an electrical signal from each detector and sensor attached to the panel. These devices are located throughout a protected area. The control panel will display the connection status for each attached device, report alarm conditions to the monitoring agency or department, and activate programmed alarm signals. These alarm signals alert occupants in the protected area and monitoring personnel. Most alarm panels are required to include an emergency source of power so that the system remains operational during power failures.

Smoke detectors and carbon monoxide detectors are sometimes overlooked as warning devices. They have proven to be successful in preventing fire fatalities and property loss associated with fires. Almost every building, particularly commercial and multitenant buildings, have smoke detectors—from hotels and apartments to office complexes and industrial facilities. Carbon monoxide detectors have also saved lives by alerting occupants of the presence of this odorless, colorless gas that is highly toxic to humans and animals. Smoke detectors and carbon monoxide detectors are commonly housed within a single device. They operate independently on batteries or can be fully integrated with a residential or business alarm and power system, and can be managed by a monitoring company or by the user via the web, a mobile device. They are inexpensive, and many nonprofit and first responder agencies provide units to local residents free of charge. The key to the effectiveness of these devices is to test them regularly, at least twice a year, and to replace the batteries at least once a year. The failure to test regularly means the device becomes a useless piece of hardware. A unit without power is a unit unable to provide an early warning to occupants.

Voicemail and Voice Systems

In the past when you thought of voice systems, the term was equated to the traditional landline sitting on your desk. Today, a voice system is a huge umbrella that includes satellite and cellular phones, translation services by voice, podcasting, or managed telecommunication services such as Google Voice. Voice systems, which also include speech recognition and medical transcription services, are a convergence of solutions for communications. We assume that they will always work when we are ready, particularly in an emergency.

A tool available to most businesses with multiple employees, many homes, as well as cellular phone users are voicemail boxes for receiving voice messages. This is a good tool that does not generally incur any additional cost to use for emergency and advisory communications. A single message can be delivered to multiple voicemail boxes with a few clicks and in under a minute or two. However, it has limited use in emergencies requiring immediate action since the person is required to be at his or her desk or telephone to retrieve and listen to the phone message. Another limitation of voicemail is that not all phones or employees within a business have voicemail services, and people are prone to forgetting their password to access their mailbox when needed most.

There is also the possibility that a voicemail box is full and the system is unable to capture and save the message. When this occurs, the user may not know that there was an emergency message for them. Another concern is that some do not routinely check their voicemail messages. This could mean that the message is buried in the middle of other messages or near the end. Depending on the type of retrieval process available, a user may have to listen to messages in the order they are recorded, therefore causing a delay in getting to an urgent message. An added issue is that, on many systems, it is easy to accidentally delete a message that you have not heard. So, even though the message was successfully received and recorded, there is no guarantee that the user was able to listen to the message or that the message was intelligible. Voicemail does not reach the hearing impaired unless the telephone is equipped with some type of visual aid to let the caller know there is a message waiting to be heard. Voicemail is a good tool but should not be the only tool used to relay emergency information. It should always be coupled with a second solution for emergency communications.

E-Mail and Instant Messaging

Electronic mail, called *e-mail*, is a method for exchanging digital messages from an author to one or multiple recipients (Figure 10.2 E-mail). E-mail is a very effective tool for communicating advisories using bulk mail lists. Since most organizations have an e-mail system in place, the use is relatively inexpensive per message. For users who have an existing

FIGURE 10.2 E-mail.

e-mail account, no additional costs are incurred. Like voicemail, the message can be targeted to a specific e-mail group and delivered in five minutes or less. E-mail has the advantage of length in that the sender can provide detailed information that you may not do when leaving a voicemail, using a video display, or in sending a text message. (See Chapter 9, "Social Media," for more information on text messaging.)

It is strongly recommended that individuals maintain an alternate e-mail address for nonbusiness communications and emergency communications. If an organization's network operations are disrupted, access to the company-hosted e-mail system may not be possible. Through planning, alternate e-mail addresses can be gathered and established to relay emergency information using any of the free e-mail services available. For example, Google's Gmail, Yahoo! Mail, or Windows Live (MSN Hotmail) are three popular solutions. All integrate with a calendar and messenger. These systems are effective alternatives to an organization's hosted e-mail system.

Yahoo is a popular free e-mail service since it does not have mailbox size limits that Gmail and Hotmail have. Gmail is known for doing a good job in battling spam but is restrictive on free space, approximately 6.5 MB of storage. Yahoo! Mail, the second-largest web-based e-mail service with 273 million users in 2010, provides unlimited mail storage and up to 25 MB attachments. Via Window Live Hotmail you can send up to 10 GB attachments and seamless integration with other Microsoft products. Each of these e-mail systems is compatible with Internet Explorer, Firefox, Safari, Camino, and other Gecko-based browsers.

In the planning process, you can develop an e-mail address naming convention for use in creating alternate e-mail addresses of employees. For example, you can establish a naming convention using any of the free e-mail systems mentioned, in the same manner as you would your internal e-mail address:

E-MAIL ADDRESS NAMING CONVENTION

Internal E-mail Address:

FirstName.MiddleInitial.LastName@College.edu

Alternate E-mail Address:

College_EDU_FirstInitial_MiddleInitial_LastName @yahoo.com

E-mail is not very effective for emergency communications, particularly those requiring immediate attention. The individual must be in front of a computer or otherwise using e-mail to know there is a message.

A user must subscribe to the e-mail service to receive the message. Customization of groups is required if the message is to be tailored to a specific group or location. Unless there is a pop-up screen or other visual aids to alert an individual of an emergency e-mail, the visually impaired are not well served. In addition to an alert, the visually impaired individual may need some type of adaptive technology to convert the text message to audio so that it may be heard, or increase the print size so that it is legible.

Instant Messaging

Instant messaging (IM) is a real-time text-based system for dialogue with others. Yahoo Messenger!, BlackBerry Messenger, Windows Live Messenger, and Hotmail Messenger or AOL Messenger are free messaging tools available on the Internet, by cellular phone carriers, and through organizations that have their own internal IM solution. Most support Google Talk and Facebook Chat and other popular social media applications, and for some serve as an alternative to Twitter. Solutions enable users to share links, data, audio, photos, and HD video files and messages live while chatting. IM is a place to update your status with friends. (See Figure 10.3 SMS/MMS/IM.)

IM is an excellent tool for those who are hearing impaired and unable to access a TTY device. Some law enforcement agencies have enabled this service in working with the community for events in progress. IM works on most any platform, including Windows, BlackBerry, iPhone, and Android. It is a tool that should be added to your toolkit for emergency communications.

Flash Messages

A *flash message* is a text message you can send a person that stands out from other messages. On a computer or mobile phone, a flash message could appear as a pop-up. This is an excellent resource for immediate and emergency communications, for example, one short message to all network-attached computers redirecting users to the web portal for additional instructions. The message can be tailored and its distribution restricted to preestablished lists of groups and users as defined by a network access scheme. For the end user, there are no additional costs, and with a few clicks, the message can be delivered in less than five minutes. Flash messaging also allows for specific instructions for incidents in progress such as shelter-in-place or a lockdown.

The trade-offs in using flash messages is that tailoring a message to a specific location may require some prework in terms of preestablishing

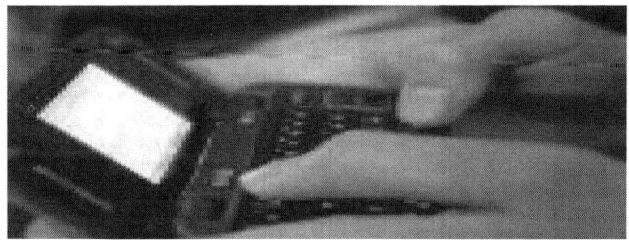

FIGURE 10.3 SMS/MMS/IM.

groups—students, staff, general public, etc. Flash messages are brief and usually instruct the user to go the web portal for more information, a second platform he or she may not have access to. The person receiving the message must be in front of a computer or mobile device. Some systems offer sound but most do not, and therefore, without some type of adaptive technology, the visually impaired may not be reachable.

Web Page

The web page of an organization is a good location to provide advisory and emergency information. The Web Content Administrator can include detailed information, provide links to additional information, including blogs and external sites, and deliver multiple messages simultaneously. Have a "light" version of your website available to launch in an emergency to avoid stress on services. Consideration could be given to having this hosted offsite by a vendor or an alternate office location that is geographically dispersed from the home site.

Web page access is subject to the availability of Internet and/or LAN service. The servers used to support the system may overload under heavy use. Many prefer Twitter and Facebook for immediate information and use the web as a resource when additional information is needed. Publishing updates may require longer lead-time to put in place, as long as an hour, before the pages have been refreshed. Managing web content requires a higher level of training or ability on the part of the person creating the message (must be able to edit web page and upload the information).

Public Media

Public media, inclusive of television, radio, and newspaper, are considered traditional media formats for providing emergency and advisory information. These formats

- Have a broad reach in making information available to the general public
- Are an inexpensive resource for getting the message out
- Can reach people in the immediate area of impact and those who may be in the general area

Public media (Figure 10.4) is losing its dominance as the first place to go for emergency and advisory information, particularly with persons 34 years old and younger. These younger age groups are mobile and depend upon their mobile technology to give them real-time information. For emergencies, many rely upon Twitter and Facebook, and only when it is convenient will they turn to a company web page and traditional media source for additional information about the event. Traditional sources generally have long delivery times—a major concern

FIGURE 10.4 Public media.

for emergency managers. Second, once the message has been delivered to the media outlet, the originator loses control of the actual message that is delivered.

The new media formats have presented new challenges for journalists who are rushing to get the next story out before all the people around them with a cell and camera are able to get their message out. Journalists are often "outside the yellow tape" while the new reporter is "inside the yellow tape," texting, uploading videos and pictures, or talking real-time. Public media sources operate in a crowded space, competing with Internet TV such as Hulu.com (a free online video service offering a selection of hit shows, clips, movies, and more; Hulu Plus is a paid service), cable, and radio (satellite, traditional), meaning that individuals have more places to go for their message. Traditional methods mean that the public must be watching TV, listening to radio, or able to access the newspaper to receive the message.

In a Classroom or Conference Room

At a workstation or in your office, access to technology is usually on your desk and at your fingertips. The same may not be true in a classroom, auditorium, or a large conference room. These are areas where the needs for emergency communication are critical. Due to what may be in progress, messages sent using speaker systems may not be heard by the occupants. Second, there are usually a large number of people with only one point of egress if an evacuation is required. This means that time is of the essence since an orderly exit may be needed. For people with disabilities and others with functional or access needs, their needs are often overlooked. This is an area where multiple solutions should be applied in any emergency, such as fire alarms, classroom speakers, and flash messages on networks attached projectors or video displays and to any computer workstation in the room.

It is a good idea for a conference to have occupants sign in or register, providing their e-mail and cell phone information. Be prepared to use this register for notification and to track who is in the area at the time of an emergency. At the start of each session, as at the beginning of every flight, a review of housekeeping and emergency procedures should start off any workshop. Ideally, the information is provided in printed format, and consideration is given in using multiple languages and pictures to provide instructions. Figure 10.5, Table of basic communication tools available in the classroom, provides a quick snapshot to use in assessing what you may already have.

Solution	Visually Impaired	Hearing Impaired	Special Building/Floor or Multiple Locations
Email	X	X	X
Web	X	X	X
Voice mail	X		X
PA system	X		X
Radio	X		
Route Alerting	X		X
Door-to-door	X		X
Video displays/TV		X	X
Classroom projectors		X	X
Text Msg/IM/Social Networking Media		X	X
TTY		X	
Computer displays		X	X
EAS msgs		X	

FIGURE 10.5 Table of basic communication tools available in the classroom.

ALERT SYSTEMS

Emergency Automated Telephone Notification System

Television is often thought of as the great communicator in getting crisis information and instructions to its audience. What many fail to consider when developing their emergency communications plans is that local broadcasting has a limited reach. Media understands this and will structure its messages to the concentrated areas it serves. Those outside these areas must filter through all the information to determine what is germane to them. Where TV stations reach their outer boundaries, it means that the communities in these outer boundaries are typically underserved when an emergency occurs. The conversion from analogue to digital television has closed the gap; it has not totally eliminated the concern. Historically, this means that word-of-mouth was too often a primary means for emergency communications such as that experienced by Jackson County, Texas, near the Texas–Louisiana state line when Hurricane Rita made landfall on September 24, 2005. To increase the capability to quickly reach the residents of this rural area, an Emergency Automated Telephone Notification System was implemented as discussed in the case study below.

CASE STUDY ... SITUATION

Hurricane Katrina had just struck New Orleans on August 28, 2005. Many Texas counties had become sheltering centers for the large number of evacuees. Less than one month later, on September 24, Hurricane Rita, a Category 3 storm, made landfall near the

Texas–Louisiana state line. Jackson County, Texas, depended upon public media and "word of mouth" to warn its residents. Large media outlets were reporting more on the Houston area, with little emphasis on exactly what to expect within Jackson County. This left Jackson County without an effective communication tool to quickly provide its resident with emergency information needed for evacuations. The lesson learned by local officials was that a better and faster method was needed to communicate emergency information.

SOLUTION

Jackson County received a grant from the FEMA Hazard Mitigation Grant Program and private fund matches to acquire and implement an emergency automated telephone notification system. In the event of an emergency, the 911 dispatcher can record a message providing residents with emergency information and instructions by neighborhood or region. The system immediately calls all registered telephone numbers within the impacted area with emergency communiqués. Emergency communiqués were expanded to include notification of prison escapes, hostage situations, dam or levee breaks, bomb threats, shelter-in-place notification, hazardous material, and flooding, in addition to severe weather and evacuation information. The system was successfully used for hurricanes Gustav and Ike in 2008, a mock drill at an elementary school, and a chemical spill at Formosa. The automated system can provide an audit trail of how many people, versus answering machines, picked up the telephone receiver to listen to the message and the number of unheard messages. The system is also better than "word of mouth" since all receive the same message as it was originally presented.[2]

Public Address Systems

In-Building Voice Announcements

Public Address (PA) Systems are great tools in delivering immediate emergency communications to those within its range. A PA system can reach the majority of persons exposed to a threat in real time. For the public, there is no special equipment or preparation required to receive a message. The sender can use a prescripted message or give specific instructions for an incident in progress, and the message can be location specific. The newer system can connect to other systems such as fire alarm panels and mass notification panels via a network or the Internet. Newer PA system can use IP-addressable speakers that are low voltage and can attach to a Network drop to obtain power over the Ethernet wiring (PoE).

PA System do require an expensive infrastructure; however, the cost can be reduced if it is a part of a fire alarm systems since its wiring is installed in parallel or as a part of the fire alarm system. A huge disadvantage of this technology is that it does not reach those who are hearing impaired or deaf. The costs of a classroom with just a couple of speakers vary widely with an average of $200 to $225 per classroom. Speakers are also needed in the common areas of a building. The ideal system will provide coverage within each building and gathering areas outside such as athletic fields, parking lots, and recreational areas.

Outdoor Sirens and Speaker Arrays (Including Voice Warning)

As in the case of their indoor counterparts, outdoor sirens and speaker arrays enable the immediate delivery of emergency messages to those who are outdoors and within range (see Figures 10.6 Exerior emergency telephone/speaker/light tower and 10.7 Outdoor speaker array). Newer system can be wireless, eliminating the need for expensive tunneling through parking lots and sidewalks. On behalf of the public, there is no special equipment or preparation required.

Although an effective tool, a speaker array can also be expensive in terms of its infrastructure requirements. Unless integrated with other

FIGURE 10.6 Exterior emergency telephone/speaker/light tower.

FIGURE 10.7 Outdoor speaker array.

systems, such as lighting or fire alarm, delivery of instructions is by voice. Instructions specific to a location by building or area is difficult since the clarity of the message diminishes significantly the further away you are from the speaker. Another disadvantage is that the hearing-impaired population is not serviced by this solution. They would be required to have an alternate system available or someone close by to interpret the message.

Community Outdoor Warning Sirens (COWS) and Community Activated Lifesaving Voice Emergency Systems (CALVES)

Over 40 years ago, 31 residents of Brandenburg, Kentucky, a community of 85,000, were killed by tornadoes. More loss of life from tornadoes occurred in 1976, 1986, and 1994. Eventually, COWS were purchased through a grant that were capable of warning via sirens all Bowling Green residents, the county's biggest city, and 80% of the residents in the entire county. Residents were educated to turn on their televisions and radios for instructions. The system was later enhanced with CALVES for warning of populations with functional needs and access such as nursing homes, day cares, school, and emergency centers. CALVES have audio and voice instructions. With the implementation of COWS and CALVES and an aggressive campaign to promote its use, what to do became second nature to the residents as discussed in the following case study.

> **CASE STUDY: SITUATION—WARREN COUNTY, KENTUCKY**
> **COWS AND CALVES PROGRAMS**
>
> Warren County, Kentucky, a rural community, faces a variety of natural hazards, such as tornadoes and other severe weather, chemical spills, flash flooding, landslides, earthquakes, and forest fires. They have a warning system that alerts residents of impending danger, enabling them to take the necessary precautions to protect their lives and property. A number of residents in the city of Brandenburg had been killed by tornadoes and severe weather. The loss of life was attributed to the fact that citizens did not receive warning of the impending storms.
>
> **SOLUTION**
>
> Warren County obtained funding through FEMA's Hazard Mitigation Grant Program (HMGP) to install 12 COWS. The sirens have the capacity to warn 100% of the residents of the county's biggest city, and 80% of residents in the entire county. When COWS are activated, residents know that they should turn on their televisions or radios for further instructions. The County later installed 250 indoor CALVES. The system is designed to warn those who are indoors or not close to a siren, such as in schools, nursing homes, and indoor sporting arenas. The system uses a series of beeps followed by a voice message from the activating agency to alert residents of an emergency. Installation was followed up by an extensive public awareness campaign to educate residents on the warning systems with the instructions "When you hear the COWS mmoooove indoors. When you hear the CALVES protect your herd!" The warning system has been credited with helping to save lives since its implementation.[3]

Speakers and Video Surveillance Emergency Phone Towers

Emergency telephone towers are highly visible and can be seen in many parking lots and parks across the country. They are usually brightly colored so that they are visible from a distance. These devices can reach the outdoor population to send a warning using audiovisuals (sirens, horns, and voice messaging via speaker for wide-area emergency broadcasting capabilities, and an attention-getting blue light), and the outdoor population can reach you (through video surveillance mounted within the unit, a panic button, or auto/manual dialing). These devices are rugged, hardened devices made to withstand severe weather conditions and are Americans with Disabilities Act of 1990 (ADA)-compliant emergency phones. They are excellent for immediate and emergency communiqués. They can also be an Internet Protocol (IP)-based system or can use analog connectivity.

These external emergency phone towers can be expensive, particularly the infrastructure needed for these units to function. A single unit can cost approximately $7,500 to $10,000 or more per phone when you begin to add camera and speakers and power supply. Units can be battery operated, work using solar energy, and use wireless technology to minimize the need for tunneling through concrete. There is the ongoing cost relative to the number of times it may be used throughout the year. There have been many organizations reporting that, on average, they may have two to five users a year, and the number continues to decline as more people have cell phones. For those organizations that retain these devices, the benefits of helping just one individual avoid danger makes it worth the cost.

Billboards, Video Displays, Digital Signage, and Community Access Television (CATV)

Billboards are large outdoor structures used for posting information in a high-traffic area such as along a busy road or walkway. Bulletins are the largest and most impactful standard-size billboards. Billboards are a "no-tech" solution, meaning that you can quickly write something on a piece of paper, and then, using push pins or tape, post it to the bulletin board. The larger the bulletin and billboard, the larger the number of people who can view the message. There are all kinds of billboards—painted, neon, inflatable, on the back of vehicles, 3D, human, and digital. Billboards can quickly capture your attention, yet some consider them to be visual pollution, particularly if the message is perceived as negative, it has been defaced, or through aging it becomes an eyesore (visual pollution). Billboards for drivers are sometimes considered a hazard because they can be distracting and lead to accidents. They are an effective tool in getting emergency communications out to those who can view the signs.

There are increasing numbers of digital billboards and video displays used indoors and outdoors—in conference rooms, recreational areas, in registration areas, lobbies, and even in our homes, places of worship, schools, and other public gathering places. These devices can provide the audience with multiple communications on a single screen—from tickers scrolling across the bottom of the screen providing the latest news, weather, sports and financial information, to multiple screens providing general television broadcasts of internal communiqués. Digital billboards are devices that can reach the hearing and visually impaired using written and audible notifications. Closed captioning and use of multiple languages are also available using this type of equipment.

Video displays can carry television UHF, VHF, HDTV, and CATV (community access television) signals. CATV is an increasingly popular way to interact with the web and other new forms of multimedia information and entertainment services, including Internet radio and television, Microsoft's Active Directory or Novell Directory Services (NDS) for targeting messaging based on location and user groups. Active Directory and NDS are centralized and standardized systems that automate network management of user data, security, and distributed resources, and enable interoperations with other directories using Lightweight Directory Access Protocol (LDAP).

Digital signage such as digital marquees, which are much like video displays, are a good communication tool for delivering immediate, emergency, and advisory communications for the population in the immediate area. The units today are programmable, networkable, and can be viewed from distances on an average of 200 feet. Digital marquees have been around for quite a while and are effective in reaching populations in the immediate area of the device. They can be analogue or IP-based systems similar to video displays. As the displays become larger, the viewing distance increases. The messages on marquees can be displayed using different colors, include animation and use of symbols, and move in different directions—left to right, right to left, up and down, zigzag, and other alternate directions. Exterior units are hardened to withstand severe weather conditions.

Exterior units may fall under the scope of local ordinances that may vary for on-premise and off-premise units. Brightness, noise, light pollution and other potentially distracting characteristics, and content play a role in the use of digital signs outdoors. Residential and nonresidential zoning may affect your selection as well as roadway type. Residential zoning affects uses along local streets, the type of building structure, and occupants of the area, and few exceptions are typically granted. Nonresidential zoning issues are building-structure types, whether you are near a freeway or a major commercial intersection, and a maximum sign size. Another stipulation is whether the sign is freestanding or attached to a building. The height of the sign and, if there are other signs around, how close the sign is to any frontage road or corner lot, and if there is a master plan in place by a developer for an area that has its own restrictions, are also to be considered. If the thought is to place a sign in a historic district, conservation district, or arts district, certain restrictions may apply if they can be displayed at all. Some areas recognize that signs are a given, preferring a well-maintained digital sign over temporary signs that can quickly fall into disrepair or be easily defaced with graffiti. For example, some may prohibit

- Intermittent or flashing illumination
- Changes in brightness, intensity or color
- Motion
- Environmentally unfriendly (high power consumption)

Common types of signs include monochrome, grayscale LED, color RGB 64K, and HD Color RGB 16–33 m. Information regarding each type is as follows:

- Monochrome
    - Single color (red or amber)
    - Used for text messages (e.g., time and temperature), corporate logos, and some graphics
    - Most common category on the market
    - Low cost
- Grayscale LED
    - Newer technology
    - Single color—multiple shades similar to a black and white TV
    - Less in cost than a color unit
    - Full-motion graphics
- Color RGB 64K
    - Entry-level color technology
    - 64,000 RGB colors
    - Can display looping animation and an advertising schedule with different images
- HD Color RGB 16–33 m
    - Professional quality—Times Square, Hollywood
    - 16 to 33 Million RGB colors
    - Full-motion, high-resolution animated graphics
    - Live video capabilities
    - Has looping playback and advertising scheduling of different images or video

Marquees on a web page, an HTML marquee, let you add scrolling text and images to your website. If using a style sheet, the information can be written once and then published on other pages on your website. These are decreasing in use today. They consume precious real estate on a web page. Considerations in using HTML marquees include

- *Background Color:* the color for the background, two-tone, a single color, what color
- *Behavior:* type of scrolling, for example, jumping, speed (fast, slow), or when to start or stop

- *Direction:* which direction the message should scroll, such as left to right, right to left, up and down, zigzag, and other alternate directions
- *Height:* how tall the marquee should be
- *Horizontal Spacing:* how much space to leave around the marquee from the edge
- *Loop:* how many times to loop or have the message repeat or set to scroll continuously
- *Scroll Amount:* how far to jump, for example, half the width of the sign or small hops
- *Scroll Delay:* how long to delay between each jump
- *Vertical Spacing:* how much space to leave around the marquee from the top to the bottom
- *What:* what the message will say
- *Who:* whom are you trying to reach
- *Width:* how wide the marquee will be

The thought behind the reduced use of HTML marquees is if you want users to see the information, it must be placed on the page so that it is more visible. The infrastructure to support a single marquee or multiple marquees can be expensive to acquire and maintain given the size of the audience it serves. Depending upon the type of unit, the costs can range from a few hundred for very small units to a couple of thousand for larger displays. Video displays are simply "nice" televisions, which make them prone to theft where unattended.

Two-Way Radios

Two-way radios are excellent and dependable resources for almost any type of emergency communications—immediate, emergency, and advisory. Unlike a broadcast receiver, such as a television, a two-way radio can transmit and receive content. They come in many different configurations—mobile, stationary, and handheld portable devices. Sprint, using Nextel pushbutton technology, offers a mobile phone with two-way radio capabilities. Two-way radios using 700 MHz are key resources for law enforcement agencies. They offer broad coverage and have longer battery life than traditional cell phones and smartphones. 700 and 800 MHz radios do require specialized programming by the local jurisdiction (Figure 10.8 Two-way radios).

Rebanding (reconfiguration) of the 800 MHz band was scheduled for completion by April 2011 to resolve interference between commercial and public safety systems in the band. It was necessitated by public safety being assigned a block of frequencies while Nextel (Sprint's

FIGURE 10.8 Two-way radios.

extended specialized mobile radio service—[ESMR]), public safety services, and community repeaters were assigned the remainder of the frequencies. ESMR looks and operates much like a cell phone; however, it is a digital trunking business radio with telephone interconnect within the 800 MHz spectrum. The double duty occurring within the 800 MHz spectrum has led to a large number of complaints of interference to public safety. Rebanding will give public safety and critical infrastructure exclusive use of 851–854 MHz.

There are ongoing discussions among industry representatives and consumer group representatives regarding market issues related to 700 MHz equipment. 700 MHz is currently being offered or developed for consumers in general and for public safety uses. This spectrum was occupied by TV broadcasters on Channels 52–69, who have since vacated the spectrum as they migrated to digital broadcast technology. The focus has been to shift from TV broadcasting to the 700 MHz band to be used for public safety and in establishing a nationwide, interoperable broadband communications network benefiting state and local public safety use.

The 700 MHz spectrum will give public safety broadband technologies that support high-speed data transmission across long distances access to video, mapping, *GPS (global positioning system)* applications and more. Topics of interest include the status of the current and near-term market for 700 MHz devices, the ability of small and regional providers to obtain devices at a competitive cost, and the effect of interoperability on promoting the public interest, including promoting competition, access to broadband, public safety, and the widespread availability of service in rural areas.

The 400 and 900 MHz bands are two radio bands in the UHF range. Cordless phones use the 900 MHz band, with the radio system remaining more private. It is commonly used for police scanners. As part of the 400 MHz band is covered by a radio receiver, more people are able to listen in. The 900 MHz range typically receives less interference and offers better sound, while the 400 MHz range offers a greater transmission range.

The 400 MHz range is getting very congested in urban areas. The 400–450 MHz UHF range covers fixed, mobile, satellite, navigation, weather aids, amateur, medical device radio-communication services, military (secondary basis), and government; (450–470 MHz) private land mobile, maritime, and rural radio service; and (470–500 MHz) fixed, broadcasting, land mobile, and core UHF TV band channels 14–18. It is a common emergency frequency used by some smaller jurisdictions such as school campuses for communications among essential employees—administrators, public works, and food services. These radios usually have a range of 1 to 2 miles with the typical unit having a 16-hour battery life. The greatest drawbacks are that these radios do not interface with radios used today by public safety agencies, and their limited reach.

The focus of emergency communications today is at the state and local levels. The national plan was established in 2007. When responders cannot communicate with their home agencies, response efforts are compromised and lives are jeopardized. The military says, "if you can't communicate you can't respond." Remember, in a crisis, no news is bad news.

Land mobile radio (LMR) will be here. Its primary mission is critical voice technology. Standardization will continue as the LMR system in the VHF and UHF bands migrate to narrowband before 2013. Broadband is not available today as a mission-critical voice technology. Broadband enables the delivery of information and video to public safety responders. Technology is a big piece of the pie, about 40%, but the larger portion of the pie that remains, 60%, covers governance, standard operating procedures (SOPs), training and exercises, and usage. Communications is a core capability. You have to use what you already have and add to it.

Weather Radios

Weather Radio Distribution Ensuring Adequate Hazard Warning

Having a radio with a weather band on it and able to run on batteries or windup/hand-cranked power has been recommended for years for every home and business, including schools and nursing homes. National Oceanic and Atmospheric Administration (NOAA) publishes weather advisories and alerts for every region of the county and into Canada, Mexico, the Caribbean on the Western hemisphere, and Saudi Arabia, Romania, Russia, the Peoples Republic of China, Korea, and Australia in the Eastern hemisphere. Through the NOAA Weather Radio all hazards public warning system, broadcasts of weather forecasts, warnings, and other emergency information is provided around the clock directly to the public. All hazards messages include natural, technological, AMBER alerts, and terrorist attacks. The alerts are given when public safety is involved, the message is from an official government source, and time is critical.

NOAA weather radios are low-cost devices and, ideally, radios are in every home and business, particularly in areas such as Tornado Alley that includes most of Texas, Oklahoma, Kansas, Nebraska, Iowa, Minnesota, South Dakota, and North Dakota, where minutes matter. Weather radios are an effective device to warn people of potential severe weather so that they can take the appropriate action quickly. Weather radio alerts also include AMBER and Silver Alerts, as well as any National Warnings or Alerts. With a small grant, Wayne County, Michigan, located in southeast Michigan, was able to purchase 860 for residents in the community. Unfortunately, it took a dangerous tornado to move through the area before getting the funding needed, as discussed in the following case study.

Case Study: Wayne County, Michigan NOAA Weather Radios

Located in southeastern Michigan, Wayne County frequently experiences severe weather and tornadoes. In 1997, a dangerous tornado moved through parts of Detroit and the surrounding suburbs of Highland Park and Hamtramck injuring 90 persons. It was the most costly tornado the state had experienced, with damages estimated at $90 million. The tornado traveled nearly 5 miles and was 2,500 yards wide. It was part of an outbreak of 13 tornadoes in southeastern Michigan, the largest number for a single day since records have been kept. With over 2 million residents, the county needed effective mitigation measures to warn adequately people of the potentially severe weather.

SOLUTION

The Hazard Mitigation Grant Program (HMGP) provided funds for the county to purchase distribute, and install 860 NOAA weather radios in every school, hospital, and nursing care facility in the county and provided tornado shelter/spotter workshop for employees of those facilities. This project helped to ensure that adequate warning time is more readily available to residents in the county. The benefit of early warnings is a reduction in the loss of life and the extent of injuries to persons in an impacted facility.[4]

Weather Radios—Enabling People to Hear and Heed Severe Weather Warnings

CASE STUDY: PORTAGE COUNTY, WISCONSIN NOAA WEATHER RADIOS

Portage County, Wisconsin, has a population of approximately 70,000 residents living in an area that is 62% urban and 38% rural. Nearly 12% of the population is 65 years old or older. There was a concern that the elderly and those living in rural areas may be unable to hear the county's severe weather warning system. The county's emergency management coordinator (EMC) indicated a need to purchase weather radios for those who are significantly at risk—those that do not always hear the warnings.

SOLUTION

Using federal grant, the county purchased 150 National Oceanic and Atmospheric Administration (NOAA) All Hazards Weather Radios.[5]

It is estimated that over 85% of the population now resides within the service area of at least one transmitter. NOAA Weather Radio is a service of the U.S. Department of Commerce, NOAA division. As the "Voice of the National Weather Service," this service provides continuous broadcasts of the latest weather information from local National Weather Service (NWS) offices.

Forecasters can add special signals to warnings that trigger "alerting" features of specially equipped receivers. This is known as the *tone alert* feature and operates much like a smoke detector in that it will send an alarm, when necessary, to warn of an impending hazard. In the past, all receivers equipped with the tone alert feature within the listening area would send an alarm anytime a warning was issued. Today, these alerts can be managed using the Specific Area Message Encoding (SAME) technology available on newer receivers to alarm only if a warning broadcast pertains to a particular location. The newer receivers allow individuals to choose the warning locations the receiver will target.[6]

Evacuation Warning System—Weather Warning System—CodeRED

Hurricanes are events for which you have some lead-time to notify affected residents—enough time to evacuate to a safe area. Lead-time can be many hours to several days. Conversely, severe weather events such as tornadoes, severe thunderstorms, and flash flooding will often give you minutes to warn an affected population to take immediate action. There is, thus, no time to evacuate. The tornado weather across the southern United States in spring 2011 took lives and devastated communities. Floyd County, Georgia, near Atlanta, is an area that frequently experiences severe weather events. In 2008, Floyd County did not have an advance warning system for severe weather in place. Through a FEMA Mitigation Grant, the county was able to acquire an early weather warning system as discussed in the following case study.

Case Study: Floyd County, Georgia CodeRED Weather Warning System

Floyd County, Georgia, is located near the Georgia–Alabama state line and northwest of Atlanta. It is an area that frequently experiences severe weather events—tornadoes, snow, flooding, and severe thunderstorms. Saturday, March 15, 2008, around noontime, an Enhanced Fujita Scale EF3 tornado struck the town. The tornado, with estimated winds of 150 miles per hour, touched three counties—Polk, Floyd, and Bartow—leaving in its path one fatality, two injuries, and serious damage to 17 homes. Floyd County did not have an advance warning system for severe weather in place at the time. It has had seven presidential disaster declarations since 1990 due to severe weather. Over a 10-year period, Floyd County had experienced 67 severe thunderstorm events, 20 flood events, and 16 severe winter weather events. In the last 50 years, the county has survived the destruction caused by 12 tornadoes.

SOLUTION

Using the FEMA Hazard Mitigation Grant Program (HMGP), Floyd County acquired the CodeRED Weather Warning system, an advanced mass alert system to mitigate future risks for the area's residents. The system is a geographically based notification system that notifies registered users of approaching dangers by calling their phones. The system was implemented and activated in March 2009. Within 5 months, more than 9,000 Floyd County phone numbers were registered. The system has been activated more than 125 times, primarily for severe weather and tornado warnings. According to Scotty Hancock, the Floyd County Emergency Management Director, residents were appreciative of the warning calls to their phones.[7]

REGIONAL RESOURCES

Highway Alerting

Highway alerts can offer a district-to-resident notification system when integrated with other systems such as Blackboard Connect-CTY service as used by Warren Township Road District in Illinois. Text messages are broadcasted on highway message signs for emergencies (see Figure 10.9 Highway alert). Messages can be delivered within 1 to 5 minutes and can reach populations enroute. The devices are increasing in number along Interstate highways and area roadways to give traffic and weather alerts, silver and AMBER alerts, such as in the State of Florida, when a law enforcement officer is killed or seriously wounded. These dynamic messaging signs alert motorists of critical vehicle description information

FIGURE 10.9 Highway alert.

such as license plates, vehicle description, and contact information that can help identify and apprehend suspects and survivors.

Silver alerts are designed to help find missing adults, ages 60 and over, with major illnesses such as Alzheimer's or dementia. They are being used more often than AMBER alerts, which are designed to help find missing children. In the State of California, over 100 children have been successfully recovered since the inception of the AMBER alert program in 2002.[8]

A major concern for public safety officials is that drivers over time will become desensitized to the alerts with the increasing number of alerts constantly on the screen. Other concerns regarding the system are that access to the signs may not be possible or requires an intermediary to service. The system can only provide a very short message per screen. This may require the use of multiple screens to provide enough information to act on. If multiple screens are needed for the message, it further limits the reach to those who are driving on Interstate highways and roadways. For motorists, other traffic may block some of the message, and they lose the opportunity to view the complete message. To help reduce the total cost of ownership, many organizations will rent air time from the owner leading to the potential of more messages. Also, if on a public thoroughfare, a special permit and restrictions may apply to the owner and user.

Dynamic message signs along major thoroughfares are typically handled by a state law enforcement agency or a regional transportation representative. The criteria used to activate an alert must be satisfied before it is displayed, causing a delay. The message will continue to display until the offenders have been captured or the threat has diminished to the point that other forms of communication are being used.

Reverse 9-1-1

When 9-1-1 was first introduced, it was a one-way messaging tool. A person would call and provide information to the Dispatch Operator on what the nature of the emergency was. Reverse 9-1-1 is a form of two-way emergency communication tools. The Dispatch Operator is not limited to "pulling" information from the caller but can now "push" information out, providing specific instructions for incidents in progress such as a tornado warning. A single voice message can be delivered in 1 to 5 minutes by geographic location.

The trade-off of the reverse 9-1-1 system is that the receiver must be in a position to receive and hear the information on the call. During major emergencies, congestion on the network could prevent completion of the call. Reverse 9-1-1 infrastructure is relatively expensive, and it requires a database for tracking all the telephones within scope. Without adaptive technologies, the call does not reach the hearing impaired.

Earthquake and Tsunami Monitoring

The first seismoscope was invented by the Chinese philosopher Chang Heng in 132 AD. It was a large urn, having, on the outside, eight dragonheads facing the eight principal directions of the compass. Under each dragonhead was a toad with its mouth opened toward the dragon. When an earthquake struck, one or more of the eight dragon mouths would release a ball into the open mouth of the toad sitting under a dragon head. The direction of the shaking determined which of the dragons released its ball and detection of an earthquake up to 400 miles away. Earthquake monitoring and early warning systems are now appearing. The federal government has shown an expressed interest in seeing the United States doing more in early earthquake warnings.

The USGS offers the Earthquake Notification Service (ENS), a free service that sends automated notification e-mails when earthquakes happen in your area. This is not an early warning system; it will notify you when an earthquake has occurred. Earthquakes are one of the few natural disasters where a credible early warning system is not in use.

Earthquake Monitoring Project TriNet

In 1997, the California Institute of Technology, the California Division of Mines and Geology, and the U.S. Geological Survey (USGS) were provided with FEMA funds to upgrade the existing earthquake monitoring systems for Southern California. This project was called TriNet, a cooperative effort to provide timely and accurate information on earthquake occurrences in the region. By 2002, 150 broadband seismometers and 450 strong-motion sensors were installed. This project spawned a new product, namely, the "ShakeMap."

Earthquake shaking is strongly affected by local geology and soil conditions, and the pattern of the intensity of shaking does not fall in concentric circles about the epicenter. With data from new seismometers available in "real time," ShakeMap (available at http://earthquake.usgs.gov/earthquakes/shakemap/) are contour maps showing the severity and distribution of ground shaking within minutes of an earthquake. Whenever an earthquake occurs in Southern California, there is a regional map available on the web that shows the shaking pattern. From this project, the Advanced National Seismic System (ANSS) was launched to expand TriNet capabilities to other urban centers in areas of high-to-moderate seismic risk, including the San Francisco Bay region, the Puget

Sound region, Salt Lake City, Reno, Anchorage, and the Memphis and St. Louis areas. The information is also used in the design and construction of earthquake resistant buildings and critical facilities.[9]

Measuring Earthquake Potential with GPS

In 2001, earthquake scientists unveiled the Southern California Integrated GPS Network (SCIGN), a ground-motion-monitoring network. Unlike other instrument networks that record shaking, SCIGN tracks the slow motion of the Earth's plates using GPS. With SCIGN, the link between the motions of the plates that make up the Earth's crust and the resulting earthquakes are observed by an array of GPS stations operating in Southern California and Baja California, one of the world's most seismically active and highly populated areas. Scientists of the Southern California Earthquake Center (SCEC) design and manage SCIGN. NASA's Jet Propulsion Laboratory, the Scripps Institution of Oceanography at the University of California at San Diego, and the United States Geological Survey are the principal SCEC partners in SCIGN, and all data from the array are openly available on the Internet.

For earthquakes outside the United States, the National Earthquake Information Center (NEIC) notifies the State Department Operations Center and will send alerts to staff at American embassies and consulates in the affected countries, to the International Red Cross, the UN Department of Humanitarian Affairs, and other recipients who have arranged to receive alerts. The USGS Earthquake Notification Service (ENS) is responsible for delivery of these messages. Users of the service can specify the regions of interest, establish notification thresholds of earthquake magnitude, and designate whether they wish to receive notification of aftershocks, and set different magnitude thresholds for daytime or nighttime to trigger a notification.

Japan's early warning system for major earthquakes and tsunamis saved thousands of lives when the March 11, 2011, magnitude 9.0 quake rumbled off the nation's east coast. The quake was followed by a 30-foot high tsunami that inundated many coastal communities. Japan, atop active seismic zones, has an interconnected warning system linking public and private infrastructure that is capable of slowing high-speed trains, stopping elevators, and delivering warning alerts to multiple devices. A comparable earthquake warning system is not yet operable in the United States.

There is a new earthquake-monitoring solution, a prototype that was being tested in the Coachella Valley of Southern California in early 2011. This warning system can provide advance notice of an earthquake, as much as 15 to 20 seconds prior. This is enough time to get the doors up so that emergency vehicles can begin exiting the garages before they are potentially trapped. Smartphones can be used as sensors with data points in a disaster. The *accelerometers* found in newer mobile devices could be used to detect earthquake motion. The data could be integrated in a *crowdsourcing application* that automatically triggers the earthquake/tsunami warning systems. An accelerometer measures the tilt and motion in mobile devices to perform different functions such as activating autoscreen rotation, muting an incoming call, or pausing the music play by turning the device face down on mobile devices. A crowdsourcing application channels data in a manner to solve a problem and freely share the answer with everyone—a distributed problem-solving and production model. Social media and smartphones will need to be a part of any national warning system adopted according to FEMA Administrator Craig Fugate.

OTHER SOLUTIONS TO CONSIDER

Combined Alerting System (Software to Allow Delivery to Multiple Platforms)

As we are learning from the Japan earthquake, the need for redundancy is real, and backup power is vital to recovery and continuing operations. The damage reported by the FCC in May 2011, less than 2 months after the earthquake, on Japan's wireless connectivity infrastructure was staggering:

- 65,000 telephone poles were destroyed.
- 1 million wire line connections were lost.
- 500,000 broadband lines were knocked out.
- 3 of 7 undersea broadband pipes were severely damaged.
- 2/3 of the base stations in the impacted area were out.
- Thousands of roads were washed out.

Combined alerting systems are becoming more of the norm. Most organizations already have e-mail, telephones, and Internet access. They want to reach more of their audiences in an emergency, and need an open-architecture structure. Tools such as InformaCast, REACT! System, or BlackBoard have plug-in architecture that enables an organization to integrate the combined alerting system with their network notification software that can include inbound and outbound, e-mail, Twitter, Facebook, RSS, SchoolMessenger, and other solutions.

Combined alerting systems are good for emergencies, advisories, and immediate notifications. Using multiple tools to deliver a message, an emergency communiqué can be delivered within 1 to 5 minutes, assuming there are sufficient phone lines to the system, which can allow specific instructions for an incident in progress. The greatest advantage is that a combined alerting system is not reliant upon a single technology to reach end-users such as text messaging, e-mail, twitter, or telephones. For the sender, a combined alerting system simplifies the steps a person needs to take to initiate a message by allowing entry to the system to generate multiple deliveries.

The trade-offs are diminishing as more companies enter into the market. Early versions required a subscription where the user was charged per message and for the number of end points registered in the database. There is usually a high annual fee to retain the service. Most organizations budget for only a few messages a year for a block of receivers. This model is good for the vendor, but not good for the user if outsourced. A single incident could lead to multiple messages being distributed, quickly exceeding any budgetary number used for planning and justification. Second, these systems require someone to maintain the database of individuals who should receive a message. Third, these systems are prone to congested phone services. The system requires redundancy in telephone services to increase the probability that the call will be completed. Adding GETS or WPR to the dialing sequence for emergencies may be a resource that can be used to help mitigate this concern.

3-D Meeting Software

Many mass notification efforts that are multijurisdictional require that teams that are geographically dispersed share technology and information, exchange knowledge, and collaborate virtually. Use of 3D meeting solutions can enable teams to collaborate virtually. The scientist working to shut down the Japan nuclear power plant was able to leverage this technology. ProtoSphere, IBM, and EON, among others, offer 3D meeting software subscription-type services. 3D virtual meetings have taken hold and will grow in popularity. Like Second World, attendees use avatars, which can approach and listen to a presenter and speak while reviewing the material. In the 3D world, attendees have an area where they gather and chat without interrupting presenters and, if desired, engage in private conversation. Attendees communicate using VoIP and text chat.

As a new generation enters the workplace, they bring with them the tools they have used at home, school, and for leisure. Those who use 3D applications recreationally are sometimes more comfortable approaching senior-level staff and executives as an avatar than having real face-to-face dialogue. It remains in its infancy in terms of a mass notification tool; however, it presents a venue for training and exercises that is safe. Using software for training and exercises can reduce travel costs and minimize the dead time of commuting. 3D software can maintain as much as possible the real-life meeting format (from simulating the material to reproducing the lighting, furniture, and floor plans of the actual event). It has been reported by several vendors, but not yet proven, that the more closely the session mirrors real life, the more likely it is adopted by the user.

Other Crisis Communications Solutions

A one-size-fits-all model is the least effective approach. The idea that just e-mail or just your fire alarm is enough is a formula for failure. It is important to reassess annually what technology you have and how it may be applied in an emergency to notify others. Other solutions to consider include

- *VoIP Phone System with paging options.* Many office telephones have a paging option much like an intercom system. It means that you can go to one phone and send a message over a loudspeaker or other speaker system that can be accessed by the phone system to send an emergency message.
- *Two-way emergency phones in rooms.* Many hospitals have emergency telephones or intercoms in each patient's room that are attached to the nurses' station. By pressing a button on the bed, on the wall, or a phone in the room, a call can be placed automatically to the nurses' station. The same solution is applied in some classrooms across America, enabling an instructor to pick up the handset, and it will automatically dial campus security. The same system can be used by a central dispatch or nurses' station to push out a message to a single phone or multiple phones. In most medical or school classroom, if the phone is used and the caller does not respond, hangs up prematurely, or sounds as though there may be questionable activities in progress, security personnel will respond to confirm the situation and take appropriate action. The greatest barrier to implementation is the cost associated with installing the infrastructure and purchasing the equipment.
- *Mass Notification System integration with fire alarm system.* This is an ideal solution. Having the capability to override the fire alarm with an audio command is defined in the 2010 National Fire Protection Association (NFPA) standards. In many areas, new building construction is required by ordinance to install such a system. The advantage this offers occupants is

that emergency personnel can override an audible alarm with voice commands that could give occupants a different set of instructions. For example, a terrorist could pull a fire alarm so that people evacuate a building, and the terrorist now has a number of targets to choose from to harm. A voice command could override and instruct occupants to return or stay indoors, lock all doors, and take cover.

- *Centralize monitoring of life safety and security systems.* Could you imaging having a different monitoring company for every building you have on a campus? An organization can migrate to a single monitoring service, whether internal or external, and significantly reduce costs. The key to centralized monitoring is to test regularly. Testing should occur during every fire or evacuation, shelter-in-place, or earthquake drill and exercise. The test should measure how long the monitoring service took to receive and respond by notifying first responders and the building owner. A review of all call down lists and lists of occupants with mobile impairments—requiring the use of a mechanical device to exit a building—will be needed.
- Have *dedicated telephone landlines available* to avoid circuit overload. VoIP systems are useful. For emergency use, a few landlines and a cluster of VoIP telephone numbers should be available. The VoIP system may experience network congestion or outages, cell phone systems may be stressed, and satellite coverage may be weak in hardened areas where some EOCs are located. Upon activation, an area can quickly swell with first responders needing telephone services.
- *Stock a mobile media center.* Mobile media centers are portable carts or trailers with camera and video equipment, projectors, lights, monitors, computers, cell phones, and other equipment used to generate video and provide network connectivity. This is beneficial when setting up a media staging area or in creating and disseminating prerecorded or live video.
- Consider use of a *computer lab to serve as your EOC.* Many organizations think that they cannot afford to have a room full of computers that are network attached, large erasable writing surfaces, and places where people can work. This would be an expensive endeavor if this were the only purpose for the area. A way to satisfy this need within many schools, libraries, and other learning areas are computer labs or classrooms, student centers, and library computer areas. Almost everything needed is available on demand. In an emergency, this is an excellent place to

start without incurring significant costs. At the close of an event or activation, the area can be returned to its normal use.
- *Look at the little things for solutions*—a child's toy, baby monitor, or hobby items. For example, at spykids.com, you can buy a $300 helicopter with a camera. This helicopter could be used to view into a window on an upper floor, that is, to see if a hostage event is in progress or to check to determine if there is anyone left in the area.

COMMON OPERATING PICTURE (COP)

There are many communication technologies and platforms, old and new, that do not communicate or talk to each other well. For first responders this causes blind spots that limit communications and hinder response activities. This matters at times of disasters where there are multiple responding agencies or multiple jurisdictions involved. This can even be an issue within an agency where the different departments use different communication tools. An example is a school campus that has a public safety department that uses an 800 MHz radio system to communicate among themselves, and other external law enforcement agencies; and the maintenance personnel who use 400 MHz two-way radios to communicate within their team. An emergency may require police to manage security and to coordinate their efforts with maintenance personnel to block off areas that students and the public should avoid. Response could be delayed as these two departments use an intermediary to relay information such as calls placed using personal cell phones or going through a dispatch center that can relay the information. In major disasters such as wildfires in Texas or tornados in Arkansas, it matters.

How we resolve these issues sounds simple at a high level, yet some effort is needed to make this a reality. You integrate existing technologies into a single COP using a *common alerting protocol* (CAP). COP can integrate visual analytics using your audiovisual systems with your mass notification system. It can provide users with real-time information and true interoperability. Users can operate this system to disseminate mission-critical information to multiple users and devices simultaneously, such as information sent from a 9-1-1 dispatch center.

A common operating picture can also incorporate mapping solutions that can put first responders at the forefront of real-time situational awareness. The capabilities of COP can reduce response and recovery efforts through improved acquisition of information. It is important that any system used enables the secure sharing of information and decision support using CAP, the National Information Exchange Model (NIEM), and NIMS. A COP can serve as the transporter of uniform data in common formats. External applications are used to visualize consumed data inside the customized user interface such as sensors, incident logs, or mobile video feeds.

Anyone with a smartphone is a potential reporter in today's around-the-clock monitoring of all locations. Having enabled or live feeds, and a spatial-temporal reasoning engine is helpful. Efficient communications using this functionality provide location-based message streams that can be programmed to automatically detect events of interest real-time within streams. The ideal system can then take the detected event and send an automated alert to multiple devices, enabling information sharing real-time. Indianapolis, Indiana, is a good example of a jurisdiction having a number of major events where COP has been used or is planned. Events include the NCAA Big Ten Championship Game, the Indy 500, the 2012 Super Bowl, and day-to-day activities.

The City of Indianapolis has smartphone and table applications that leverage its IP multimedia subsystems. The hooks among these devices and applications promote interoperability by advancing mobile technology. The IP multimedia system (IMS) is available to all cellular providers (AT&T LTE, Sprint 4G WiMax, Verizon LTE, T-Mobile HSPA+). AT&T will be the only carrier that offers two layers of network technology delivering 4G speeds—HSPA+ and LTE—in 2011 as it enhances its *backhaul*. Backhaul as used in telecommunications involves the transport of traffic between access points and a centralized point of presence.

IMS allows the convergence of voice and data regardless of access type. Converged networks provide a valuable first step toward the vision of an all-IP environment using Android, IPhone, Microsoft 7, and BlackBerry platforms. IMS combines media such as voice, text, pictures, and video while giving users the tools to personalize their communications. IMS-based services enriches communications by combining the various media to share events in real time. Adding presence information, users can better control how, where, and by whom they can be contacted.

We begin by using the City of Indianapolis, again, as an example. Security personnel with a cell phone responsible for a large event such as the NCAA Big Ten games can draw a geo-fence, a trigger point, of when you are automatically notified of events within the geo-fence using a unique ring back tone. The ring back tone can be of a different tone than other calls received on the same device. Geo-fencing could also be activated when you are in the perimeter identified by the geo-fence, and deactivated when you are outside the geo-fence using a smartphone. This process would be effective for someone who is on duty while in

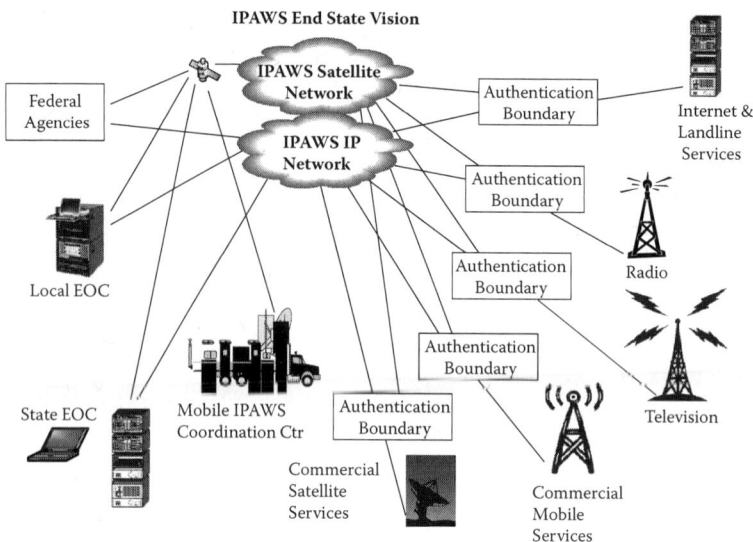

FIGURE 10.10 IPAWS end state vision. (From Sandia National Laboratories, www.sandia.gov.)

the geo-fence and off duty when outside it. The information could be available in real time using a smartphone or the web, or integrated with other technologies. BRS Labs video analytics 2.0 solution combines artificial intelligence (AI) and behavioral analytics that can integrate to provide alerts based on unusual behavior caught on video for a given area.

Any COP should be CAP compliant using the IPAWS architecture (Figure 10.10 IPAWS end state vision). This should be thought of in the design phase. Failure to begin with the end in mind, which includes compliance, can lead to additional costs associated with redesign of the system. COP can also present bandwidth issues, particularly at a time of heavy congestion on systems. Providers can minimize bandwidth issues by using *quality of service* (QoS) to increase services to first responders. QoS is a term used to define the prioritization of network traffic so that the most important data get through the network first.

IMS is useful in promoting Next Generation 9-1-1 (NG911). The University of Maryland is studying, and the City of Indianapolis is beta testing, the use of IMS with NG911. Other systems that are enhanced by a COP are a computer-aided dispatch (CAD) system that is often used by agencies to dispatch personnel to a location. COP allows a user to overlay information from different systems. Critical Incident Management Systems (CIMS), visual analytics, and notification systems can also be enhanced by COP.

SUMMARY

There are a number of solutions available to you today for use in a crisis—no-tech to high-tech. No-tech is the use of pen and paper, walking around, flyers, and posters. Medium-tech solutions are e-mail, voicemail, and other technologies that most of us use every day. High-tech solutions require the aid of specialist to use and administer, such as sonic buoys and highway alert signs. In a crisis, you begin with what you have and work your way toward what you need. Most of us already have e-mail, cell phones, voicemail, public media, text messaging, smoke detectors, and fire alarms. Alerting systems that are out there occupy the range of emergency telephone notification systems, PA systems, and emergency telephone towers, among others. The desired goal for mass notifications systems are combined alerting systems and creative ways to integrate and use the systems you already have.

There are many despairing communication technologies and platforms, old and new, that do not communicate or talk well with each other. This creates blind spots that limit communications and hinder response activities. The trend, going forward, is emergency communications for work or home that will provide targeted messages so that you and the person sitting next to you will receive a different set of instructions based on your profile for the same event. This can be achieved by integrating existing technologies into a single common operating picture using a common alerting protocol.

ENDNOTES

1. FEMA. (2007). Mitigation Briefing. http://www.fema.gov/mitigationbp/brief.do?mitssId=4566 (accessed March 27, 2011).
2. FEMA.
3. FEMA. (2003). http://www.fema.gov/mitigationbp/brief.do?mitssId=2549 (accessed March 16, 2011).
4. FEMA. (2002). Mitigation Brief. http://www.fema.gov/mitigationbp/brief.do?mitssId=11 (accessed March 12, 2011).
5. FEMA. (2008). Mitigation Briefing. http://www.fema.gov/mitigationbp/brief.do?mitssId=8012 (accessed March 12, 2011).
6. FEMA. (2008). Mitigation Briefing. http://www.fema.gov/mitigationbp/brief.do?mitssId=8012 (accessed March 15, 2011).
7. FEMA. (2006). Mitigation Briefing. http://www.fema.gov/mitigationbp/brief.do?mitssId=6930 (accessed March 15, 2011).
8. California Highway Patrol. Help save a child's life. http://www.chp.ca.gov/amber/index.html (accessed April 30, 2011).
9. FEMA. (2011). Mitgation Briefing. http://www.fema.gov/mitigationbp/brief.do?mitssId=2870 (accessed March 10, 2011).

CHAPTER 11

Learning Your Systems Requirements

> Precision of communication is important, more important than ever, in an era of hair trigger balances, when a false or misunderstood word may create as much disaster as a thoughtless act.
>
> James Thurber, 1960
> *American author*

THE CHALLENGE

Crisis communications is about dealing with natural, technological, and other man-made events. Hurricanes and tsunamis will happen, hazardous material spills and fires will occur, and cyber-attacks and active shooter will affect your organization and community at some point. The challenge is developing and implementing effective crisis communication systems to advise of an event as early as possible, reaching as many affected individuals as possible.

Case Study: Hurricane Ike and Campus Closure

In 2008, Hurricane Ike was one of the largest storms to affect the U.S. coastline, becoming the third most costly storm of all time. Its path was unpredictable. Was it going to make landfall near Corpus Christi, Texas or New Orleans, Louisiana, a span of 570 miles? As the storm neared the Texas coastline, at H-120 it was projected to make landfall near the Texas and Mexico border. At H-72, the projections shifted north and so did all the resources that were staged and ready to restore normalcy to the affected communities. At H-48, the projects again shifted farther north toward Freeport, Texas. The Texas

State Operations Center (SOC) was faced with relocating the staged resources farther north, while at the same time needing to adhere to downtime requirements for those transporting emergency vehicles. Where to stand up shelters was another concern. When and who needs to evacuate was the greatest challenge as the projections came closer to the Galveston/Houston region. For colleges the questions were when do we close and when can we reopen? Lone Star College System (LSCS), with more than 15 locations, closed all campuses at the same time and advised the media. Reopening campuses was not as easy since each came online at different times over a two-week period due to sustained power outages in the region. The media would report the information provided incorrectly, stating that either Lone Star College was opened or closed, failing to report specifically which campuses were opened or closed, confusing the community.

SOLUTION

A reliance on e-mail, the home page on the web, text messaging, and a prerecorded message on the main telephone number was used to keep the Lone Star community updated and keeping the number of calls to a reasonable level regarding the status of each campus. This was also supplemented by completing Incident Command System (ICS) forms and forwarding them to the Harris County Office of Homeland Security and Emergency Management (HCOHSEM) Joint Information Center (JIC). The forms were used to help ensure that accurate information was communicated across the airways. HCOHSEM and LSCS became trusted sources for information and the place where the community would go for current and accurate information regarding LSCS. As power was restored to each campus, the campus would reopen. The last two remaining campuses were opened by the end of Week 2, post-Ike.

The challenge in this situation was effective communications to a large community with different communications requirements and stakeholders who used different communications tools to stay abreast of the situation. The lesson learned from this experience is that crisis communications is very different from everyday communications. People take, process, and react to the information differently.

The business challenge is finding the appropriate balance between information "underload" and "overload." Information *underload* is when you are unable to obtain the information needed to make informed decisions in a timely manner. Information *overload* is all the information coming at you and not having the analytical capability to analyze it all to distinguish between good and irrelevant information. These business challenges are further exacerbated by inefficient processes for capturing data and the quality of the data captured (integrity of the data).

Learning Your Systems Requirements

FIGURE 11.1 Business challenges.

The goal is to find a happy medium in the amount of data available and what is needed to address your business needs through technology.[1] (See Figure 11.1 Business challenges.)

FUSION CENTER AND THE EOC

Emergency Operations Centers (EOCs), public information officers and the information and intelligence section under National Incident Management System (NIMS) operate much like a fusion center without the criticism they sometimes receive. *Fusion centers* gather information and share the information with partners. Since it is critical to relay emergency information to the affected populations in a timely manner and accurately, the use of best practices for intake, analysis, and dissemination of information and intelligence through improved information sharing and security is needed. Within the EOC, people, processes, and technology from this viewpoint must be addressed by:

- Creating a collaborative environment for intelligence and information sharing.
- Identifying, developing and using a platform that accesses existing disjointed databases and systems to maximize information sharing both internally and externally.
- Using the National Information Exchange Model (NIEM) to integrate existing systems for seamless communications.
- Implementing and enforcing the use of security measures that help to ensure that access to networks, systems, and information is controlled and monitored.
- Integrating technology systems and people that enable jurisdictions to effectively share critical information in emergencies, as well as support the day-to-day operations of agencies.

An EOC with operations similar to a fusion center requires attention in developing the list of requirements for the center and the technology used for gathering and disseminating emergency communications. The general design begins with a concept, documentation of the current state, and identification of gaps. Armed with this information, you can begin to build the future state with prioritization of your requirements and a review of what is available on the market, and mapping this to your organization's strategic vision. In previous chapters we reviewed the technology you may already have. The new platform should use these systems and their associated databases, then augment to extend your reach to as many stakeholders as possible.

SOLUTION SELECTION FOR YOUR ORGANIZATION

An industry norm is to use the systems or software development life cycle (SDLC). (See Figure 11.2 Software system development life cycle.) SDLC is the process of creating or altering systems, including the models and methods used in developing these systems. SDLC is composed of five key modules: analysis, design, implementation, testing, and evaluation. You need not use all components of the SDLC to arrive at the best solution for your organization. There are key attributes to any system that should be considered, and these will be reviewed in this chapter.

Analysis

The first matter of order is to understand why you need to even consider a mass notification system for your organization. Is it the right thing to do for your stakeholders? Do you find that, when you need to send a message quickly, you spend more time on the phone identifying someone to send the message to the different systems you have? You conclude that there is a need to at least investigate further what your needs are, what you can afford, and when or if you should move forward.

Stakeholders and Target Audiences

The first step is to interview your stakeholders. You need only a few representatives from each group. To gather this information, it is common to develop a short list of questions to ask. Stakeholders include your members from your target audiences as illustrated in Figure 11.3 (Target audiences). The interview will provide information on business processes, perceived needs, and business rules from their perspective. In an attempt to solicit the most information possible, open-ended questions are suggested. Questionnaires and Likert-scaled surveys usually contain what

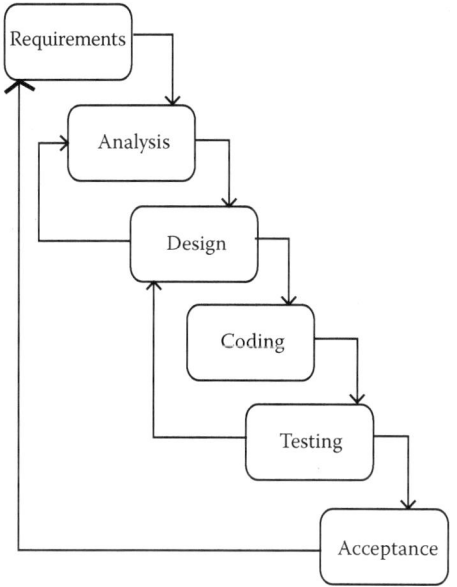

The waterfall model (Systems Development Life Cycle)

FIGURE 11.2 Software system development life cycle.

you already know and limit interaction with stakeholders and learning new information.

The questions to ask your stakeholders are

- What systems do you know to be available for emergency communications within the organization?
- Are these enough?
  - Can you reach your target audience based on your hazard profile using at least three to five systems?
  - Do these systems accommodate people with disabilities and others with functional and access needs?
  - Can these systems send messages to all the devices you need to or want to?
- If not, what do we need (gaps)?
- What can we afford?
- What systems are most important to you for emergency communications in order of preference?

At the close of this process you will have valuable feedback in balancing the needs of the organization with the perceived needs of stakeholders. Often, this is one of the first places that gaps become readily available. The next step is to begin a technical review of the systems you have.

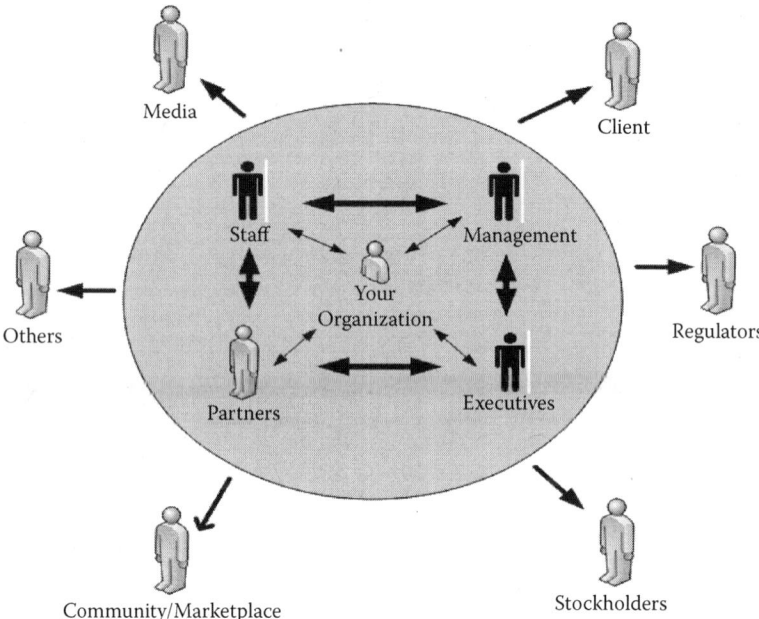

FIGURE 11.3   Target audiences.

Begin with the end in mind by having a list to follow when interviewing the stakeholder. Your Information Technology (IT) department can help to validate the list you have. Next, define your target audiences—internal and external groups—and what technologies are available to reach them.

From the previous chapters we have listed commonly available solutions that are available to you when you are sitting in front of a computer, by the telephone, watching television, or listening to the radio. We also defined technologies for those who are mobile—smartphones with mobile apps, highway alerts, digital displays, speakers, sirens, and horns, to mention a few. We reviewed technologies that are available for people with disabilities and others with functional and access needs. This list can supplement the list you have to identify gaps.

When defining your target audience, you also need to ask the following questions:

- How quickly do you need to reach them based on your hazard profile?
- How many different groups?
- Do these groups need to be prioritized? If so, how?
- What access is needed to any systems you have by group?

- Where are they typically located (geographically)
  - During normal work hours?
  - After hours (nights, weekends, holidays)?
- Do you have alternate contact information for each?

Emergency Communication Needs

The next area to review is communications. Given your target audience and hazard profiles, how many messages will you need to generate dynamically, and what can you prescript (develop beforehand)? What is the bandwidth limitation of your existing systems for messages (sending and receiving?) The threshold size for messages must be quantified. Your IT or Telecommunications department can assist you. Your local telecommunications provider (telco) can also be of assistance.

To obtain a rough estimate of your appropriate threshold, examine the products used for messaging and assess the bandwidth and the delay (latency) for a given point-to-point connection. For example, send a test e-mail from your work e-mail system while sitting at your desk to your personal e-mail account. Conduct this test during peak (normal work hours) and nonpeak times (after work hours). With each message, request a delivery confirmation. This is usually an option in formatting an e-mail. Next, send a text message to your cell phone and possibly a few others in the office that live in different parts of the town or elsewhere in the country. Ask for their carrier. The third step is to test the same in reverse. Gather data from these tests and document the delay (latency) for a given point-to-point connection. The *delay* is the difference between when the message was sent and when it was received. The accuracy of your findings will increase as more data are gathered. Regardless, use whatever information you have.

A breakdown of call types and call counts is an area that can provide valuable information. Only a small subset is needed, again, using point-to-point communication functions. Your IT or telco provider can assist you in gathering as much of this data as possible. If you have a call center or a department that has a high volume of calls, these would be good starting areas for data collection. The functions to review are

- Messages and calls outbound (*Isend*)
- Messages and calls inbound (*Ireceive*)
- Messages and calls that are placed on hold (*Wait*)
- Message and calls dropped or lost (*Drop*)
- The number of broadcast calls or messages (sent to a group of users) (*BCast*)
- Percentage of point-to-point calls (*%PTPC*)
- Bandwidth utilization (maximum, average) (*BDWU*)
- Point-to-point buffer sizes

TABLE 11.1 Call and Messaging Activity

Function	Number of Calls/Month	Statistics[a]
Isend	2,900	29.00%
Ireceive	2,530	25.30%
Wait	3,910	39.10%
Drop	65	6.5%
BCast	0	0%
% Point-to-point calls	9,800	98.0%
Bandwidth utilization (maximum, average)		6,5
Point-to-point buffer sizes		2.0%

[a] Denotes that this column of data represents fictional data.

Each of the call and message activity variables have been captured in Table 11.1 (Call and Messaging Activity). If you are unable to gather all of these variables, make a minimum attempt to gather the number of calls inbound and outbound for your organization. This information is helpful in determining how robust a system is needed for emergency communications using peak demand activities. The next step is to set goals.

DESIGN BRIEF

Goal Setting

SMART Goals

A *goal* is a statement regarding what you want to accomplish. Goal setting is a bottom-up activity when using an existing system and striving to minimize costs. When looking for a strategic solution, goal setting becomes a top-down activity. A review of assumptions and constraints help to build goals that are SMART. *SMART* is the acronym for specific, measureable, attainable, realistic, and timely.

A specific goal has a greater likelihood of success than a general goal. Specific goals answer

- Who—who is involved
- What—what we need to accomplish
- Where—the location to be affected
- When—the time frame to complete
- Which—the requirements and constraints
- Why—the purpose or benefits for accomplishing the goal

SMART goals are measurable. They establish the criteria to use in measuring progress toward attainment. *Attainment* is how you will make accomplishment a reality. To be realistic, the goal should be one you are able and willing to work on. The goal should have an end point. A goal without a time frame or key milestones associated with it is one that is perceived to have no sense of urgency.

An example of goal that is SMART:

> Select a product that offers at least five multiple layers of notification for each of the primary target audiences defined in the Emergency Communication Plan. This product must be selected using existing systems and infrastructure. This system should be fully implemented and operational by June 1, 201X. This system will enable ABC Company emergency management team personnel to gather information and intelligence using a secure system.

Problem Statement

Assumptions

The project should have assumptions clearly stated that are associated with the development and implementation of the system. Assumptions should include future situations and the current state that are beyond the control of the project. The outcomes from these assumptions are known to influence the success of a project. Examples of assumptions are

- Pending legislations or changes in industry standards
- Any pending litigation related to the project
- Hardware or software availability
- Changes in technology
- Future trends in demographic changes of the target audience
- Significant organizational changes.

Constraints

The development of a system will have constraints. *Constraints* are conditions that are outside the control of the project that limit design alternatives. They exist because of real business conditions or technical limitations. Constraints are broadly categorized as either technical or nontechnical. Examples of constraints are

- Nontechnical
  - Government regulations
  - Strategic decisions

- Technical
  - Technical standards imposed on the solution (the use of a specific Database Management System).
  - The application must use a layered architecture that can only access the layer directly below it.
  - Implementation—all software components of the application shall be programmed using Java.
  - Testing—the application shall include built-in self-test software that automatically and continuously test the component while it is in operations.

Context

The concept of system is provided using as simple a diagram as possible. If you review many U.S. patents, you will find that most are submitted and approved using simple handdrawn or Visio-style diagrams. The *context* of a system depicts the connections and relationships between the system and the outside world as illustrated in Figure 11.4 (Context diagram).

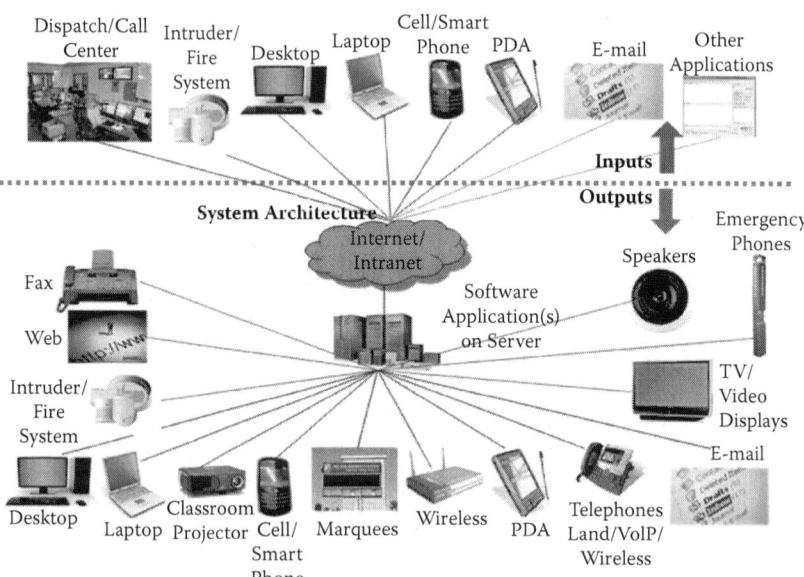

FIGURE 11.4 Context diagram.

Budget

When moving forward with a project, it is important to consider the costs associated with it. This is true even if you are using only existing resources to develop your organization's mass notification system. People are pulled away from their regular assignments to integrate or develop processes for your system. This costs money.

The decision to invest in new technologies requires decision making. The market for emergency communications is dynamic, and changes rapidly as experience, new entrants, intangible or unquantifiable factors, and new information emerges. In developing a budget, three factors influences it—project risks, solutions' fit with operations, and other factors. Other factors include strategic issues, the need to maintain expertise, and employee morale.

There may also be a need to consider the return that the project will or should produce. Accounting or financial groups may question whether the proposed investment and associated expenses can be recovered over time. The second question is if the return is sufficiently attractive in view of risks involved and potential alternative solutions. Since patterns of investment and cash flow can be quite different from project to project, there is no single method for performing a project analysis that is ideal for all cases. Methods commonly used are

- Selecting between alternative designs for the right solution
- Choosing between an outsourced, in-house, or hybrid solutions

The steps involved in developing the budget is to make a decision that first recognizes and formulates the problem, considers alternatives, analyses what is available, and then allocates funds based on the information in hand.

In addition to economic factors, noneconomic factors need to be included in the analysis. Some of the most common nonmonetary or intangible factors to consider are

- Ethical and social values
- User likes and dislikes
- Target audiences
- Environmental impact
- Government regulations

Emergency communications solutions for organizations are selected focused on containing the costs to implement and maintain over maximizing profits. Factors that usually lead to cost containment are

- Meeting community expectations
- Maintenance of a desired public image
- Improvement of safety in operations
- Reduction in emergency communication delay
- Maintaining the flexibility to use new technology

Variables to include in the budget are

- Software License
    - Pricing varies widely by vendor.
    - More functional modules (e.g., distribution and a high number of end users logging onto the system). You can expect to be at the higher end of the cost range.
    - Multiple locations tend to increase this cost.
- Database License
    - Usually based on the number of end users that log into the system.
    - More users, you will typically pay more.
    - If you already own a license for the database that the selected software runs on, you probably can eliminate this line item.
    - If you do not have your own license that you can plug in to your budget consider this as a starting point (Tier 1 usually offers the most features, serves a larger set of users; Tier 5 is usually the basic offering with a minimum amount of service and a small group of users):
        - Tier 1 ($75K)
        - Tier 2 ($50K)
        - Tier 3 ($20K)
        - Tier 4 ($10K)
        - Tier 5 (None)
- Hardware
    - Possibility of new hardware for the server, any client machines, and infrastructure upgrades (scanners, wiring, etc.).
    - If your hardware is current and select a system that is compatible with your current infrastructure.
    - If you are moving to a new server platform, there will be a larger impact to project costs.
    - Add the costs associated with redundancy and uptime requirements needed (such as mirrored servers or mirrored drives within a server).
        - Tier 1 ($100K)
        - Tier 2 ($50K)
        - Tier 3 ($30K)

- Tier 4 ($10K)
- Tier 5 ($2K)
- Implementation Services
  - The biggest component of your software budget
  - Difficult to estimate. Factors that influence costs are
    - Project management
    - Consultant's expertise
    - The amount of business process reengineering required
    - Hourly rates (the faster you implement, the lower the cost)
  - For early budget estimates, use a ratio to the software license cost. For every dollar of software license cost, you can expect to pay $X for implementation. Adjust the ratio for the complexity of your situation and the length of time you anticipate it will take to implement.
    - Tiers 1 and 2 (2:1 Ratio)
    - Tiers 3 and 4 (1:1 Ratio)
    - Tier 5 (None, if self-implemented)
- Training
  - Various training options (Computer Based, Train-the-Trainer, Off- and On-Site). Your selection will affect your budget.
  - For early budget estimates, use a percentage of the estimated software license.
    - Tiers 1 and 2 (15% of Software License)
    - Tiers 3 and 4 (10% of Software License)
    - Tier 5 (None if self-trained)
- Annual Maintenance
  - An annual fee you pay the software vendor.
  - May include upgrades and high-level technical support.
  - Understand the services that are included—you may need to purchase additional support.
  - For planning purposes, use 20%; expect to pay 18%–25% of the software license cost annually.
  - If Tier 5, you will only purchase updates as needed.
  - Plan to start paying annual maintenance in the first year unless you negotiate differently.
- Internal Costs
  - Will vary by organization.
  - Calculate an estimated billing rate and the number of hours allocated for the project.
  - Other factors you can include are
    - Duplicate work effort (e.g., data entry during testing)
    - Meetings
    - Lost productivity during the project

- Facilities—temporary work space
- Travel expenses
- Subcontractors and temporary staff
- Overtime, bonuses, or other incentives

Time

The associated timeline should reflect what the budget has allocated. A short list of questions in addition to those mentioned for the budget is offered. The timeline should include a review asking:

- How much time it will take to accomplish project duration?
- Have you identified all the resources needed?
- Will their existing work be reassigned to other resources or deferred?
- Will the project need to allocate time to release staff for the project?
- Are there any anticipated delays in the receipt of any items ordered or services to be delivered?
- Are there any government regulations, internal processes, or industry requirements that could require time to validate?

The timeline should include at least the following milestones:

- Project Start activities (selecting/assigning team members)
- Project Kickoff Meeting
- System Concept Development Phase
- Planning Phase
- Requirements Analysis Phase
- Design Phase
- Development Phase
- Integration and Test Phase
- Implementation Phase
  - User Acceptance
  - Go-live Date
- Operations and Maintenance Phase
- Project closeout activities (documenting lessons learned, hand-offs, project control book sign-off, file)

Solution Analysis

Prototypes

If you are considering the use of proprietary systems or beta systems, then building small models of elements and gradually expanding them for user experiences may be required. If you are using technology that has a sound track record within your industry, is off-the-shelf, and uses open-architectures that are compatible with your existing systems, then a small pilot group and a test environment may suffice.

Prototyping is favored when the following applies:

- Users do not have a grasp of the information or system capabilities required.
- User needs are rapidly changing.
- There is little experience among team members in delivering the type of system sought.
- There is a high risk in delivering the wrong system or a system that does not work.
- A high level of user involvement is required for success.
- There are multiple solutions and design strategies that must be tested.

Use Cases

Developing use cases is a time-consuming effort and can be shortened when a pilot group is selected. This is an iterative process where you work and refine your cases. Start with a small group of basic users of the system. Test sending alert messages to this group and then obtain feedback. If this is successful, then you can add more actors to your testing process from other target groups. An actor can be a system since a system plays another role in the context of your new system. The system will have interactions with other actors as your use cases evolve.

Use cases that are most likely going to occur when the system operates as planned are called *Sunny Day* use cases. These are your primary use cases and your starting point. From these you can develop *Rainy Day* use cases where you develop scenarios that may introduce problems with the system.

An alternative to use cases, which are time consuming and may be more than required, is *stress testing*. *Stress testing* an application measures whether the system will give consistent or satisfactory performance under extreme and unfavorable conditions. It may involve using the system while experiencing heavy network traffic, heavy processes load, working under maximum requests for resource utilization or in the system or in interfacing with other devices, etc. Stress testing can save you significant time in moving forward with the selection of a solution.

Functional Requirements

Functional requirements paint a picture of the core functionality of the application. This section includes the data and functional process requirements, and your specified or derived requirements. Process requirements describe what the application must do. These are the attributes from the data gathered from your stakeholders. The requirements are phrased with the definite word "shall" and will have a unique number assigned to it for reference.

> Example: 5.1 *The system shall provide authorized users access to project data*

Other requirements are considered "*design features.*" These are features requested by stakeholders but not a core requirement of the system. These requirements usually include the phrases "desired" or "if available."

Interface Requirements Interface requirements are the system functions that need to interact with users, hardware, software, and communication. *User interfaces* are those that are to be implemented by the system. *Hardware interfaces* are those to be supported by the system (logical, physical, and expected behavior). *Communication interfaces* define any interfaces to other systems or devices (LAN, WAN, mobile devices, etc.). *Software interfaces* are the applications the system must interface. Documentation of software interfaces includes

- The name of the application
- Application owner
- Details of the interface

Data Requirements *Data requirements* define the business data needed by the application system, not the database or fields within a database. Data elements include data entities, their decomposition, and definitions. A data dictionary may be required showing each of the data elements.

Operational Requirements This is the section where operational requirements, describing how the system will run and communicate with other systems, target audiences, and operations personnel, are defined. How these requirements will be satisfied is covered later. Areas to address in this section are

- *Security*—the need to control access to data
    - Consequences of a breach
        - Erasure or contamination of application data
        - Disclosure of sensitive or private information
        - Disclosure of intellectual property
    - Security requirements
        - If there is a need to control access to the facility housing the application

- Control access by class of user
  Control access by data attribute
- Control access based on system function
- A need for accreditation of the security measures adopted for this application
- *Audit trail* requirements of the system—list of activities that will be recorded in the applications audit trail
- *Data currency*—a measure of how recent data are
- *Reliability*—the probability that the system will be able to process all work correctly and completely without aborting
- *Recoverability*—the ability to restore function and data in the event of a failure
- *System availability*—time when the application must be available for use
- *Fault tolerance*—the ability to remain partially operational during a failure
- *General performance*—response time to queries and updates, throughput, and the expected rate of user activity (i.e., number of transactions per hour)
- *Capacity*—required capacities and expected volumes of data in business terms (5,000 e-mails per minute, 10,000 SMS within 5 minutes across the United States)
- *Data Retention*—the length of time the data must be retained

System Requirements Specification

We are now ready to develop a list of requirements. Good requirements are

- *Achievable*—with budget, time, and to expectations
- *Appropriate*—within the scope of work
- *Complete*—contains all information needed to elicit a complete response
- *Concise*—describes a single need
- *Consistent*—does not conflict with other requirements
- *Correct*—adequately reflects the desired function or performance
- *Implementation independent*—can be satisfied by more than one design and implementation, eliminating the use of specific hardware or software
- *Ranked for importance*—identifier to distinguish, that is, order, rating, weighting
- *Unambiguous*—having only one interpretation
- *Verifiable*—tested and confirmed

Requirements Specification

Most IT departments have established operating environment standards (hardware and software) and a process to follow to determine when changes in the standards should occur. These standards should be included in your list of requirement that a system must use or integrate with.

Any new application software should conform to

- Software will be tested according to a test plan and will conform to the change control process before it is implemented.
- Back-out and recovery procedures should be provided.
- The software should conform to defined
  - File allocation and naming conventions
  - Job execution class
  - Forms standards
  - Accounting fields
  - Job name standards

Server standards are ideally reviewed annually and the hardware standard configurations published. They should include the hardware configuration and operating system (OS) software.

The desktop standards within the organizations should also be included and reviewed annually. These cover hardware and software, including any security related applications. Most products are commodities with short life cycles. Any solution should take advantage of technological enhancements and feature of efficiencies. Consideration should also be given to when vendor support or maintenance has reached end-of-life or sunset and there is need to make updates.

Mass Notification Requirements Specifications—Numbered List The following is a detailed list of requirements used in selecting a solution that is a good fit for your organization. We have only listed those that are most common. Remember all those using the term "shall" are required; all others may be required or desired depending upon the needs of your organization.

1. *Security*
   1.1 *Group Structure*
      1.1.1 System allows users to be organized into groups.
      1.1.2 System to allow multiple groups to be combined into higher-level groups.
      1.1.3 Allow simultaneous operation of groups with no channel-busy delays.
      1.1.4 Device ID/user must be integrated into the system.

1.2 *Dynamic grouping*
- 1.2.1 All system devices must be capable of being reassigned to a new group dynamically.
- 1.2.2 Groups must be able to be established quickly and easily.
- 1.2.3 System administrators must be able to control and configure their groups within their departments' responsibility.
- 1.2.4 System administrators must be able to configure the attributes of the members of user groups under their control.
- 1.2.5 User configurations/profiles should be transferable between systems.

1.3 *Partners*
- 1.3.1 System must allow for interagency access (public and private agencies and partners wherever feasible).
- 1.3.2 System must allow for interagency communications (public and private agencies and partners wherever feasible).
- 1.3.3 System must be able to interface with partner communications system while maintaining constant communications and full user functionality.

1.4 *General*
- 1.4.1 System must provide for the secure (e.g., encrypted) transmission of sensitive communications and data.
- 1.4.2 Encryption key groups need to be user defined.
- 1.4.3 The ability to disable encryption keys, a desired feature.
- 1.4.4 System should be capable of passive/active attack monitoring and defense deployment.

1.5 *Authentication*
- 1.5.1 System shall require that user/device be authenticated before use of network resources.
- 1.5.2 System must require that user/device be authenticated before the use of specific network resources.
- 1.5.3 Capability shall exist to search for a user/device on the system and disable the user/device immediately if necessary.
- 1.5.4 Lead system administrators must have the capability to disable a user/device immediately.
- 1.5.5 User/device must be able to be authenticated and authorized from anywhere on the network.
- 1.5.6 User/device authorization must be tied to a role-based access control method.

1.5.7 System must be immune to attacks against the integrity of network and communications traffic.
1.5.8 System shall provide safeguards to detect and prevent unauthorized access, reading, modification, or destruction of data.
1.5.9 System shall be resistant to cyber-attacks where possible.
2. *Audit Trail*
    2.1 System shall provide selectable audible and visual out-of-coverage indications for messages sent.
    2.2 System shall have audit trails, logs, and reports.
        2.2.1 All transactions shall capture the author, date, and time.
        2.2.2 The system shall generate reports that can be exported into CVS file format.
        2.2.3 Verification of receipt by recipients.
        2.2.4 Reporting.
        2.2.5 Current status.
        2.2.6 Alert tracking.
        2.2.7 User and system activities.
3. *Data Currency*
    3.1 Data Repositories for vulnerable people, employees, students, minors
    3.2 Access to data in real time
    3.3 Access to active historical data that is at least 3 years old
4. *Reliability*
    4.1 System must provide reliable coverage with at least 99.9% uptime when portable.
    4.2 System must be able to communicate and operate regardless of location (home office, remote locations [domestic and international]).
    4.3 System must provide constant, reliable communications.
    4.4 System shall provide constant, reliable, and readable data.
    4.5 System shall provide constant and reliable access.
        4.5.1 Mean time to repair
    4.6 Routine system maintenance and testing must be able to be performed while the system is online without any noticeable degradation on the system.
    4.7 Networking software shall provide for online network expansion without changing application software.
    4.8 Battery life of backup systems must be adequate.
5. *Recoverability*
    5.1 System shall have the capability to automatically self-diagnose, detect faults, and alert, log, and report events, and realign.
    5.2 System shall incorporate state-of-the-art system management capability.

6. *System Availability*
    6.1 *Prioritization*
        6.1.1 Must assign communication priority by groups.
        6.1.2 Highest priority given to an emergency button/switch activation.
        6.1.3 System allows higher priority users to access seats, data, systems used by lower priority users when there are no vacant seats available.
        6.1.4 System must be capable of assigning and managing an adequate number of levels of priority.
        6.1.5 Priority levels shall be able to be set for each device type.
        6.1.6 Priority levels shall be able to be set for each user group.
        6.1.7 If system becomes busy, requests for service shall be placed in queue according to priority level.
        6.1.8 Requests in queue are assigned on a priority basis as seats become available.
        6.1.9 System shall provide a call-back feature when a unit receives a busy queue.
        6.1.10 If the system becomes busy, there must be a provision for user preemption, allowing the user's activities to be interrupted for an emergency.
        6.1.11 Selective alerting shall be provided as a means for alerting individual users or groups.
        6.1.12 System shall be capable of assigning seats in real time automatically.
    6.2 System shall provide for continuous availability and integrity without human intervention.
7. *Fault Tolerance*
    7.1 The system must provide for reliable networking and communications during a primary system failure.
        7.1.1 The mean time between failures (MTBF) must be at a minimum for the system.
    7.2 System design should include redundancy and fault tolerance.
    7.3 In the event of a system failure, the system shall provide automatic rerouting without operator intervention.
        7.3.1 System restarts and failovers must be accomplished in minimal time.
    7.4 In the event of infrastructure failure, network/system administrators must be capable of setting up an ad hoc network for emergency communications.
    7.5 System must operate through commercial power failures.

7.6 System must be capable of surviving and maintaining sustained operations during times of natural disaster.
7.7 Any hardware must be durable, meet industry standards, and take into account the climate in this area.
7.8 Vendor must guarantee the availability of equipment, parts, and software support for a reasonable period of time.
7.9 System must be capable of operating without harmful interference, desensitization, or degradation of performance in hostile environments (systems operating on backup networks).

8. *General Performance*
   8.1 System must integrate with current organizational systems—e-mail, voicemail, digital displays, and networked attached computers and projectors.
   8.2 Existing infrastructure should be taken into account, but not be the sole factor in planning where to locate hardware/software.
   8.3 The system must support GPS that are accurate within 1 meter.
   8.4 System must be capable of analyzing a received wireless signal to derive user location when no location data is available within the transmission content.
   8.5 System must be capable of supporting a variety of passive and active sensors that transmit data burst at periodic and nonperiodic intervals.
   8.6 System shall support satellite communications.
   8.7 System shall support Voice over IP (VoIP).
   8.8 System shall support broadband.
   8.9 System shall support connectivity to other communication networks.
   8.10 System shall be able to send alerts using multiple channels for mass notification.
      8.10.1 Network attached computers
      8.10.2 Digital displays
      8.10.3 Phones (landline, mobile, satellite, cable, VoIP, Internet)
      8.10.4 Fax
      8.10.5 Voicemail
      8.10.6 E-mail
      8.10.7 SMS, MMS
      8.10.8 Flash messages
      8.10.9 Website updates

- 8.10.10 Video
- 8.10.11 Two-way radios, including walkie-talkie
- 8.10.12 Public media (Television, Newspapers)
- 8.10.13 Public address systems

8.11 *System core features*
- 8.11.1 Automated alerts, instructions, and advisories
  - 8.11.1.1 Automated data-driven call solution
  - 8.11.1.2 Manual
- 8.11.2 Targeted notification and delivery by users, groups, and location
- 8.11.3 Centralized and distributed control
- 8.11.4 Control via a secure connection over the Internet
- 8.11.5 Automated or dynamic triggers for ongoing communications
- 8.11.6 Support for multiple languages
- 8.11.7 Global reach
- 8.11.8 Web-based incident management tool that contains all decisions, action plans, pictures, etc.
- 8.11.9 Incident call summaries that are sent via e-mail and can be delayed in their delivery

8.12 *Training*
- 8.12.1 Training to include train-the-trainer.
- 8.12.2 Training shall be provided for all personnel.
- 8.12.3 Training video or podcast shall be available for all users.
- 8.12.4 User training manual shall be available for all users online.

8.13 *Alert types*
- 8.13.1 Location-specific evacuation instructions
- 8.13.2 Lockdown instructions
- 8.13.3 Hazardous material and fire alerts
- 8.13.4 First responder activation
- 8.13.5 Severe weather
- 8.13.6 "Code" alerts (medical, crime in progress, AMBER, SILVER, etc.)
- 8.13.7 IT alerts for virus warning, cyber-attacks, e-mail outages
- 8.13.8 Business continuity updates for critical systems and power outages
- 8.13.9 Testing of systems
- 8.13.10 False alarms
- 8.13.11 Emergency road closures and traffic information
- 8.13.12 Dynamic and easily customizable alerts

9. *Capacity*
   9.1 System must be capable of assigning and managing an adequate number of levels of priority.
   9.2 Growth
      9.2.1 System assignments must be adequate to meet coordinate user needs.
      9.2.2 System must be capable of future groups (e.g., expand system capacity to meet peak period workload requirements).
      9.2.3 System must be flexible enough to meet organization needs and unusual occurrence needs.
   9.3 *Partners/Division*
      9.3.1 System must allow for establishing as many private subgroups necessary for admin or tactical operational needs.
   9.4 *Operations*
      9.4.1 Upon activation of a Unified Command structure, system must allow users from partner agency groups to be established "on the fly."
   9.5 System must have adequate capacity and bandwidth to support the anticipated amount and types of communications.
10. *Data Retention*
    10.1 Information must be available upon request.
    10.2 Critical data must be available to push to specified users.
    10.3 The useful life of data and records shall be at least 3 years.
11. *Other*
    11.1 Critical dead spot areas, to include critical coverage of buildings, must be addressed in new system design
    11.2 System must be able to integrate with systems, providing reliable in/out radio spectrum coverage in previously identified densely populated or critical activity areas of the organization/community.
    11.3 System access must extend to the maximum extent possible (land, sea, air) to access mobile devices in these areas.
    11.4 Alphanumeric paging is a desired feature.
    11.5 Selective alerting (medium).
    11.6 *Ease of Use*
       11.6.1 Operation must be simple and convenient for the user.
       11.6.2 User must be able to operate the system when using adaptive technology such as screen readers, text-to-voice commands, etc.
       11.6.3 Ability to enable LED/alert tones needs that are programmable or user selectable.

11.7 The useful life of the system shall be at least 15 years.
11.8 At a minimum, the system shall have
    11.8.1 An adequate number of modes
    11.8.2 Be capable of digital/analog operations
    11.8.3 Groups
    11.8.4 Group alerts
    11.8.5 Alert encode/decode
    11.8.6 Dynamic regrouping
    11.8.7 Point-to-point IDs
    11.8.8 Emergency alarm
    11.8.9 Out-of range indicator (audible and visual) for messages sent and communications
    11.8.10 Can integrate with touch-screen technology
    11.8.11 Encryption on/off for selected devices
11.9 System shall not be proprietary.
11.10 *Compatibility*
    11.10.1 Seamless migration from old systems must be possible.
    11.10.2 System must provide as much backward compatibility with prior implements as is cost effective and feasibly efficient.
11.11 *Laws, Regulations, Rules*
    11.11.1 All systems and associated hardware/software must meet or exceed all applicable federal and state rules and regulations and standards.
11.12 *Easy and rapid deployment*
11.13 *Easy updates*
11.14 *Devices*
    11.14.1 Single-button alerts activated from embedded alarm systems, desktops, phones, wireless devices
    11.14.2 Multiple channels of contact and communications
    11.14.3 Automated alerts, instructions, and advisories triggered by other systems
    11.14.4 Instant notification and targeted delivery
    11.14.5 Each alert can contain a different message for separate groups based on role/location

EFFECTIVE GOVERNANCE OF THE PROJECT

This section describes the processes for providing effective governance of the system to support the business' operations and strategy for emergency communications. The governance model defines the relationship between the vendor and the business at strategic, functional, and operational levels. It is intended to accomplish

- Strategic
  - Set IT goals.
  - Review the business and vendor relationship.
  - Provide a forum for senior executive exchange.
- Functional
  - Develop and review the technology plan for the business as it relates to emergency communications.
  - Review regularly published status and performance.
  - Define the direction going forward for the solution.
  - Conduct project reviews.
- Operational
  - Support emergency response teams, EOC, public information offices, and the gathering of information and intelligence.
  - Review and resolve problems at least weekly.
  - Review changes and change status weekly.
  - Manage ongoing change requests.
  - Manage expectations of solution.

One of the first steps is to define who the members of the governance team are. Members should include representatives from management and assume the following responsibilities:

- Manage the prioritization of business and IT requirements.
- Act as the liaison between the business and the vendor.
- Own the budget.
- Mediate and resolve conflicting business and IT priorities.
- Approve any changes to the corporate technology plan that affect the ongoing delivery of the solution.
- Communicate and gain commitment of end users to support IT plans, including requirements definitions, table setup, testing, education, and data clear or conversion.
- Own the business resources needed to successfully implement and maintain the solution.

SOLUTION ANALYSIS

Risks and Benefits

There is a close connection between decision and risk analysis. Risk analysis should make a valuable contribution to decision making for the project. Dimensions of risks are

- *Risk identification*—what are the most critical variables?
- *The inability to forecast project performance*—what chances are there of a very unfavorable outcome?
- *Risk varies with the level of the organization*—risk needs to be interpreted in relation to the project and the total business context.

The process to use for risk determination is to use a logical sequence of steps for handling risks, such as

- *Risk identification*—develop an understanding of the nature and impact of risk on the current and potential future activities of the organization.
- *Risk measurement*—the assessment and classification of risk situations.
- *Risk evaluation*—the judgment about actions to handle risk and the possible need to reevaluate your options.

There are various types of risks to consider for IT-related projects that make risk assessment difficult. Risk types include

- *Business risk*—a function of the organization's normal operations
- *Financial risk*—a function of the capital structure and the buying decision, such as outsource, in-house, or hybrid
  - *Outsourcing or Hybrid Benefits*
    - Inability of organization to invest in capital needed for the project (equipment, resources, facilities)
    - Fixed rate
    - Payment from pretax rather than after-tax earnings
    - Hedge against inflation (future costs to be paid in inflated dollars)
    - Payments can be coordinated with cash flow
    - Longer terms
    - Convenience
    - Level payment
    - 100% financing
    - Capital acquisitions can be amortized
    - Budget limitations
  - *Outsourcing or Hybrid Risks*
    - Overall cost
    - Prestige of ownership
    - Flexibility
- *IT risk*—decision to select a new solution almost invariably to be made on the basis of insufficient information
- *Implementation risk*—associated with the successful implementation and completion of the project with the prevention of

budget overruns, longer-than-planned implementation timeline, or having a solution that is below expectations.
- *Natural disaster risk*—risk associated with natural disasters such as earthquakes, floods, or severe weather that could lead to unforeseen delays.
- *Human judgment risk*—selection includes use of previous experience; however, due to rapid changes in the industry, much of the previous experience becomes less meaningful.
- *Other risk factors*—commercial, political, and geographical.

As there is a close connection between decision and risk analysis, the same holds true for decision and benefit analysis. What may be a risk in one organization is a benefit to another. For example, the flexibility of an existing public safety radio system since it is using 700 MHz is a benefit, or the inflexibility of the existing radio system since it is using 800 MHz is a risk since 800 MHz has been reassigned. Other benefits could be

- Having a single platform for all emergency communications and notification
- Ability to quickly integrate external partners into the system "on-the-fly"
- Increased likelihood of stakeholders' needs being met in terms of successful emergency communications
- Activity easier to track, including audit logs of all messages sent and when received, users activating messages, and tracking time intervals associated with crafting and sending communications
- Updating of crisis communications plans as lesson learned are incorporated
- Maintaining image of the organization

The list will be unique to your organization and project. Working with the project team, asking each member to provide his or her top three benefits for the project will help you to develop a comprehensive list for the project and the Design Brief.

SUMMARY

Your goal is to have an effective crisis communication system that is able to advise the target audience quickly. You have an EOC operation and a team able to work collaboratively. The model you want to follow mirrors, where applicable, the NIEM so that you are in alignment with your many partners. You also seek a system that strikes a balance between information "underload" and "overload," the need for efficient processes, and quality data. You start with interviewing stakeholders to learn what their perceived needs are. Next, you define your target audiences and what each needs from your organization when an emergency occurs. Working with IT, you begin to learn what your current state is in terms of your communication requirements. There is an emphasis on bandwidth limitations, latency, and your current call activity. This helps in knowing what your existing systems can handle in terms of call volumes and messaging activities. The design brief is the start of documenting your goals, needs, and expected outcomes. It includes your list of assumptions, constraints, drawings, the project's budget, and timeline. The next area of the design brief covers any prototypes and use cases you select, functional requirements, and system requirements specification. Almost every project will have bumps along the way. To keep it at a minimum, an effective governance model is recommended, which outlines an agreed-upon approach for managing change and conflict. The solution analysis concludes with the risks and benefits associated with the project. You are ready to review all the possible solutions available to you and narrow your choices down to just two or three.

ENDNOTE

1. Bill Gates. (2006). Beyond Business Intelligence: Delivering a Comprehensive Approach to Enterprise Information Management (http://www.microsoft.com/mscorp/execmail/2006/05-17eim.mspx) (accessed May 1, 2011).

CHAPTER 12

Picking a Solution That Fits

> When budget dollars are tight ... leaders must look to technology to make the utilization of those dollars more efficient and more productive.
>
> —Tom Pauken
> *Chairman, Texas Workforce Commission, 2010*

You are looking to succeed in selecting a solution that fits your budget, meets expectations, and can be delivered on time. This is a tall order yet achievable. Taking the time to gather your requirements, conducting adequate testing, and effective management of the project from start to finish can improve your chances of success.

Bob Lawhorn of CA, Inc., gives presentations on software failure. In March 2010, he cited these startling statistics when businesses face project issues:

- 66% project failure rate is contributed to poorly defined applications because of miscommunication between business and IT (Forrester Research).
- $30 billion or more every year is what U.S. businesses lose to poorly defined applications cost (Forrester Research).
- 60% to 80% of project failures can be attributed directly to poor requirements gathering, analysis, and management (Meta Group).
- 50% are rolled back out of production (Gartner).
- 40% of problems are found by end users (Gartner).
- 25% to 40% of all spending on projects is wasted as a result of rework (Carnegie Mellon).
- Up to 80% of budgets are consumed fixing self-inflicted problems (Dynamic Markets Limited 2007 Study).

This chapter will present information assuming you are selecting a combined alerting system—software enabling delivery to multiple platforms. The primary reasons for this assumption are

- Best practices indicate a need for multiple communication medias.
- Technology may fail when you need it most.
- Motivation to use commercial-off-the-shelf (COTS) solutions over in-house developed solutions. COTS includes all development efforts required to interconnect different products.
- The need to accommodate uncertainty.
- You have time to select the right solution given your requirements.
- Reduces overhead, risk, work, confusion, and lowers the total cost of ownership (TCO).
- According to Gartner, TCO incorporates a holistic view of an IT environment, including all elements that affect the IT environment whether obvious or subtle. TCO tracks tangible assets and resources.[1]

FINANCIAL CONSIDERATIONS

Financial consideration is an area worth further discussion. Many organizations select a product based on best price as their primary factor. In the case of commodity items such as desktop PCs or printers, this may be an effective approach for your organization. Some solutions should have other factors possessing a greater influence or at least weighted higher during the selection process. Other factors to place above financial considerations are

- Public health and safety and the protection of property goals
- Quality of product and availability
- Vendor guarantees
- Domain coverage

Financial considerations should not be a barrier to the selection of the right solution for your organization. Two products being equal, then, yes, the solution with the lower TCO is probably the best choice. However, if there is a significant difference between the solution desired and the one your project budget can buy, the temptation is to forgo the better solution you think you can afford. Be careful. It is so important to review the cost of a product for its life cycle—beyond your cost to buy, implement, and use for the first year or two.

A better deal could be in hand through negotiations. Negotiations will vary from person to person and with each situation. The goal of negotiations is getting to a win–win outcome. A win–win is obtaining an agreement that meets the needs and expectation of all parties involved. A few tips in negotiating a better deal are

- Ask if you can serve as a referenceable account. This is an account the vendor can use to tell others how you value their service. Possibly, you are used as a case study they can publish, they can have prospective clients call you for a good reference, or they can refer to your organization in their literature as a customer using their product.
- Consider becoming a beta or test sight. This approach could be risky, as a tightly defined agreement is suggested regarding what is included in any testing and what will be covered if other systems or operations are disrupted.
- Financial incentives. If the product is delivered on time, to documented expectations and within budget, you may offer a percentage or two above contract; conversely, there are no extra dollars or penalties if one or more deliverables are missed.
- Use an RFP process. The process should include an overview session that all interested vendors are required to participate in or at least given extra points for participating in the overview. This gives vendors an opportunity to see who else is interested in your business within their industry and may adjust their price downward to win your business.
- Work with the vendor to develop a solution that you could jointly market to others in your industry or other organizations in your area.
- Consider a consortium strategy. Agree to buy and share the same solution. Collectively, organizations can buy; individually, they could not afford to.
- Consider lease options rather than buying.
- Some vendors will set their price high knowing they are more than likely going to accept a lower price. Tell the vendor the price is higher than you expected. A good salesperson will probably explain the features in more depth, covering the value of the product, stating you are receiving a good price, and the solution they are providing is worth more than what they are asking. In turn, explain that these are the reasons you have selected their solution, yet you have budgetary constraints. Anticipate that the salespersons may walk away—that is OK. Remember that most

salespersons have a limit, if any, to reduce the price "on the spot." They usually consult with their supervisor before making a counteroffer or electing to stay with the price. Many times, they return with a lower offer or will make their best attempt to have you raise your offer, and will be open to continuing the process. You may want to consider increasing your offer up to half of what you are willing to pay. By continuing this process and remaining patient, persistent, and polite, you are proving that you are willing to pay a fair price, and most likely you will get a good deal.

SELECTING A PRODUCT

As discussed in previous chapters, a requirements-based approach is adhering to best practices. It simplifies and adapts the technology to the business rather than the business adapting to technology. This is important when you factor in

- The constant changes in technology
- The risks associated with system or operations failures
- The requirement for new knowledge and skills
- New laws, regulations, and standards
- Budget constraints in a global economy

A requirements-based approach usually increases the amount of time needed to evaluate solutions. As the number of requirements increase, the amount of time needed increases. If you are unable to devote much time to selecting a product, the number of products you can acquire and implement are reduced as illustrated in Figure 12.1 (Solution rejection rate).

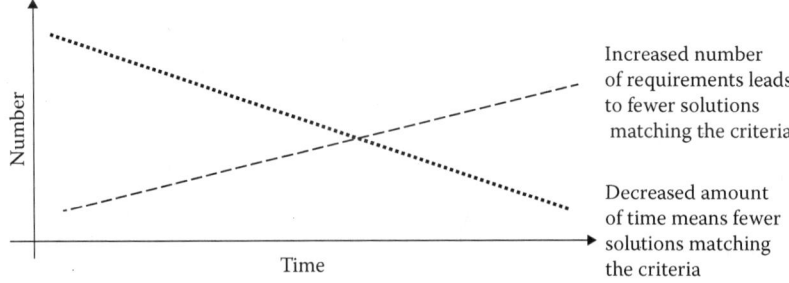

FIGURE 12.1 Solution rejection rate—time versus number.

Evaluation Process

Multiple techniques are available to assist with the decision-making process. Three common scoring techniques used are

- *Nonweighted*: All variables carry the same weight (the product offers the feature or it does not). A score of "1" is given if the product has the feature, and a "0" if it does not. The summation of all 1s is the score for the product.
  - *Benefit:* It is easy to use.
  - *Concern:* Accuracy is low.
- *Weighted*: (Each variable is prioritized first as High = 3, Medium = 2, and Low = 1). If the product has the feature, and the feature has been prioritized as "High," the score is "3"; if the product has the feature, and the feature has been prioritized as "Medium," the score is "2"; if the product has the feature, and the feature has been prioritized as "Low," the score is "1"; and if the product does not have the requirement, the score is "0." The summation of all "1s", "2s", and "3s" is the score for the product.
  - *Benefit:* It is simple to use.
  - *Concern:* Provides more confidence than the nonweighted approach and takes more time.
- *Enhanced weighted*: Each variable is prioritized first as High = 3, Medium = 2, and Low = 1. Next, each variable is assessed, indicating whether the solution was able to demonstrate (exceeds = 5, fully satisfies = 4, partially satisfies = 3, did not satisfy = 2, and not observed = 1) the stated requirement. The score for each requirement is measured as PRIORITY times ASSESSMENT equals the score. The highest possible score for a solutions is 15 (HIGH × EXCEEDS), and if the product does not have the requirement, the score is "0." The total score is the sum of scores for all variables.

$$\text{PRIORITY} \times \text{ASSESSMENT} = \text{SCORE}$$

  - *Benefit:* More confidence in decision than the previous scoring techniques.
  - *Concern:* More complex, and takes significantly more time, skill, and knowledge.

Using any of these scoring techniques for each product under consideration will provide an objective approach to product selection. The top two or three products should be provided an opportunity for further consideration. It is suggested that you do not stop with only the product having the top score. As you go through the acceptance process and continue price negotiations, you may find that issues or risks not previously identified may occur and require further review. Legal issues with vendor agreements are often the first place where issues may occur in the acceptance phase.

Acceptance

The next step before making your final selection is a review of all licenses and negotiation of legal issues with the vendor. Each licensing situation is unique, but there is core information any technology license should specify. These are

- License grant—extent of rights
- Payments—commercial and financial considerations
- Jurisdictions
- License ownership—who owns the product
- Risks and liabilities assumed by each party
- General considerations
    - Disputes
    - Government regulations
    - Warranties
    - Waivers
    - Licensor/licensee obligations
    - Force majeure
- Privacy, confidentiality, and secrecy

Technology is owned by someone who has the intellectual property rights unless it has become part of the public domain and freely available for use. The acceptance and use of technology in that part of the public domain does present inherent security and other unknown risks. Any organization should carefully consider if this is the right approach for your organization. Many social media tools such as Facebook, Google Mail, Yahoo IM, and Twitter are free, used extensively, and riddled with security issues and constant changes that could have dire consequences for your organization.

At this stage in the process an external resource is needed. An organization needs a competent professional with licensing expertise. Ideally, this person is a lawyer experienced in intellectual property rights and technology. Licensing provides an income source for the vendor, and you the right to use the product within the scope of the license agreement. ***Never sign what you have not thoroughly read and understood.*** Ask questions and modify the agreement. If the vendor is unwilling to make any modifications, it may be time for you to reconsider the other alternatives assessed.

Use Cases

Development of use cases should start when you have narrowed your list to no more than two or three finalists. A *use case* describes how users will perform tasks and activities by using a product. The two primary components of a use case are

- The actions a user will take to accomplish a specific task a product states it can perform
  - A sequence of interactions between the user and the product
  - Does not state the user interface
- The way the product should respond to the actions of the user

Each use case has an actor, interaction, and a desired outcome. The *actor* is the individual using the product. The *interaction* states what the user wants to do. The *desired outcome* is the goal of the user. As you go through your use cases, your list should end with one finalist and one alternate product. The alternate product is considered further if the finalist cannot address the financial and legal considerations.

Developing the use case is similar to writing a play or a program—an easy-to-understand narrative. It can be as simple or complex as needed. Engaging members of the team who were involved in the requirements development are excellent resources to build your use cases. The steps in developing use cases as stated by Edwin Kinworthy in 1997 are

- Identify those who will use the product (actors).
- Pick one to be your actor.
- Define what the actor will do using the product.
  - Each step the actor performs using the product becomes a use case.
- For each use case, determine the normal course of events when the actor is using the product.

- Describe the basic course in the description of the use case. Describe in terms of what the actor does and what the system does in response that the actor should be aware of.
- Consider an alternate course of events and add those to "extend" the use case. To *extend* a use case is to establish a relationship between the base use case and the extension use case, explaining the behaviors between the two.
- Look for what is common among the use cases. Reuse them where needed.
- Repeat steps 2 though 7 for all other actors.

A sample use case is narrated in the following text and illustrated in Figure 12.2 (Sample use case diagram).

Combined alerting system use case
  1. *Actors*
     a. *Public Information Officer (PIO), Incident Commander (IC), IT Department Security, Member of Target Audience*
  2. *Preconditions*
     a. *PIO has logged on to the system and selected "List of Messages" option.*
  3. *Basic course*
     a. *System retrieves PIO's account and available prescripted messages from the system database.*

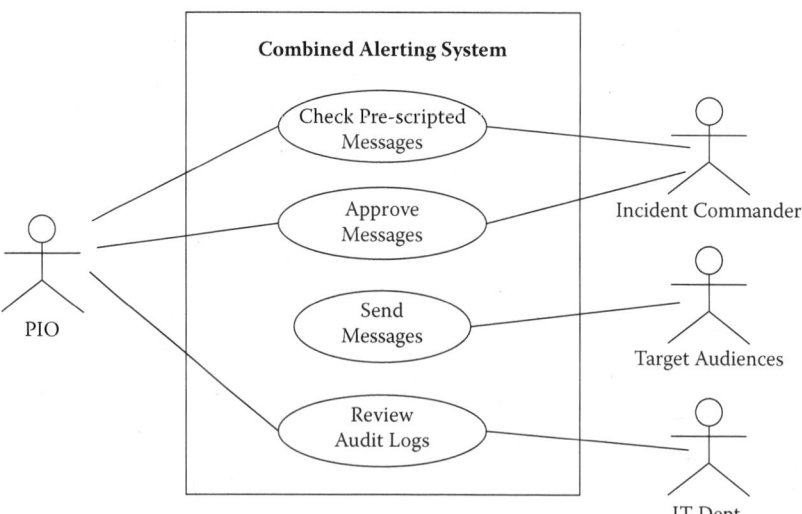

FIGURE 12.2 Sample use case diagram.

    b. *System asks PIO to select a group of messages he or she wants to change; PIO selects a message within the group.*
    c. *System asks PIO for any changes to messages, and time to send and duration; PIO provides information.*
    d. *If changes are accepted, then the system asks PIO to select the communication mediums he or she wants to use.*
    e. *If request accepted, then the system asks PIO to select target audiences.*
    f. *If request accepted, then the system asks PIO to select request for IC approval.*
    g. *When the request is approved, then the system will send the message to the target audience.*
    h. *System displays transaction summary.*
    i. *PIO requests to view the Audit Log for the message sent in the System.*
    j. *(Then other work may continue that does not affect the use case or its outcome.)*
4. *Alternative courses*
    a. *If no desired prescripted messages are available, then the system should ask if you want to create a message.*
    b. *PIO creates message, then continues, and then return to Step 3.c.*
    c. *(Same as 3.j, above.)*

Test Plan

A *test plan* defines the unit, integration, system, regression, and Client Acceptance testing approach. The test scope includes the testing of all functional, application performance, security, and use case requirements listed in your use case documents, your quality requirements, and end-to-end testing, including interfaces of all systems that interact with the product chosen.

The primary objective of system testing is to ensure that the system satisfies all requirements, including metrics using the use case scenarios. Quality of the product should be maintained throughout the testing process. Upon conclusion of testing, the users know whether the product has met expectations as defined in your Design Brief. Remember that the design brief contained your requirements, budget, concept, use cases, and other documents used in making your decision as discussed in Chapter 11. System testing also reveals issues (bugs) and risks as all areas of the system are scrutinized and bugs found are addressed appropriately.

The test plan will include assumptions about testing. Assumptions may include

- Other test types have been conducted such as unit testing and integration testing, and requirements were met.
- User acceptance testing is performed last by end users.
- Any test results are reported on a regular and frequent basis, for example, daily with any failures reported to the vendor.
- All use cases and test scripts are developed and approved.
- All significant dependencies are reported as learned.

Test Types

Your IT department typically conducts multiple test types that may be applicable to your project and before product approval. Common test types include usability, unit, and iteration/regression testing. *Usability testing* ensures that new components and features function as stated and are acceptable to the customer. The testing is conducted by simulated end users or end users themselves. They are nontesters, not a part of the IT department.

Unit testing is conducted by IT during code development. The purpose of this test is to ensure that code coverage and functionality are achieved during coding. Sign-off is needed before proceeding to the next stage of testing. Areas where unit testing is conducted are database conversions and development, and executables (*.dll, *.exe).

Iteration/regression testing is the repeated cycles of where bugs are found and new builds (bug fixes code) are introduced. This phase of testing should contain two or three cycles or iterations after any new builds. A debriefing is conducted after each iteration. This testing would involve the testing of any installation instructions under different conditions (manual install, automated install—packaged and pushed out across the network), and any documentation that comes with the product (training manuals, administrator manuals, user guides).

User acceptance testing is also called *final release testing*. This phase of testing comes after the successful completion of all other testing and is conducted by the test team and a group of end users. A successful user acceptance test means that the product is acceptable to the customer, known issues have been addressed, and the product is ready for release and installation.

Test Deliverables

Testing documentation, test cases with any bug write-ups, and reports are the deliverables from the system test plan. Figure 12.3 (Testing dependencies) is a diagram illustrating the dependencies among these deliverables and progression as you move through the process.

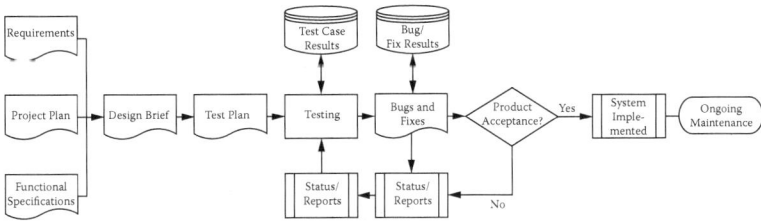

FIGURE 12.3   Testing dependencies.

Testing the Environment and Resources

The testing toolkit contains a few items in addition to what the IT department and testers already have. They are

- Testing resources:
    - Tracking tools, such as a tracker, are used to enter bugs and issues and track them throughout the testing process.
    - Configuration management is the management of evolving software systems. It includes the identification and definitions of source code, test cases, and requirements specifications. It includes managing changes during the software life cycle and tracks to status on all items including change requests.
- Testing environment:
    - Hardware
        - The minimum hardware requirements needed to use the application.
        - Access to a server in a test environment separated from the production environment.
        - An adequate number of PC workstations having different configurations using the information contained in the Design Brief.
    - Software
        - The minimum software is the standard base load on the computers in your operations. This usually includes an operating system (OS), office tools, e-mail, testing tool server software, and other products designed by your IT department. 2

SYSTEM INSTALLATION AND IMPLEMENTATION

System installation and implementation is the process of moving from an old system to a new one. It requires the conversion of data from the

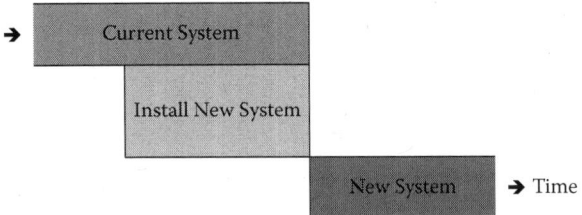

FIGURE 12.4 Direct installation.

current system for use in the new system. At this time, your IT department has received all of the appropriate approvals and has begun scheduling the install. Those involved in the selection of the solution who are not a member of the IT department are kept informed of the progress through project reports and members. IT and the vendor are working together to deliver the solution so that you can begin to use it.

Multiple approaches may be used to install a system and then implement or "go live" with the system. There are four defined approaches used to complete the installation process: direct, parallel, single location, and phased installation.

Direct installation is when you cut over from the current system to the new system with no overlap. The new system is installed, while the current system is still live (in production). This approach to installation is called the *big bang* since the old system goes away once the new system is live (see Figure 12.4 Direct installation). This approach presents the most risks yet is the least expensive. This type of installation is conducted when the same hardware will be used or when a project has identified that the cost savings outweigh the risks.

Parallel installation occurs when the current system and the new system run in parallel, at the same time, for a period. The new system is installed, while the existing system is live and in production. At the end of this time, the current system is deinstalled or decommissioned (see Figure 12.5 Parallel installation). A parallel installation is less risky than a direct installation. Conversely, it is more expensive since it may require additional resources to operate, and maintenance may still be required for the current system while it remains live and in production. This approach is also confusing for users since the temptation to use what is familiar remains on the system, and, therefore, they may use it more than the new system or they may have to perform tasks on both systems for a period of time.

The third approach is the *single location installation*. The product is installed in one location using the parallel installation model and then, once cut over, the team installs it in a second location using the same

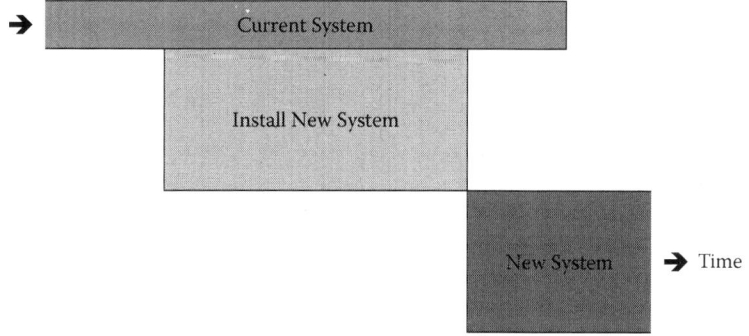

FIGURE 12.5 Parallel installation.

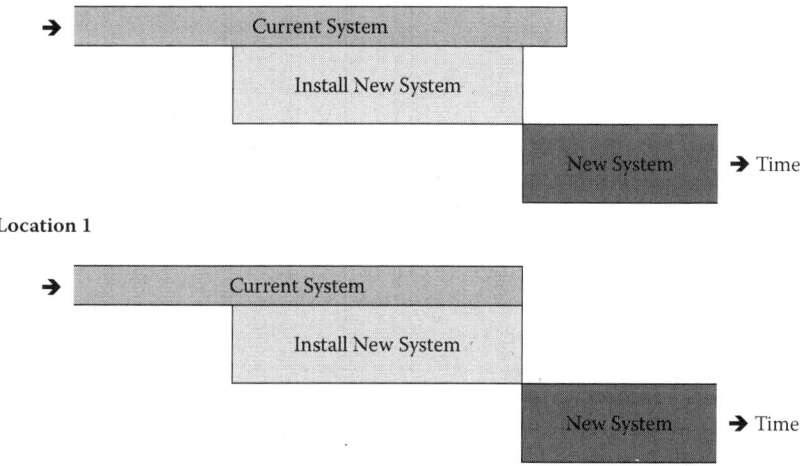

FIGURE 12.6 Single-location installation.

approach. This process is followed until all locations have the product installed (see Figure 12.6 Single-location installation). Installation of a new system in a single location is more manageable when a system must be installed in more than one location. This process will require careful version control to ensure that each location is using the appropriate version of the application while the conversion is in progress.

The fourth approach is a *phased installation*. A phased approach is a blended approach when multiple locations will have a new system installed (see Figure 12.7 Phased installation). Location 1 has the new system installed. The current system runs in parallel for a period before it is turned off or uninstalled in Location 1. The current system remains

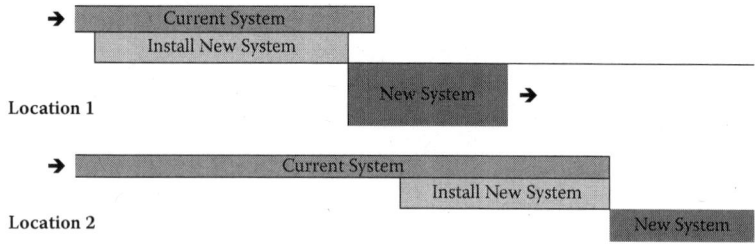

FIGURE 12.7   Phased installation.

active in all other locations. As the new installation is done at the second location, the current system remains live for a period after it has been installed and runs in parallel with the new system for a defined period. This process continues with the current system running until all locations have completed the new installs and run in parallel. At the close of the last location, the current system is decommissioned or uninstalled. This process may appear on the surface to be the most expensive; it presents the least amount of risk when installing an application at multiple locations.

Implementation Success or Failure

Unfortunately, not all projects run smoothly, and some fail. Bob Lawhorn of CAI reported that a 66% project failure rate is due to poorly defined applications. Some would say this occurs because management support and user involvement is lacking. The reasons could be many; however, a study conducted in the United Kingdom as reported in the *International Journal of Information Management* found that the main cause of implementation failures was change. These changes occurred because of

- New business/strategic development
- New technology
- Organizational changes
- Original specification is inadequate
- Government/legal changes
- External factors, for example, suppliers, customers
- New policies, for example, security, financial cutbacks
- Original specification is not property implemented
- Personnel changes[3]

To turn around a failed project and make it successful, it may be time to take a "pause." This is an opportunity to review all that you have and to identify gaps. Start with the design brief and look at the changes that have occurred since it was initially developed. Address the question "how have you adapted to these changes?" The second must be "how do we overcome the issues?"

Tips to use in fixing a failed implementation are

- Learn what the priorities are for the organization. They may have changed. Have these changes affected your project?
- Revisit the requirements documentation and design brief. Did you identify all of the functionalities you needed in the application? Did the review process correctly capture the features and functions of the product? Was scoring applied based solely on an abbreviated product overview or a detailed review of the product against the matrix?
- Are you having problems that a "quick fix" could buy you time to develop a detailed resolution? "Quick fixes" are not recommended with technology because they can sometimes introduce new problems, such as processing errors. If you can apply a quick fix, then work diligently on a long-term solution and update your project plans.
- Was user training adequate? For someone who is computer savvy, user training may not have been a requirement as would be for others. A manual and a computer can get the job done. For others, they learn best with classroom instructions. Consider offering additional user training modifying the materials user, extending the instructional time of the class, or a new trainer that understands your business.
- Did workflow processes change for employees? If so, were employees given adequate time to adjust to the change? Maybe you used the direct installation approach to stay within budget, but the risk was that the old system was no longer available to employees. A learning curve was not considered. Consider modifying the installation process.
- Were the issues due to personnel problems within or among departments? Were there significant unplanned absences, overworked or stressed staff, political battles, turf wars, or hidden agendas? Meeting with all involved, conducting a confidential online survey and used effective communications may help. Maybe additional resources or rotating the resources you have may help. In the world of "you're fired," some are afraid to say there were problems early on because they did not want the blame. Tread slowly in terminating an employee who was on the project as it may affect the morale of the team. Did they have all the resources they needed

to be successful? Find where there is common ground and work toward team performance through team-building efforts.
- Did you take a commercial-off-the-shelf (COTS) and attempt to customize it? Maybe the customization was a stretch for the technology selected.

At some point, the project team needs to give solutions to the organization that will bring back the trust needed to proceed. This is not the time to overcommit and underdeliver. As you move forward, keep senior management informed and provide your users with regular updates.

A simple report for management covers 80 hours (2 workweeks) in bullet format for the following three areas:

- What was accomplished?
- What is planned over the next two weeks, including any planned absences?
- Hurdles faced and unplanned absences.

Have everyone on the team complete this short report and turn it in on the last day of the workweek. The project manager can consolidate the reports, meet with the team, and then share them with management. Key points should be shared with end users.

The second document is a snapshot on the health of the project. A simple one-page document containing the following is suggested:

- Stop light chart—one line for the project
  - Green (on target; no issues)
  - Yellow (slightly behind schedule or minor issues)
  - Red (significantly behind schedule or major issues)
- Issues/risk lists
  - Severity in order of high to low (1 = high, 2 = medium, 3 = low)
  - Status in order of unresolved to resolved (unresolved, ongoing, resolved)
  - Owner, date opened
  - If resolved, the resolution

User Documentation

Your product has been thoroughly tested and has successfully completed testing. The system has been installed, but wait! Where is the documentation? The installation process must include the documentation you need to operate the technology and use for end-user training. There are two audiences for the final documentation—IT and end users. The documentation needed is system and user documentation. *System documentation* is one used by those responsible for the administration and maintenance of the system. It includes the design brief, system design specification, system functionality, and internal workings. *User documentation* is records about the system, how it works, and how to use it such as your user guides and procedures manuals. It should be easy to read and follow.

Training for Users and Administrator

Training allows system administrators and end users to learn how to operate and use the technology provided. Training has many benefits, including

- Improved efficiency by minimizing downtime and errors
- Increased productivity for your organization and worker confidence

Multiple training methods should be deployed to enhance the learning experience. Formats that are effective are

- Online training courses
- Classroom sessions
- Open lab where a trainer is available to answer questions (Hybrid—online and face-to-face)
- Tailored on-site sessions specific to the department or location

When it comes to emergency communications, this is an area you do want to take for granted—that your staff knows how to use the equipment. In many areas, training is mandated, and it is not an option.

> **CASE STUDY: BILL REQUIRING MINIMUM TRAINING FOR 9-1-1 OPERATORS SEEKING SUPPORT**
>
> State Rep. Uvalde Lindsey of Arkansas in late 2010 was seeking support for a bill that would require minimum training standards for 9-1-1 operators across the state. Lindsey said, "We think everyone in every county ought to be assured that when they call 9-1-1, they're going to get someone who's trained and certified." Currently 9-1-1 operators aren't required to receive formal training.[4]

An effective crisis communications training program should prepare those assigned the responsibility of crisis communication to handle most any situation, develop real-time crisis management and messaging skills, learn to use the technology assigned to this process, and teach the fundamentals of media interviews. The use of technology begins with what each button does. That is only a beginning. For the crisis communications team, this is orientation. The technology should be included in each exercise.

This training can help the spokesperson to be better prepared. It should be interspersed with media training. This will help the spokesperson who is on the computer sending messages and then having to get in front of the mic to maintain his or her composure before the media whenever unexpected—or unwelcome—circumstances arise. Ideally, the training should also include assistance with messaging, imaging, strategy, and assessment of the risks and scope of a crisis quickly through practice. If coaching sessions are possible, they should help in focusing the response strategy on getting the message out right the first time while balancing the need to protect the organization's reputation by identifying the best response. Effective communications in a crisis can help build credibility and goodwill and turn a negative into a positive in the long term.

SUMMARY

To succeed in selecting a solution that fits your budget, meets expectations, and is delivered on time is a tall order yet achievable. Taking the time to gather your requirements, conducting adequate testing, and effective management of the project from start to finish can improve your chances of success.

Many organizations select a product based on best price as their primary factor. Financial considerations should not be a barrier to the selection of the right solution. A requirements-based approach is recommended because this method simplifies and adapts the technology to the business rather than the business adapting to technology. As you go through the acceptance process and continue price negotiations, you may find that issues or risks not previously identified may occur and require further review. Legal issues with vendor agreements are often the first place where issues may occur in the acceptance phase.

Development of use cases should start when you have narrowed your list to no more than two or three finalists. It is similar to narrating a play. Testing involves all functional, application performance, security, and use cases requirements listed in your use case documents, your quality requirements, and end-to-end testing, including interfaces of all systems that interact with the product chosen. Once tested, system installation and implementation will move you from an old system to a new one.

Unfortunately, not all projects run smoothly; some fail. In attempting turn around a failed project and make it successful, it may be time to take a "pause." In closing, process documentation is created for the system on how to operate the technology and for training. Training allows system administrators and end users to learn how to operate and use the technology provided.

ENDNOTES

1. Gartner. (2002). *Product Champion: Government Seat Management and Total Cost of Ownership.* http://www.gartner.com/4_decision_tools/measurement/measure_it_articles/2002_10/gov_tco.jsp (accessed May 1, 2011).
2. National Institute of Health. (2007). *Test Plan Template.* cabig.nci.nih.gov/workspaces/CTMS/Templates/Test... (accessed May 2, 2011).
3. Fitzgerald, G., Philippides, A., and Probert, S. (1999). "Information Systems Development, Maintenance, and Enhancement: Findings from a UK study." *International Journal of Information Management,* 19: 319–328.
4. http://www.nwaonline.com/news/2010/dec/20/emergency-service-calls-bill-targets-911-training/12/10/2010 (accessed April 5, 2011).

CHAPTER 13

Effective Communication Plans

> President Barack Obama challenged the nation to not just hang in there but rather to see the hard times as a chance to "discover great opportunity in the midst of great crisis."[1]

News cycles measured in minutes rather than hours are also 24/7. Traditional media sources are only one of several places people look to for news and instructions on what to do in an emergency. A rapid response to an event is essential to managing the event. A communication plan that includes the management of a crisis is necessary and potentially dangerous. Without a plan, indecision freezes organizations because no one for sure knows where to begin. A plan is potentially dangerous because it can give an illusion that the organization is fully prepared. Having a plan is not a guarantee that you will recover from a crisis.

WHAT THE PLAN IS AND WHY YOU SHOULD HAVE ONE

The objectives of a communications plan are to gain public confidence and keep the community calm. The communications plan gains this confidence by providing information that is (ACCEPTed):

- Accurate
- Caring
- Credible
- Empathetic
- Pertinent
- Timely

Following a good plan can help to keep stakeholders calm by providing information and instructions that

- Acknowledge an uncertainty
- Explain the process for finding answers
- Are transparent and indicate when you do not have the answers
- Give people specific actions to do
- Recognize the fears people have
- Take care not to over-reassure
- Ask more of people—to share the risk when needed

The core elements of a communication plan are

- A spokesperson selected based on order of preference by management and by hazard type
- The communication team members
- Message development
- Identification of key media contacts and resources
- Training and exercising the team members
- The process for updates to the plan

The plan is a blueprint for your organization and its reputation. With a communication team selected, the collective knowledge and experience fortifies the communication plan continuously.

The communication plan is a living document needing constant attention. People change jobs, organizations change, and new intelligence is obtained on media contacts. Constantly, new information is introduced to analyze and act on. As you incorporate crisis management into the communication plan, it becomes a strategic document with a wealth of tactical details. The communication plan's effectiveness goes beyond telling you who to go to; it also states how to ask the right questions, provides the rules of engagement with each media contact, how to manage leaks and any attackers, and what online resources are available to you.

In the light of recent natural and human-made events, political unrest, terrorist activities, earthquakes, severe weather, and even highly publicized situations of senior officials and company executives, it is important to have a plan in place. The media will flock to the story and quickly nationalize it, and even international coverage could happen. When do you make your first statement?

There are countless examples of figures waiting too long to give a credible statement. Tony Hayward, former CEO of BP U.S. operations, repeatedly stumbled during the 2010 oil spill, alienating federal and state officials and Gulf Coast residents. The spill started in April 2010. Robert Dudley replaced Hayward as CEO three months later. Tiger Woods' delays in responding to allegations he was having extramarital affairs led to him giving multiple apologies and losing major product endorsements. He first responded with brief statements and days later made a public apology. By then it was too little, too late. Former New York governor Eliot Spitzer's sex scandal story was another fumble. He was late to his own press conference by more than 30 minutes. This gave the news media waiting time to speculate and provide impromptu comments to the public while waiting. The comments were negative. He resigned shortly thereafter, following only 13 months of service. Where were their communication plans, especially their crisis communication plans?

It is important to be prepared. You never know when an event will lead to a crisis that could threaten the health and safety of members in your community or threaten the integrity or reputation of the organization. You begin with a communication plan that includes crisis communications and your day-to-day operations. Expanding on the tenets of a communication plan previously mentioned, there are seven steps to an effective plan. These steps are

Step 1: Plan, then detail your plan—its purpose and objectives
Step 2: Put your team in place
Step 3: Train and exercise
Step 4: Prepare to manage the message and the media
Step 5: Communicate early and often while maintaining transparency
Step 6: Prioritize your target audiences and tailor the message to them
Step 7: Review and update at least quarterly

PLAN, THEN DETAIL YOUR PLAN— ITS PURPOSE AND OBJECTIVES

A good offense is the best defense—anticipate a crisis and be prepared. You want to take an all-hazards approach in the development of your plan and scour resources in your industry for plans that may already exist. Many government agencies, educational and research institutions, and nonprofits publish their plans online. There is no cost to review these. Find organizations that are similar in reach as your own, for instance, yours may be an organization with one location providing a single product to consumers. Many educational institutions also have a single location providing educational service to students. Maybe yours is an industrial or manufacturing facility using hazardous products or materials as a part of a process. Many research institutions have distinctly different approaches. Review their plans. These are good places to find a template to use. You will be able to determine if you need to hire a consultant to help or if your resources can manage this phase of the process. The key is to start now—do not wait until the last minute.

Sample language for this section:

> COMPANY's crisis communications plan outlines the roles, responsibilities, and procedures that will guide the company in promptly sharing information with all of COMPANY's audiences during a crisis or emergency. The audiences for this plan include our customers, employees, and the communities we serve.
>
> This plan is a part of the company's overall Emergency Management Plan, coordinated through COMPANY's Office of Emergency Management. The communications plan will be carried out in a manner that aligns messages and operations, promoting effective communications across the community we serve.
>
> Our guiding principle will be to communicate facts as quickly as possible, updating information as circumstances change, to ensure the safety of our employees, customers, and the community, and the continued operation of essential services. Our efforts to be accurate and timely may mean that some communications are incomplete. We accept this, knowing that how we communicate in an emergency or a crisis will affect public perceptions of COMPANY. Honesty, speed, and transparency are the most effective means to avoid lasting damage to a company and widespread second-guessing by the public, which expects immediate access to accurate information. We recognize that in a crisis, people will likely expect us to have more information than we may have. It is paramount to speak with accuracy about what we know and not to speculate about details we do not know.[2]
>
> We will use multiple mediums to reach as many people as possible with accurate and timely information. In the first hours and days of an emergency or a crisis, this becomes vitally important. Our goal is to be open, accountable, and accessible to all audiences while mindful of legal and privacy concerns.
>
> For the purposes of this plan, a crisis is defined as a significant event that prompts significant and, at times, sustained media coverage and public scrutiny. The crisis has the potential to damage the reputation, image, or financial stability of COMPANY. A crisis could be affected by an emergency or a controversy. An emergency is a natural or human-made occurrence that represents an ongoing threat or other event that involves a response from emergency personnel. A controversy could be employee misconduct or protest against new legislation.[3]

PUT YOUR TEAM IN PLACE

Select members from within your organization with diverse backgrounds and representative of your stakeholders. Internal resources are members from different departments that interface with your stakeholders. At a minimum, you need the CEO or president, vice president, external affairs, public relations, and the public information officer (PIO). A representative from marketing, emergency management, IT, your webmaster, your social media guru, and legal department should be included as primary resources. Additional resources to reference are your media contacts and a multilingual resource (someone that fluent in at least the two primary languages spoken in your region).

As the head of the organization, the CEO or president sets the tone and lends creditability to messages with his or her name on it. The vice president is important. This person is usually the second highest-ranking individual within an organization and often an officer of the company. They give you the depth you need when the president is not available. It is a good idea to work on their itinerary with plans of limiting their travels together, particularly in the same vehicle.

Public affairs communicates the priorities, goals, and objectives of an organization to its stakeholders by disseminating information, media relations, publications, and other services related to requests from the media about the organization. They often serve as the PIO within the Emergency Operations Center (EOC) when activate and may manage the Joint Information Center (JIC), if it is activated. Public affairs is often responsible for interfacing with the media, providing communications on behalf of the organization on community issues, and speech writing support to key executives.

External affairs advances the goals, objectives, and priorities of the organization by building and maintaining relationships with key stakeholders, communicating the organization's strategic messages, assisting and interface with elected officials, and for some organizations, assuming responsibility for soliciting private financial support. These two areas, public affairs and external affairs, know those who can make a difference for your organization and think in terms of long-term relationships. They are often the first to represent your organization in times of a crisis until it is appropriate for the president or vice president to speak. Like your executives, they are the face of the organization, particularly in times of crisis.

A representative from marketing can assist you in constructing messages. Their role is to promote the business, so you can grow the business. When a public affairs or external affairs representative is not available, this is a superior alternative. They understand the marketplace in which you operate and have developed their own contacts, which you may need to leverage in a crisis. This increases in importance when a crisis is related to a product or service you offer.

The *emergency management coordinator* (EMC) is a good resource on the team. This person can assist you in developing message templates for specific scenarios that are highly probable in your area. They are often the maestro to ensure that the communications team is prepared to deliver and can assist in designing exercises and drills using your plan. The EMC has a list of contacts to assist in an emergency. You may need to reach out to one or more of them in a crisis, depending upon the type of event. For example, the fire chief is the Incident Commander (IC) for a fire. As IC, they are the spokesperson until the situation has been normalized. Your communications team may assist the fire chief but will not have the expertise the fire chief has in dealing with fire suppression, hazardous materials, search and rescue operations, emergency medical services, and fire prevention. He and his team have details that you most likely do not, in the early stage. The EMC can assist in coordination of a JIC to help with communications. It is critical to have the EMC available as a key resource as you develop and prepare to test your plan.

The *Information Technology Department* (IT) is often thought of as a utility service to the organization such as electricity and water. They are often excluded until there is a problem with the "utility." You have a message but how can you leverage technology to get the message out quickly to as many as possible? Quite possibly, the IT disaster recovery plan must be activated and services moved and restarted at a backup location. This could require a different way to access the "utility." The IT department can assist. They are a key resource to assist with the technology components of your plan—they know who to go to for an e-mail blast, a flash message on all your computers, or your internal TV system, for example. They can also assist you in determining what technology limitations you may face during peak performance or an outage. It is important that your plan include both "tech" and "no tech" methods for communications. IT needs to know what service levels are expected of them if your plan is activated.

The *webmaster* is the one person typically responsible for publishing information on your website. The webmaster may be a member of IT or Public Relations and usually works closely with both. This individual may be responsible for shifting to a "web lite" version of your website that you may want to activate in a crisis. The webmaster's role should be addressed in your communication plan.

Your *social media guru* is a very important resource to have as a member of your team or a key resource. This individual can assist you with posting to different sites. Equally important, they can assist you in gathering intelligence from online sources and analyze what is being said about your organization and the event online. Remember, social media sites such as Twitter and Facebook are often the first place people go to for information.

The legal department is another key resource to have for developing your communication plan. They can review and provide input into what you have developed from a legal perspective. It is important to strike a balance. They are not those you want in front of the camera or mic. The plan should include when they are involved in creating messages dynamically. Having a legal resource is a benefit; they can also slow the delivery in the dissemination of needed information as they review communiqués from an "avoid litigation" perspective.

Additional resources to reference are your media contacts. These are secondary resources. They can share with information on blunders made and pitfalls to avoid. They can also advise you on how to approach them for information and to assist you in getting your message out. Because there is no such thing as "off the record," it is best to develop a trusting relationship over time.

It is wise to have multilingual resources, someone that fluent in at least the two primary languages spoken in your area, available to you. As you craft your message, they can assist with translating it into another language. In addition to a multilingual resource, having a sign language interpreter available to assist with dissemination of your message is of great benefit. These resources can assist you in reaching communities that are often overlooked. It is wise to have an agreement in place with a translation service provider and a list of available sign language interpreters. They can assist you on-demand with the translation of your message into other languages.

Sample language for this section:

Crisis communication team members (by role):

- President/CEO
- Vice President
- Head of Public Affairs
- Head of External Affairs
- Emergency Management Coordinator or Public Information Officer
- Webmaster/Manager
- Any others appropriate for your organization (see Figure 13.1 Sample crisis communications team organizational chart)

The President, Vice President, or designee will add other team members, as appropriate given the circumstances. The President,

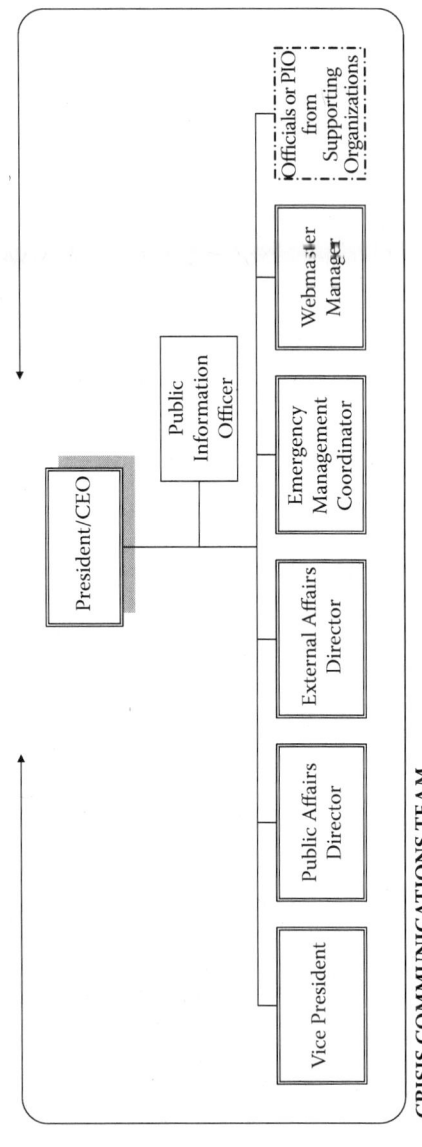

FIGURE 13.1 Sample crisis communications team organizational chart.

Vice President, or designee will contact each member by phone and e-mail to convene immediately.

The Crisis Communications Team headquarters for most crises will be the Public Affairs Conference Room, [LOCATION]. The conference room is equipped with satellite and cell phones with batteries, five dedicated landlines, laptop computers, video displays, electronic board, television, conference-call capabilities for members who cannot attend meetings, press packets, press identification badges and parking passes, and copies of this plan.

The back up location will be the second floor training room. Any training in progress may move if feasible or reschedule. Special arrangements have been made with [XYZ CONFERENCE CENTER] at [ADDRESS] if there are technical problems at the locations mentioned or need to be in proximity to the news media if they are set up on or near COMPANY property.

PREPARE TO MANAGE THE MESSAGE AND THE MEDIA

Here is where the rubber hits the road. Preparations mean answering the "W's:"

- *What* is the problem?
- *What* is the nature of the crisis?

Here is where you create the message and keep it clear and consistent. The team can assist in determining what areas you can have prescripted messages or templates to use. For examples, you have launched a version of your product that is radically different from your competitors. Because it is new, it may be prone to a higher rate of defects reported by early adopters. Going through "what-if" scenarios of how such a situation should be addressed, you can begin defining your message. When working with external stakeholders such as the media, you are at the mercy to meet their needs and requirements to have your information disseminated.

Sample language for this section:

> Our Crisis Communications Team will convene when the OEM or designee officially declares an emergency. In the event of an ongoing threat, COMPANY policy and ANY LAWS require COMPANY to make timely notifications to the community. Our goal is to make that notification within [30 MINUTES OR LESS] minutes. Depending on the nature of the emergency or crisis, it may not be possible for Emergency Response Team, the Emergency Management Coordinator, or designee to convene or make quick decisions. Given the urgency of rapid communications, the President, Vice President, or designee has the authority to begin

taking action immediately until a broader decision can be made about how COMPANY should proceed. Also the President, Vice President, or designee may identify a potential crisis or controversy that is not an immediate emergency and assemble the Crisis Communications Team to prepare a communications strategy as part of a coordinated COMPANY response. The execution of the plan will be adjusted accordingly after senior management meets and has an opportunity to determine the nature of the event.

CRISIS COMMUNICATIONS TEAM EXPECTATIONS

It is important to be proactive; a crisis will not go away if you hide. Hiding or avoiding the situation could make it significantly worse. You want to communicate early, often, and maintain transparency. When executing the plan, you should also active the team. The team convenes and designates a spokesperson. In an emergency or crisis, it is critical for a high-ranking leader of the organization—in most cases, the president—to be the organization's public face and take the lead in communicating key messages and answering questions. The IC or PIO may deliver the first message until it is appropriate for the president to step in. This demonstrates that the situation is under control and that efforts are being made to address any questions that have arisen. It also serves to calm various audiences. As the situation evolves, the senior leader acting as the key spokesperson may change. You do not want your president in front of the mic ill prepared.

The team should also select a *secretary* or *recorder*. This individual is responsible for note taking and records keeping. As the team assesses the situation, they begin developing *fact sheets*. Fact sheets are used to help determine the communication mediums to use. These are used to update websites, e-mails, news releases, and other communication channels. They will also help guide the team's overall strategy as events unfold. During a crisis, it is important to maintain an organized log of interview requests so that requests are returned promptly. It is a missed opportunity if the members of the media are unaware of your key messages and facts as understood by the organization.

The plan should include a media staging area, preferably a location that has some technological infrastructure, and place to provide food, bathing, and resting facilities, and parking for news crews and their vehicles. Members of the media can set up in the designated area with access to an interview room. It is recommended that security be assigned to prevent unauthorized access to other parts of the building or area.

It is critical that you stay aware of how your organization is being portrayed in early and ongoing coverage. The information gained can be used to adjust the communications response for rumor control, error correction, and to maintain confidence in the organization. Collectively, this step and the steps of this section determine the communication methods to use to get your first message(s) out.

Sample language for this section of the plan:

> The Crisis Communications Team will implement some, or all, of the steps defined below, given the circumstances and in coordination with the OEM. Throughout a crisis, the team will meet frequently to review updated information, assess whether key messages are reaching target audiences, and determine whether strategies need to change. Success of this plan rests on open and frequent communications between the Crisis Communications Team and the company's EOC and OEM.
>
> In an emergency, our goal is to issue our first communication to key audiences within [30 MINUTES OR LESS] minutes of notification of the event, with regular updates as needed. Some situations may require even faster initial communications.
>
> The team will do the following activities:
>
> 1. Designate a secretary that will maintain notes, to-do lists, information files, and other items throughout the event.
> 2. Review and document facts whether released or not released to the public. The information will be used to determine whether a response is appropriate and for which target audiences. The facts gathered will be used to fill in templates for news releases, text messages, and other items that have already been developed. A list of potential crisis, a list of audiences by crisis type, and who will be responsible for coordinating communications to each is included in the plan's appendix. It is critical that updated information be used for new fact sheets as the situation changes.
> 3. Develop several prescripted key messages to be included in all company communications. One message will address what COMPANY is doing to ensure the health and safety of employees, customers, and the community. A second message is forward-looking and addresses what you are doing to make sure the crisis or a problem with our response do not reoccur. All of the messages should evolve as circumstances change. All messages should aim to restore and maintain confidence and calm, balancing a sense of concern with resolve and action. Sample messages are included in the approved templates.
> 4. Determine the spokesperson, a senior leader. The senior leader will serve as the public face for the company. A second team member will run briefings and handle media questions between the formal press gatherings. A third member of the team is responsible for communicating

key messages and emerging facts to the spokesperson and addresses any last-minute media training. The senior leadership of COMPANY should share copies of the most recent news releases and other messages, so everyone is aware of what is being shared with the public.

5. Assign Crisis Communications Team members responsibilities to communicate the facts of the situation and our response to key audience. Each member will use approved messages and templates for this effort. When a crisis declaration occurs, employees assigned to the Crisis Communications Team will be relieved of their normally assigned duties to help execute this plan. It may also be necessary to have additional help. The Vice President has the authority to enlist the help of communicators from across COMPANY and assign them as needed to the crisis response.

6. Do a web response. COMPANY webmaster is responsible for the following updates to the website and will use XXX as the primary source of updates, linking from there to other pages with detailed information. COMPANY webmaster may also create web pages or sites about a situation, linking to the home page of COMPANY. A "website lite" with a "lite" homepage, and limited navigation and features has been developed and may be used as needed. The webmaster may also remove Flash features, take down images, and obtain backup website support. This support is to ensure that web effectiveness including download speeds is acceptable during peak demands.

7. Assign a communicator as needed to handle calls using a script developed from the key messages and facts the Crisis Communications Team has developed. When additional resources are needed, the Communicator(s) should reach out to other units accustomed to customer service or heavy call volumes. These employees will monitor and update the recorded message on the company's hotline. A separate log will be maintained to record all calls and interview request from the media. The Communicator and team will be responsible for ensuring that all calls are returned.

8. Develop communications for the President, as appropriate. It may be necessary for the President to communicate to the community about the emergency. Public Affairs will assign a communicator to draft presidential correspondence and any other written materials, such as speeches or talking points.

9. Approvals of outgoing information during normal operations require multiple approvals before it is distributed. COMPANY recognizes the need for rapid decision making

during a crisis to enable timely and accurate communications. We have developed templates, approved in advance by senior management and legal to expedite the approval process during a crisis. Final approval for all communications rests with the Vice President or, in his/her absence, Public Affairs or the highest-ranking member available on the Team.

10. Open the media center and determine whether press conference(s) should be held. Interviews of designated COMPANY representatives for news events and by news reporters, as well as photography, should occur in the Media Staging Area. The designated Media Staging Area may be moved to an alternate location, as needed. Public Affairs will be responsible for opening this facility, bringing needed supplies (identification badges, equipment, tables, chairs, parking passes, media guidelines, press packets), determining a staffing schedule, and coordinating press conferences and related media advisories as needed. If the emergency requires the opening of a media staging area, COMPANY will handle this as a priority. Some emergencies may require the media staging area to remain open 24 hours for an extended period. The head of Public Affairs has the authority to use communicators from other departments within the company for staffing. Shifts should be limited to 12 hours, followed by a formal handoff to relief personnel. Staff should be granted adequate time to rest and relieve stress between shifts.
11. Assign a resource to monitor media coverage to anticipate issues with the flow of information to the media. A second resource is recommended to monitor print coverage, broadcast media, and social networking postings. Daily summaries of relevant media coverage will be provided to the Crisis Communications Team.
12. Determine how we should report on the situation for internal audiences. The Vice President will be responsible for coordinating print, video and audio coverage, and working with Public Affairs of events for internal notification.
13. Assess how to help our community recover, return to normal and, if needed, regain faith in the company after the crisis is over, in coordination with senior leadership. This may include the need for community town hall meetings, letters from the President expressing sympathy, detailed plans to prevent another such crisis, etc.
14. By the tenth day after the end of the event, assess the effectiveness of this plan, update the plan as needed, and recognize the work of partners whose help was valuable.

15. The end of the crisis will be determined by the Emergency Management Coordinator or President and normal communications processes resumed. The decision to declare the emergency over will trigger an after-actions review of how the crisis was handled and how communications can improve.[4]

PRIORITIZE YOUR TARGET AUDIENCE AND TAILOR THE MESSAGE TO THEM

The process of prioritizing your target audience begins with identification of those who are internal and external, that is, directly affected, from those who may have been affected indirectly. Each audience must be addressed with specific information regarding the crisis and the appropriate plan of action put in place. Whenever possible, the first audiences to inform are internal—employees and any others on-site. Others could be dignitaries, students, contractors, vendors, or other guests. This would depend on the nature of your business. The next audiences to address are external. This may be a broad range of groups such as the media, community leaders, parents, customers, government agencies, first responders, and nongovernment organizations to name a few. The communication method used to communicate to each may be different. This will require a detailed plan to reach these audiences.

Sample language to use in this section is as follows:

> Whenever possible, the first audiences to inform are internal audiences directly affected, such as employees, [OTHER AUDIENCES] and others on-site. The next groups are external, and typically would include community leaders, [OTHER AUDIENCES], and the media. The Crisis Communications Team, depending on the circumstances, may identify other audiences and assign responsibility for them.
>
> The communication method used to communicate to each may vary by audience. A team member should be assigned to develop detailed plans to reach the designated audience. Below is a list to reference:
>
> - **Customers** (Public Relations or designee)
> - **Employees** (Vice President or designee)
> - **Contractors, Vendors** (Department Heads)
> - **Senior government leaders** (External Affairs)
> - **Community Leaders** (External Affairs)
> - **Local law enforcement** (Head of Security or Public Safety)
> - **Media** (Public Affairs)

- Federal and state leaders, agencies, and contacts (PIO)
- Visitors (Public Affairs)

Appendix B: Sample Mass Notification System Activation and Criteria Guidance Sheet includes a list of activation levels, suggestions for who has decision authority, and activation criteria.

TRAINING AND EXERCISES

The communication plan should incorporate a section for educating team members and other employees who may become involved in the process. Different forums can be used to facilitate the process. An *orientation seminar* is a low-stress informal group discussion involving little or no simulation. Additional seminars should be conducted for new employees and as membership changes within the Crisis Communications Team. This is an opportunity to introduce team members to the plan. After plan orientation, how to use the plan and fine-tuning it can be reinforced through exercises. Conducting exercises is the next phase of training.

Exercises are conducted to evaluate the team's capability to execute one or more portions of the crisis communication plan. By conducting exercises on a regular basis, the organization can

- Test and evaluate plan and the associated procedures
- Identify weaknesses that may be in your plan
- Reveal resource gaps that may be present
- Improve individual performance, organizational communication, and coordination
- Train personnel on the plan
- Clarify roles and responsibilities
- Satisfy regulatory requirements

Exercises ideally should become progressively complex with each exercise building on the previous until exercises are as close to reality as possible. Involve your external partners such as the fire department, law enforcement, local public health, and Red Cross when possible. These exercises should be carefully planned to achieve one or more identified goals. Tabletop, functional exercises and full-scale exercises are different types of exercises used.

A *tabletop exercise* is the type of exercise recommended after an orientation seminar. A TTX, is a low-cost facilitated group analysis of an emergency in an informal, stress-free environment. The TTX is designed to examine the plan from an operational perspective, to identify problems, perform in-depth problem solving, and examine operational plans. TTX should be conducted at least twice a year using different scenarios. At least one TTX should involve the use of no technology. A TTX should be conducted regardless of whether you have recently activated your plan (within the last 6 months). A TTX should also be used when there have been significant changes in the plan or significant organizational changes occur.

A *functional exercise* is a fully simulated and interactive exercise, testing the organization's capability to respond to a simulated event. Events to consider are product defects causing health issues, employee misconduct that left consumers with losses, or any type of event or hazard your organization could face. A functional exercise focuses on the coordination of multiple functions or organizations and usually takes place in an EOC. This exercise type strives for realism without the actual deployment of equipment and personnel. A functional exercise should be conducted at least once a year or within 12 months of plan activation.

Short of an actual event where activation of your plan has occurred, the *Full-Scale Exercise* (FSE) is a simulated emergency event. A FSE is as close to reality as possible. It would use all functions related to the plan and requires full deployment of equipment and personnel. This exercise, in a business setting, should include your external partners such as fire, law enforcement, emergency management, and other organizations. Actors should be considered to fulfill roles such as the media, survivors, and fatalities or other roles as identified in the scenario. Actors could be employees, volunteers, or others as you have identified. FSEs should be conducted when there are significant organizational changes or if it has been more than two years since activation of your plan.[5]

In addition to testing the plan, it is wise to provide media training to those who may serve as your spokesperson, senior leadership, and the PIO.

Sample language to use in this section of your plan:

> The head of Public Affairs will take the lead in educating our community about when and how members would get messages from your company in a crisis.
>
> COMPANY will conduct annual exercises of the crisis communication plan with emergency management and members of the Emergency Leadership Team. Members of the Crisis Communications Team will test the crisis communications plan at these times with participation.
>
> Public Affairs will schedule media training sessions for senior leadership and key team members. After the initial training session, additional sessions will be scheduled annually for employees who are new to the Emergency Leadership Team, other employees whose role may need to speak to the media in the event of an emergency,

or the Crisis Communications Team. Every two years, all members will attend a refresher course in media training. Public Affairs will develop a list of these employees to track and ensure they have completed training.

PLAN REVIEWS AND UPDATES

Having a plan, training employees on it, conducting exercises, and maintaining currency will help you to be prepared. It helps the organization protect itself when a crisis occurs. Trained employees provide a better response because they are familiar with its content. Ideally, the plan is reviewed and updated as follows:

- Quarterly or after every exercise or plan activations
- When there have been significant organizational changes
- When there have been significant changes in technology
- As required by law, regulations, or company policy

This section of the plan clearly states the review and update process, who is responsible that this occurs and who approves any changes.

Sample language to use in this section:

> The communication plan will be reviewed twice a year. Appendices covering key resources and contact information for key resources and external resources will be reviewed and updated quarterly.
>
> Public Affairs will update contact lists for members of the Emergency Leadership Team, OEM, and the Crisis Communications Team. Public Affairs will oversee updates of media lists and fact sheets. Public Affairs, working with the Information Technology Department, will oversee updates and improvements to e-mail lists for internal audiences. Public Affairs, External Affairs, and OEM will update contact information for external resources.

PLAN APPENDICES

In the appendices, you add information that will require updates on a regular basis. This would include a listing of personnel and groups, essential personnel, external resources, and their contact information (names, physical and e-mail addresses, phone numbers, and other information needed for each). Another appendix would include your templates and any prescribed message by event type and target audience, such as early closure, significant flooding, disgruntled employee's actions, fatality, or armed suspect at work. A third appendix should include a listing of all technology and communication mediums used to support this process and contact information for troubleshooting. Communications media should be listed by target audience. A fourth appendix should include an equipment and materials lists, where to obtain the items, and suggested lead-time. Other appendices may cover a list of products or services you offer, a detail reference list of materials, news conference guidelines, or media relations reminders. Two appendices warranting additional insight are news conference guidelines and media relations reminders.

News Conference Guidelines

There should be very few individuals within the organization who are authorized to notify or speak with media on behalf of the organization. Most organizations frown upon individuals within their organization who voluntarily seek out the opportunity to represent the organization without the appropriate authorization. That face becomes your organization. This is an area for all on the team to have a heightened sense of awareness. When you notify the media, try to anticipate the size of a crowd. When establishing your media staging area, consider the different formats you may use in advance among other factors reviewed in the sample as follows:

- When notifying the media of news conferences, press availabilities, or press briefings, ensure you define what kind of event you are having. *News conferences* are held to make an announcement about an event for the first time. *Press availabilities* are held to make individuals available to answer questions or demonstrate something. *Press briefings* are held to provide updates and possibly answer questions.
- Limit news conferences, press availabilities, or press briefings to when changes occur or the event remains in progress for an extended period. Avoid unnecessary news conferences/availabilities. If it is not worth their time, the media may become angered.
- If holding a news conference, attempt to advise the media beforehand about some of the details of what you will be announcing.
- Try to anticipate the size of your crowd carefully when establishing your media staging area. It is better to have too much than too little space. Make sure equipment, seating, lighting, and water are in place at least 30 minutes prior to the event.
- Decide upon a format in advance—who will introduce speakers, who decides when the question-and-answer period ends, and other details.

- Determine in advance whether handouts are needed. If the speaker is giving a talk for which there is a text, you may want to wait and handout material after the talk so media will stay and listen. However, it is advisable to tell the media you will provide a text of the speech, so they are not irritated by having to take unnecessary notes.
- Check to see what else is happening in the community and within your organization before scheduling a press conference.
- Consider if other organizations and agencies need to know you are having a news conference and consider inviting them.
- Decide who will manage the news conference, where cameras will be set up, and seating arrangements.
- Try to plan the length of the news conference; however, remain flexible.
- Be mindful of the time of the news conference. If you want to make the noon, 6 p.m., or 11 p.m. TV and radio news, you need to allow time for crews to travel and edit tape.
- If you are setting restrictions on an event such as limited photo access, try to put the restrictions in writing and communicate to the media at least 24 hours in advance.[6]

Media Relations Reminders

Media relations are to work with different media to help keep the public informed of the organization's mission and practices consistently, in a positive and credible manner. It is a relationship managed by Public Affairs; however, every member of the Crisis Communications Team should be aware of a few principles of media relations. You always want to return their calls, avoid antagonizing them, consider how information is released, be proactive in releasing information, and give credit to other agencies where it is deserved. A sample of items you may want to consider for this section:

- Always return calls from the media. Having a cooperative spirit is better.
- Make communications two-ways with the media—talk to them as well as listen to them. During crisis time, you may learn a great deal from the media that can be useful to your organization in handling the crisis.
- Avoid antagonizing the media. Your tone of voice and body language at a press conference, during a phone call, or elsewhere are monitored just as carefully as what you say and can affect

your future relationship with an individual or other media who may hear about the conversation.
- Consider a dedicated call-in phone line to offer information to the media or others. Information on news conferences, rumor control information, and newly acquired information can be recorded and updated as needed. This is useful when regular phone lines are congested.
- Consider how the information released to the media may affect other organizations or individuals. If something you say may result in the media calling other agencies, call those agencies in advance to advise them of possible calls.
- When speaking with the media, ensure you give credit to others working on the crisis, including your staff.
- Try to be proactive with new information even in the midst of chaos. When you acquire new information regarding the crisis, reach out to the media rather than waiting for them.
- Monitor information flow and public response for any information shared with the media and adjust your message and approach accordingly.[7]

SUMMARY

The purpose of writing a crisis communication plan is to use the resources available to you and understanding that communications is a tool to help solve problems. Organizations with a crisis communication plan are usually better prepared. They have begun thinking of what to say, why it should be said, and who is needed to say it, knowing that each factor could make the difference between keeping your employees or closing your doors. You can write your own crisis communication plan for your organization using the materials covered in this section and reviewing the crisis communications plans of other organizations online who are representing your industry. Identify target audiences and determine targeted messages per audience. Spend time in preparing your first announcement(s) for different types of situations your organization may face. Ensure your prescripted messages and templates are reviewed and approved before an event occurs that requires you to use them. Ensure it takes into consideration any regulatory requirements that are applicable to your business. Also, identify materials you may need and put a list together in an Appendix. The plan should address the possibility that technology will fail. You must have redundancy and a no-tech approach such as face-to-face communications incorporated in the plan. An effective communications plan continuously scans the environment for what information is shared and how the public responds to it.

ENDNOTES

1. MSNBC. (2009). *Obama: Crisis Is Time of 'Great Opportunity'*. http://www.msnbc.msn.com/id/29567427/ (accessed April 27, 2011).
2. Dickinson State University. (2010). Crisis Communication Plan. http://www.dickinsonstate.edu/emergency/pdf/CrisisCommunicationPlan.pdf (accessed April 27, 2011).
3. Duke University. (2010). Crisis Communications Plan. http://emergency.duke.edu/plan/ (accessed April 28, 2011).
4. Duke University. (2010). Crisis Communications Plan. http://emergency.duke.edu/plan/ (accessed April 28, 2011).
5. State of Washington. (2010). Exercises and Drills. http://www.emd.wa.gov/about/documents/Business_Exercises_and_Drills.doc (accessed May 2, 2011).
6. Texas Wesleyan School of Law. (2011). Crisis Communications Plan. http://www.law.txwes.edu/?tabid=60 (accessed April 28, 2011).
7. Texas Wesleyan School of Law. (2011). Crisis Communications Plan. http://www.law.txwes.edu/?tabid=60 (accessed April 28, 2011).

CHAPTER 14

Getting the Message Right the First Time

> Perception is reality. The way in which your organization is perceived, is the reality of your organization's image .... By doing a certain amount of planning in the form of "what if?" scenarios, the goal of turning a potentially negative situation into a positive is a real possibility .... It is important to remember that the degree to which your organization effectively responds to a crisis affects your relationships with all your publics.[1]

AFTER THE CRISIS

After each crisis, where the activation of the crisis communication plan occurred or whenever an emergency communication is sent, an after-action review should be conducted. The same is true whenever the plan or components of the plan have been tested or exercised. The review process evaluates what went well, mistakes made, and any best practices that can be applied to enhance or improve the plan. From these reviews, an *after-action report* (AAR) should be generated. The AAR is a record of the review, including the process used, findings, outcomes, and improvement plan. If earlier AARs exist, it is good practice to review each as a part of the analysis. At a minimum, the last two should be included as a part of the review.

An effective crisis communications planning process is partly about how you conduct the review and what you learn from it. AARs can reveal and provide an understanding of the competencies you want to retain, those to build upon in-house, and those you can turn to partners to fill.

Change management is a key process to incorporate as a part of the crisis communication plan. It is a core component of success. If there is a disruption in the process, such as a shift in the situation, it is imperative to be prepared to adjust and respond quickly. Change management is a structured approach to shifts in people, processes, or technology from a current state to a desired future state.

The individuals responsible for crisis communications must be nimble and able to adapt to changes during an event. Crisis communications team members need to known when to think beyond the prescribed and structured templates provided in the plan to achieve success. The outcome captured in many AARs is that, after the initial communiqué, the team moved too quickly without any validation of any changes, or the changes reported were insignificant. This negatively affected subsequent communications. Finding the balance often means slowing down the scheduled releases to ensure that meaningful information is provided with each communication, creating additional checkpoints and increasing internal communications.

At the other end of the spectrum, the outcome reported in some AARs is that an organization lacks balance between timely and effective crisis communications. Their internal processes for gathering or approving messages create unnecessary delays in delivery although the message is accurate and complete. When the gap is too long, the affected populations will speculate based on rumors, prior experiences, or aged data that are no longer applicable and take inappropriate action. Correction is required to obtain the balance needed by restructuring the crisis communications collection timeline and adding resources as needed to increase speed in sending messages. At the end of the day, the implementation of the crisis communications plan will yield a unique experience and opportunity to challenge the status quo with each event.

Crisis Communications Planning—Things You Need to Do Now

Other lessons learned and captured in AARs include

- Communications—Identify the most likely incidents or hazards that could occur in your area or within your organization.
- Build incident scenario templates in your notification tool and consider building groups by notification groups. The latter may create conflicts or require a great number of groups to manage.
- Select your technology based on your preparedness plans rather than attempting to force fit your preparedness plans with the technology.

- Despite temptations, do not overuse your mass notification system! Make an attempt to limit its use to emergencies and testing of the system, as you would your fire alarm. Your audiences become desensitized to your messages if sent too often and will disregard those having the greatest importance.
- Technology is important, but it does not constitute a crisis communications plan. To be prepared, you should have a plan.
- The most advanced technology in the world is useless if you do not test it.

A weakness consistently cited are mistakes communicators continue to make. This is the subject of the next section where the common mistakes are explored further and suggestions made for overcoming them.

COMMUNICATION MISTAKES COMMUNICATORS STILL MAKE

Is your emergency communication message getting lost because you are making one of these common errors? For many years, communications professionals have lamented over the way they are perceived. Terms such as "the last ones to know" and "the department of analysis paralysis" were common just a few years ago when referring to the Public Information Officer (PIO) function. This is changing—slowly, according to many emergency management leaders. As other members of the command and general staff get a better rap, challenges still exist when it comes to conveying emergency messages to internal resources and the affected community.

Learn to communicate effectively by avoiding these pitfalls:

- *Failing to convey the vision for emergency management before an event occurs.* Marcia Wright, Director of Security with a California-based aerospace firm, has been in emergency management for decades. She has become increasingly interested in emergency communications as the world became more computerized. Communicating was much simpler—you send an e-mail, pull a fire alarm, or send a prerecorded message over a speaker. Also, the threats were not as complex or as broad in reach or impact," she recalled. "Now our jobs are more complicated because we have to still deal with all the noise and threats that are automated, information sent and received from more sources, yet we also need to be prepared for the more complex and advanced method coming our way." For Wright, it means handling current threats and plans for the immediate future.

- The team is required to be forward thinking and dealing with the situation in hand simultaneously. Find the balance to avoid getting caught up in only fighting the fire rather than fighting the fire and having clear projects. This means spending time looking and updating the vision to address where are you going and your strategy for getting there. For emergency communications to reach the target audience in a timely manner, you need to work so that time is not wasted on the "cosmetics" of the emergency communications but rather focused on the essence of message in a manner that is easily accessed, understood, and trusted.
- *Neglecting to relate emergency communications to everyone.* Everyone needs to understand how emergency communications will work for them—all stakeholders. This means that, during the planning process, the PIO needs to understand how effective the existing emergency communication solutions, processes, and people are by listening to user pain points and how to continually improve with this information in mind. Questions to ask users should include
    - Was information obtained quicky?
    - Was special access required?
    - Were the targeted audiences addressed, including those with functional and access needs?
    - How can the process be made to work more easily in the future?

  If we can solve problems for the message receiver by giving them useful instructions and an accurate status of the situation, the receivers will not mind an extra step in gaining access to the information or a brief delay in the receipt of the message.
- *Failing to consider different cultural norms.* Most U.S. urban areas are heterogeneous, that is, people representing different cultures, nationalities, languages, religions, and other differences that make us unique. Some homogeneous communities have people sharing the same culture that may be different from those outside the community. When communicating it is important to consider the varying cultural differences that may exist within a community. The varying cultural norms mean that emergency communications need to be conveyed using multiple mechanisms to avoid misunderstandings. A message that is straightforward in English and then translated into Spanish using a free English–Spanish translation tool online can be viewed in an entirely different manner by the Spanish-speaking community. A one-size-fits-all approach can lead to problems and missed opportunities to reach more of your target audience.

Consider the term *"inside lane."* In the United States, the term means the lane nearest the center of the road; conversely, the same term in the United Kingdom refers to the lane nearest the side of the road. You send an alert on the highway electronic signage stating

Traffic accident on the inside lane of Hwy 59 Northbound.

If I am familiar with U.S. driving patterns, you know the inside lane is usually to the left of the driver. However, if the person reading the message is from the U.K. and unfamiliar with the U.S. driving pattern, he or she would interpret the message to mean that the "inside lane is to the right of the driver." Two people reading the same message would come to two different conclusions.

If clear text were used, the message would be written as follows:

Traffic accident in the left lane on Hwy 59 Northbound.

The message is specific, avoiding any confusion as to which side of the road the message is referencing. It is critical to view emergency communiqués from the lens of those who will receive the message so that emergency communication is audience-appropriate.

- *Failure to make the business case for emergency management preparation and planning.* The profile of emergency management within an organization and the community has risen significantly since the unheard of events of 9/11. Despite the increased attention, executives and employees look to others to give them what they need at the time of need. They often do not consider the need for training and exercises as important. Emergency management must remain present in the minds of all within the organization and the community. One way to achieve this goal is to explain to others the risk they face and the potential outcome if they are not prepared. Risks including sending poor instructions, not knowing who is authorized to approve or send a message, not reaching your target audiences—those needing the critical information.
- *Neglecting the timing of the message is paramount.* In many cases, the PIO role can become burdened with analysis paralysis, negatively affecting when a message is delivered. It is about timing. If the message is too late or too early, it will be ignored. If you give the right information to the right person at the right time, people will act. If you miss one or more of these, the

message will get lost and not received. In the greater Houston area, hurricanes occur on average once every 3 to 5 years. This event makes the headlines, and business and people have an interest in them and want to learn more and be prepared. As you move beyond 2 years after the last hurricane, getting the message out to invest in communication solutions becomes increasingly harder. It is easier to get the investment dollars right after an event that caused some harm to an entity or person; it is much harder as new competing priorities for funds begin to bubble up. Timing of the message is critical whether in an event or in preparing for the next event.

- *Forgetting the PIO's communication role changes frequently.* Executives preparing for an upcoming meeting with a client often use "Murder Boards." A *murder board* is a place where you list all the possible questions someone may ask about a situation, and you take this information and begin developing your response. The same approach can also be used in preparing the IC and command staff who may need to provide an update or a briefing. It is important to prepare mentally for these meetings by reviewing the available information and determining what role you as the PIO and others will play during this communication process.

  The PIO may be the leader up front or an adviser behind the scenes, and may be the publisher of good news or bad. The PIO must have the ability to work in any of these roles at different times given the situation. PIOs need to understand many different things so they know what to ask and whom to ask, and then relay the information in clear text or nontechnical terms so that the audience understands. With a broad range of knowledge, the outcome is more likely to yield a more effective message.

- *Failing to communicate when communication is a waste of your time.* Despite our best efforts at effective communication, what you have to say, regardless of format or method used, may become meaningless or may not make a difference. Are you the right person for the position, or do you feel that, as the PIO, you are being set up for failure? Do you have the support of senior management in planning, preparation, and sponsorship? Unless senior management has some understanding of, and offers support for, emergency management and the need for crisis communications planning, it may not matter how good you are, for it will be difficult to get anywhere with planning and preparation before an event occurs that could bring harm to the entity.

GETTING THE MESSAGE RIGHT

A crisis requires a communications strategy that prepares employees and all those affected for the transition to the next steps. An effective crisis communications strategy should permit progressive updates at regular intervals. It should ensure that there is outreach to all stakeholders affected by the event, including external stakeholders such as regulatory bodies, vendors and contractors, and customers. Your success, or failure, is about more than just the event. The challenge lies in communicating factual and timely information in such a manner that people can use the information.

You are more likely to get your messages out right the first time with a good plan, preparations, and training. Effective communications is a human issue that succeeds when all involved work together to disseminate the message. Communication is effective when the communicator stays focused on the outcome so that the outcome satisfies the need and delivers only what the target audiences wants to know. This is possible when the needs of the target audiences are clearly defined and available in writing.

Effective Communications Is a Human Issue

Requirements are the primary means of communication between the communicator and the target audience. For a crisis event where interests are high and changes occur quickly, requirements become the primary means of communication between the communicator and the target audiences for what the latter needs. The crisis communications plan can have the force of a contract, yet the needs that it articulates come from the audiences it should serve.

Writing a message in a crisis can be difficult. The drafter should have an understanding of the requirements and be able to organize these messaging requirements into a logical structure. The drafter should understand possible solutions and what to do, and should be a good writer who can produce text that is clear, concise, and complete. The drafter should consider the following when creating a message:

- Determine what the target audiences need to know
- Help organize these needs into a clear documented structure
- Structure these needs and then prioritizes them.
- When time permits,

- Include representatives from the targeted audiences to test and offer feedback.
- Conduct a review.
- Ensure that the message is evaluated after each use and updated as needed.

It is easier to understand the message requirements when using complete sentences. Phrases, single words, and a collection of acronyms and abbreviations alone are insufficient data points for a requirement. The use of complete and accurate sentences is a better approach to documenting requirements. For example, in the requirement

The students must be able to understand that classes will be cancelled.

The subject of the message requirement is *students*, and the predicate phrase is *understand that classes will be cancelled*. The phrase *must be able to* connects together the two segments of the sentence to form the entire message requirement. Each documented message requirement should be drafted this way so that those who create and later evaluate the effectiveness of the message clearly understand the requirement. The subject implies the audience that the message requirement is related to. The predicate should be an action phrase or express an activity that must be performed by, for, with, or to the subject.

The verb *to* establishes the connection between the subject and the predicate. The words *must*, *shall*, and *will* have significant meaning. The word *shall* serves as an indicator that the sentence is a requirement that must be adhered to or followed. The words *must* and *shall* indicate mandatory conditions of the requirement. Using the foregoing example, this sentence relates to what is needed by the target audience from the communicator to fulfill the message requirement. Less restrictive words are used to designate discretionary or optional requirements. An example of less restrictive words used in drafting message requirements is

The students should receive class cancellation information before the top of the hours.

The more complex a crisis, the more communiqués are needed. This results in more requirements. Those responsible for drafting and communicating can fully understand a finite number of requirements that are organized logically. As in writing a program or narrating a play, the ideal structure to use is to follow the natural pattern of use.

Use clear text and plain language to draft your message requirements. This helps to ensure that everyone can immediately understand the written requirements. Eliminating the use of jargon, acronyms, and technical terms and instead opting for everyday language is the one medium that both the communicator and the target audience share. Clear text and a structure to organize requirements are equally important.

The structure should give the message drafter and intended audience a vision of the big picture. Message requirements with various levels of detail enable readers to follow along section by section when a good structure is applied. One method is to use are universal symbols when supplementing the message. Easy-to-follow flowcharts and diagrams are effective for less complex events. When these become larger, they become difficult for the user to interpret and understand.

Most crises require more than one communiqué to keep the target audience informed of what to do and expect going forward. Tracking the status of the event and adherence to requirements is essential. The tracking process begins during the planning and preparation phases of a crisis. Tracking of message requirements should record

- The source of the message requirement including when it was created and the logic or purpose of the requirement
- Whether the message requirement has been proposed, reviewed, accepted, rejected, or modified
- The message requirements and its status even though you may elect to take no action
- If the requirement mandatory, optional, or desired
- If the requirement is high, medium, or low priority
- How the requirement will be verified (inspect, test, analyze, simulate, or confirm requester has the authority to make the requirement)
- If these are other constraints to consider (performance, safety, health, loss of property)
- Comments, action items with owners assigned, and suggested changes

High-priority and mandatory message requirements should be captured in a manner that it is easy to identify. These requirements could be considered for analysis and could possibly incorporate a narrative of what would happen if they were not satisfied. The key is to limit your mandatory high-priority requirements to those that are of value to the target audience.

After-Action Review

The purpose of an after-action review is to discover issues early enough to address them early and usually at a lower cost (tangible and intangible). The review process can be formal or informal. An informal review can be conducted by sharing the messages with the crisis communications team and representation from the target audiences. A formal review is broader and conducted in a formal meeting. The meeting is used to identify actions and decisions made on the items reviewed. Any action items should be tracked to completion.

The review process is

- The collection of comments from stakeholders
- Decisions on what to do about the comments
- Agreement on changes to the crisis communication plan

The process begins by providing the stakeholders with a copy of all messages sent, related tracking records, and a copy of the crisis communication plan for review. Ask the reviewers to write and send their comments to the review organizer prior to the meeting. The review organizer collects and sorts all comments. The organizer creates a consolidated document to send to stakeholders to use in preparing for the formal meeting. The stakeholders come to the meeting with the understanding that they are to make decisions on the suggestions offered. Knowing that having stakeholder representatives and the crisis communications team together in a meeting is expensive, the prework becomes a critical step in the process. More can be achieved within the time allotted. The purpose of the review meeting is to improve the quality of the crisis communication plan, updated or new prescripted messages, and input into future training.

Following good meeting practices is important. Have an agenda, adhere to timelines for discussions, and publish meeting notes incorporating recommendations, decisions, and the status of action items.

The review process should

- Encourage criticism to help improve the plan.
- Identify a small group who can allocate time for the review process and have a good understanding of the organization and the target audiences.
- Ensure that changes to the plan have been reviewed and accepted, rejected, or accepted with modifications. When stakeholders have completed their prework assignments, decisions on requirements can be made within a couple of minutes. Use of a timekeeper or facilitator can help with this step in the process.
- Strive to minimize interdependences within a message requirement to keep it easy to understand.
- Aim to have a formal review. Consider the use of informal reviews with a small group to minimize the time and costs of formal review meetings.

Drafting Your Message—What to Avoid

We have identified what you should do. Next, we will review what should be avoided to create a good message.

1. Avoid ambiguity ... write clearly. For example, the word *for* and other subtle errors can cause ambiguities. For example,
   a. *Message shall also be able to incorporate words and pictures for students who speak English or French.*
      i. Ambiguous:
         1. Which messages?
         2. Is this for messages that are an advisory, just a warning, or both an advisory or warning?
         3. Is the message for both those who speak English and French or a single language?
         4. If a single language, which language and what conditions apply?

2. Avoid drafting multiple requirements within a sentence. Avoid using conjunctions, words that join sentences together, which can cause confusion for the reader. A conjunction can confuse the reader who is attempting to decipher which segment of the requirement applies to them or if there is a conflict. For example,
   a. *The computer's low power warning indicator will turn red when the power available drops below 10%,* **and** *the current desktop view or open files should be saved.*

3. Avoid the use of escape clauses or let-out within a message requirement. The requirement sentence begins with a definite action yet closes with other options. This type of poorly constructed requirement leads to problems when trying to have an audience take the appropriate action. Words that can imply a let-out or an escape clause are *although, about, except, if, unless,* or *when*. For example,
   a. *The sirens shall automatically activate when the National Weather Service has issued a tornado warning, except when the campus is closed.*
   b. *The siren shall always be sounded when a tornado in the area is detected, unless the maintenance director is testing the alarm or the inspector has suppressed the alarm.*

4. Defects in writing such as long rambling sentences, references to unreachable documents, and use of jargon or other mysterious language can cause error and confusion. For example,

a. *If the specific switch going to the designated server is set correctly so the system is able to differentiate from a test and a production code so that the output will comply with the required framework of Section 4.2.8 that indicates the job has reached its desired completion state.*

5. Message requirements should specify the need and refrain from using any supporting technology system. Indicators that the focus has shifted away from the message content are when names of materials, products, or database fields, for example, are used. A sample message requirements statement indicative of a design statement is
    a. When sending this alert message, the supporting messaging system shall be capable of tracking who sent the message, using SQL as the database with access limited to the Public Affairs Director and requiring the use of strong passwords to access the system.

6. Avoid mixing different kinds of message requirements when providing instructions. Organize the instructions logically to identify overlaps and gaps in the message. This is the fast path to confusion, mixing up instructions within a message. An example,
    a. *Audiences shall be able to view the current message on LCD panels on campus and use closed-captioning on Line 21 of any analog TVs.*

7. Do not speculate or add a *wish list*—general terms about something that somebody wants but not necessarily need. Message requirements are a part of a contract between the Crisis Communications Team representing the organization and the target audience. Warning signs that the speaker has strayed from facts include generalities or vagueness about the topic or response and the use of generalization with words such as *generally, normally, often, typically,* or *usually.* An example is
    a. *Police usually want early indication of a burglary in progress so that they have a better chance of apprehending a suspect.*

8. Do not play on ambiguous requirements. The use of *or* and *unless* in a message allows each reader to interpret the message differently from the same wording. This is dangerous. It could lead to confusion, with people responding differently to your instructions. For example,
    a. *You can postpone evacuating for now.*

    Write the message clearly and for all stakeholders who need to take the stated action.

9. Do not use vague and undefinable terms. Company jargon, slang, and other words used to informally indicate or convey a message within a group are too unclear for use in a crisis message. Vague or general terms include words such as *approximately, flexible,* or *as possible.* A message using these terms are unable to be verified because there is not anything to use to confirm that your instructions have been followed. For example,
    a. *You should leave as soon as possible.*
    b. *The police will be flexible in determining whether you may return to your neighborhoods.*
    c. *There are approximately 200 people waiting in the emergency room.*

10. Avoid stating suggestions or possibilities not plainly stated as a message requirement. This information may be ignored by the audience and may lead to confusion. Use of suggestions or possibilities is indicated with terms such as *could, might, may, ought, perhaps, probably,* or *should.* Examples are
    a. *Your mobile phone is probably able to receive a call in the basement.*
    b. *You may want to tape up your windows just in case there are strong winds with the approaching storm.*

11. Avoid wishful thinking or dreaming, that is, asking for the impossible. Terms associated with this form of communications are *always safe, handle all unexpected events, never fails.* Examples of use include
    a. *The life vest is 100% safe in normal operations.*
    b. *The fire department will always come and rescue your cat from the tree.*

12. Avoid overlooking possible constraints related to your requirement that could negatively affect health and safety or loss of property. There may be other constraints resulting in a unique situation that should also be considered, such as *goodwill, financial, performance,* or *legal;* however, most are secondary to health and safety or loss of property. The message should convey
    a. How probable is each hazard.
    b. How it can be mitigated to an acceptable level.
    c. What to do if a safety breach occurs.

Other Messaging Guidelines

In addition to all that has been covered thus far, there are a few more pointers to consider when you are the sender of a message and must prepare for speaking with your external audiences. These tips should be used for every message and, in particular, when interfacing with the media. The intent of your message is to provide information about a crisis and instructions on what to do. With this in mind, there are guidelines to consider:

- *In a timely manner, deal with an event to avoid a potential crisis that could become more damaging in the future.* Regardless of how catastrophic a problem may appear, your stakeholders will measure you on how you handled the crisis, not that it occurred. A crisis can occur at any time and when you are least prepared. Ignoring and hoping it will go away is an approach you can use; however, in most instances, the situation continues to develop, sometimes rapidly. This is time lost in planning and working on resolving the issue and alerting others. Delays can have tremendous influence on how the crisis is perceived. By giving a timely response you are maintaining the confidence of your stakeholders that this kind of incident will not happen again and that you are prepared.
- *Honesty is the best policy.* In a crisis, the temptation is to hide the facts or rush to make an announcement before you have the facts. Pause and stay with your plan to get the whole story told, providing all the facts your target audiences have a right and need to know. You must balance this with your legal liability concerns and the privacy of clients, employees, and their families, and what is and is not appropriate public information must be considered.
- *Avoid guessing or "no comment." It is not an appropriate response.* Using the phase "no comment" gives the perception that you have something to hide. Guessing can lead you to providing misinformation that must be explained later. If you do not have the answer, state that you do not have an answer but that you will follow up.
- *Any written document should be considered public domain.* The saying goes " ... if it has been written, it has been said." Adhere to a clean desk policy. If you do not have one, consider creating one. What you leave on your desk—notebooks, keys, contact lists, papers, for example—can be viewed or, with a cell phone, a snapshot taken so that the party with the information can view it later or send it to others such as a reporter or have it posted on the Internet.
- *Effective communications can help contain the crisis.* Once you have released your message, it is important to monitor for misperceptions and misinformation. As a situation unfolds,

message updates containing factual information should be provided. It is also important to contain endless reporting. You do not want to feed the frenzy for related stories and emotional reactions. Limiting the message to the facts and not speculations helps to contain spin-off stories. Adhere to scheduled press conferences and news briefings. Always be on time and stay with the talking points. If the media is on your doorstep, have your designated spokesperson available and a designated area to direct the media. Media should be supervised at all times.

Also remember that anyone with a mobile device that has a camera or video is a reporter, including your employees. They are "inside the yellow tape." You have the family members and friends of employees, visitors, and witnesses, for example, who are aware, and, in the process of sharing information, facts morph into false information and rumors.

- *Follow your crisis communication plan.* If you do not have one, create one. At a minimum, the plan should have the following information:
  - Crisis communications team members. Ensure members are aware of their responsibilities.
  - List of spokespersons with their contact information. Remember all employees are spokespersons for your organization. Some have had some training, although others have not, in crisis communication management.
  - Gather the facts and assess current conditions of the crisis.
  - Identify your target audiences and how best to communicate with them. A press conference may be required for urgent or complicated issues.
  - Prepare your messages using your fact sheets and pre-defined templates.
  - At the close of the crisis, conduct an after-action review that includes an evaluation on the effectiveness of the crisis communication plan. Any changes should be documented and all affected parties notified.
- *Reestablish relationships.* After the crisis ends, it is time to reach out to reassure your stakeholders that any harmful effects have been minimized. This is whether the situation is one of your making (a product defect) or was outside your control (severe weather). Draft a press release or final communiqué that declares a return to normal operations or long-term recovery, preparations to eliminate or minimize the opportunity for a reoccurrence, a word of thanks to all who assist, and other points you may want to consider.

- *Remember your website.* Your website is just that—yours. Post press releases and other information you want available to your target audiences immediately. Routinely reference your website as the location to visit for current information. The currency of your website is important. Provide updates regularly; simultaneously remove information that is no longer relevant.
- *Create your own agenda and make your points early in your message.* Write down the main ideas you want to cover in the interview and the talking points for each. The first sentences in your message capture the most attention. Prioritize your talking points and address them in your message in order of importance. Keeping the message concise helps to keep the receiver's attention through the end of the message. Long preambles and rambling could cause you to lose your audience.
- *Use clear text.* Your message should convey empathy and caring. Company jargon, professional buzzwords, and abstractions will have you appearing clinical and uncaring.
- *Statistics tell a story.* Numbers capture our attention in a crisis since it can help paint the picture on the magnitude of the event. Make your numbers listener friendly. Rather than saying "20% of students had the flu," say instead" *one out of five students had the flu.*"
- *Stay "on the record"*; there is no such thing as "*off the record.*" If you do not want what you said in the evening news, do not say it when the media is around or in the media staging area where media equipment is present.
- *Dealing with multiple questions.* If you followed your communication plan, you most likely had a fact sheet to use for constructing your message. After your first message was distributed, you received feedback, whether solicited or unsolicited. Review all of your materials and anticipate tough questions. Prepare straightforward responses. Look for ways to bridge the gap between negative information and a positive response. Know your answers inside and out. Practice your responses aloud. Prepare written documentation the reporter can reference later. Pick out one or two questions for which you have the best response. Present your answers calmly and professionally. This helps to set the tone for the remainder of the conference. Next, ask the media what other questions they may have. The media will ask the unanswered questions.
- *Anticipate interruptions.* You do not want to appear that you are in a battle with the reporters—you stand to lose more than they do. Instead, pause until the interviewer is finished with the interruption. Then say "I'll be happy to address that in a

moment. As I was saying ..." and quickly finish your point. There are many reasons a reporter will interrupt you, ranging from you rambling to reporters to competing among themselves to ask a questions. A "pause" can reset the interview so that your message is heard.
- *Use paraphrasing for clarity.* The use of paraphrasing, repeating what was said, is a good practice. When you are presented with a question, it helps the listeners and the sender know that you heard what was said and you understand the question. For example, say *"Sir, you asked how long it took for us to send our first message. Correct?"* or *"Ms. Doe asked if it was true that it took two minutes to sound the first siren."* The use of paraphrasing is also beneficial when someone has restated your message incorrectly. A good response would be *"Mr. Smith, that's not what was said. What I said was ..."* and repeat your statement.
- *Negative introductions, inaccurate information, and false charges.* Messages and interviews that start with a negative introduction can leave a negative impression. If you remain silent, the audience will assume you agree with it. When given the first opportunity to speak calmly, state, *"there's something I'd like to correct ..."* and then politely correct the initial statement and advance to the positive point. If the information provided by others in the interview is incorrect, specifically if coming from a report, graciously correct it if you have the correct information. Do not assume it is correct. If you are unfamiliar with the information, indicate that you are not familiar with the information shared and if the reporter will share the information with you. You advise the reporter that you will be happy to review it and comment on it. If the information shared by a reporter is false, immediately correct the false assumption of guilt without restating the charge by saying, "That isn't true," and then describe your view of the situation in a positive manner.
- *Skepticism.* Throughout the interview, let your natural enthusiasm shine. It is easy for question-and-answer sessions to generate skepticism in a crisis. Longer sessions often lead to the same question being asked by different reports in a different way. This may be taken as your response being questioned. Actually, it may be that there is nothing new to add and so the same questions are being asked, only phrased differently. Stay focused and reaffirm the validity of your statements. Consider introducing a new fact supporting your position or simply restate your original position. Be consistent with your response.

- *Long pauses are OK, you do not always have to rush with a response.* Human tendency is to say what comes to mind first. In dealing with a crisis, a cool head and rational thinking are needed. You need not rush to fill lulls in conversations. You do not want to say something that will require correction later. If you need a moment to respond, take it. It could be that the reporter needs more time to ask a new question. In this case, calmly wait on the reporter or say, "*I believe I've answered your questions. Is there any anything else you wish to discuss?*" If not, you can move on to the next topic or adjourn the session.
- *Become a master of the short answer.* When you can answer in 30 seconds or less, you are less likely to give out too much information or conflicting statements. Making each response a self-contained message minimizes conflicts with information given earlier or later in the interview. Abbreviated messages are less likely to be edited, which helps in reducing the opportunity your message, or the intent of what the message, is conveying to be changed through editing.
- *Practice so you learn to speak with power and enthusiasm.* Rehearse your message aloud using a mirror and recorder to review how you convey your message. Learn to vary your pitch, rate, and volume, adding interest to your message. You want to move quickly through ideas the audience understands and slow down when new or complex information is shared. Storytelling should be given in a quick and relaxed manner. Throughout the interview, let your natural enthusiasm shine. If your behavior reflects that what you have to say deserves attention, the audience will listen.
- *Dress the part and think like a star.* Looking prepared and having the right attitude is a winning way to start the interview. This is true whether it is a job interview or press interview; you come prepared in the same manner. You start by leaving home dressed professionally and ready to have pictures taken. You want to avoid distracting fabrics, colors, and patterns such as stripes, polka dots, or fluorescent colors, and the sparkling jewelry also known as the bling. If you normally wear glasses, wear them; avoid them if they are only a prop. From the time you walk out your door until you return, assume you are on camera. Never assume you are "off the air" until you are told that you are. Best practice is to assume you are on the air until you have left the recording area.
- *Make eye contact.* Different cultures view eye contact differently among men and women. In the United States and much of North America, you want to maintain eye contact throughout the interview. You want to look at the interviewer rather than directly

into the camera unless directed otherwise. If sitting, you want to sit comfortably and lean forward slightly to appear assertive. Let your gestures flow naturally, and keep your hands in front of you. Avoid putting your hands in your pockets or locking them, and crossing your arms or leaning on them.
- *Get the details for the interview*—what the interviewer wants. Ask the interviewers to provide you with a list of questions or topics they want to discuss and who will be the intended audience. Then do your homework on the reporter so that you have a good understanding of what his or her level of expertise is in the subject. You also want to know what type of interview it is, such as if it will be broadcast live or recorded, edited, and viewed later. Also, ask the interviewer if anything is needed visually for the interview, such as company logo in the background or a blank white background.[1]

SUMMARY

Getting the message right the first time can save time, energy, and money, and it may be your only chance to get your message out. The goal is to take a negative situation, the crisis, and communicate to your target audience what is going on around them and what action they should take. The information conveyed must be clear, concise, and easy to understand. In Chapters 9 and 10 we discussed the FEMA PLAN program that will send emergency text messages that are up to 90 characters in length to cell phones. This should serve as an indicator for us all that providing facts and instructions in a clear and concise manner is what gets the job done. Having a crisis communications plan as discussed in Chapter 13 that is routinely exercised, reviewed, and updated is an effective tool. With a good plan, you are more likely to get your messages out right the first time. We have reviewed in this chapter common pitfalls in communication and guidelines to avoid them. At this point, we have reviewed technology used for emergency communications since the beginning of time, the dynamics of effective communications and solutions available in the market today, and exercising and activating of the plan. In the next chapter the subject to discuss is the continuous improvement process to maintain currency of your emergency communications program.

ENDNOTE

1. Life Services Network. (2002). *Crisis Communications Handbook.* http://www.lsni.org/linkclick.aspx?fileticket=66lijzaxusi%3d&tabid=102 (accessed January 25, 2011).

CHAPTER 15

Prepare to Prosper

> So, what must an organization focus on to ensure that it reaches the promised land? In short, communication, people, and process. That is, communicating (and communicating and communicating) your ... process to your people.[1]

KEYS TO SUCCESS

A key to success in an emergency is communications. The primary failure in emergencies is the failure to communicate. An effective approach to prospering in emergency management and crisis management, and most important for those responsible for creating, approving or sending emergency communications, is to get to know those in the field. These individuals could assist you in an event before an emergency occurs. You do not want to have to exchange business cards in a disaster. Your job is to know enough people who know enough people to get the job done. This is also important because people like to work with people they know. Learn the names of people across an organization, not just its leadership. Often, those who are not in the most senior position are your way to people to get things done.

A response to an event starts with interpersonal relationships. For example, the States of Texas, California, New York, and Illinois have robust mutual aid agreements in place that are practiced. For example, in the State of Illinois, one exercise is to have 99 emergency response vehicles and responders credentialed and staged within 90 minutes at a specific location during rush hour. Some states conduct statewide interoperability exercises among the first responder community, such as police, fire, and emergency medical services (EMS). Public health agencies as well as other organizations conduct communication exercises, sending alerts to the external partners to exercise mutual aid agreement to stand up points of distribution (PODs) to dispense prophylaxis within 24 to 72 hours, such as practiced by the Montgomery County Hospital District in Texas. The focus is to know who and what your resources are.

It is equally important for an organization to be prepared to work without technology. The "no-tech" approach uses pen and paper, face-to-face contact, and walking around for communication. Acquiring flashlights with batteries and backup batteries will work after the power goes out, but only for a short period—as long as the batteries last. Today, there are a number of flashlights that use solar power, or you can "wind up" to operate. These are products to consider adding to your emergency supplies in the Emergency Operations Center (EOC) with your pens and paper.

As a communicator you should have a list of essential contact information programmed in your cell phone and office telephone directory and loaded in your personal e-mail contact list. At least twice a year you should print out a hardcopy to keep in the trunk of your car and by your bedroom or home office phone.[2] The list should include

1. Airport Security Directors and Emergency Managers
2. American Red Cross
3. Attorney General or Deputy Attorney General
4. Chief of Staff and Deputy Mayors and County Executives
5. Department of Transportation (city, county, state)
6. DHS National Operations Center (NOC)—202-282-8101
7. EMS agencies
8. FBI
9. FEMA Regional Administrator (political) and Deputy Administrator (career)
10. Fire agencies (including suppression, hazmat, fire marshal, search and rescue)
11. GIS
12. Governor's Chief of Staff and Deputy Chief of Staff
13. Hospital administrators/Emergency Department Directors and Head Nurses
14. Hospital District Officials
15. Jurisdictional Chief Information Officer (CIO) or 24-hour Technical Support Center
16. Key elected officials for each jurisdiction
17. Law enforcement agencies
18. Mayor's Chief of Staff and Deputy Chief of Staff
19. Media (print, radio, television, and key community bloggers)
20. Port Security Directors/Emergency Managers
21. Public Health Officials
22. Public Works (city, county, state)

23. School Superintendents and Facility Directors
24. State and County Public Health Director
    a. Epidemiologist
    b. Medical Examiner
25. State Forester
26. State Homeland Security Advisor
27. State Office of Emergency Management (SOC)
    a. Deputy Officer and Watch Desk
    b. Director and Deputy Director
    c. Operations Chief
28. State Operations Center Incident Commander and Regional District Officer for each jurisdiction
29. State Patrol/Police
30. The Salvation Army
31. Top appointed officials (city managers, county administrators)
32. U.S. Secret Service
33. Utility companies' Business Continuity Staff (power, phone, wireless, wastewater, water suppliers, natural gas, cable, Internet service providers)
34. Your organization's senior management personal contact information (cell phone, home phone, personal e-mail address).

Another critical skill for communicators to have is the ability to talk to others as though they were the most important persons in the room at the time. President Clinton was considered a master at this. Other great communicators are former Apple CEO Steve Jobs, President Barack Obama, President Ronald Reagan, and Debt Detractor Suze Orman. Each could capture the target audience's attention and hold it. To start building this skill, there is no time like the present, whether at home, work, or out having fun. For example, if you are out to dinner, practice with the wait staff; they have received training, too. If you have noticed, most of them make immediate eye contact with you and have a smile and enthusiasm about their menu. The waiter or waitress may touch you on the shoulder or drop to one knee at the table when talking to patrons so that they are at eye level. Studies have shown that wait staff that practice this technique earn higher tips.

Having competency in your field and a solid reputation is another critical skill. Effective communicators are usually people others do not mind having on their team. People who are perceived as arrogant, stubborn, never wrong, and self-centered or do not ask questions or listen are often difficult to work with. These individuals may need additional training to work well as a member of a crisis communications team or serve as a spokesperson. Remember, while your day-to-day role indicates that you may be a good fit in a certain situation, other factors may indicate that you are a better fit in another crisis role. There is a role for everyone in a crisis. Identifying the role that is the best fit is the challenge.

Great communicators work constantly to build relationships. They network, network, network, and understand that relationships are a key to success. Communications in a relationship can bind people together because effective communication is a two-way street. Nonverbal clues, such as body language, can speak louder than words and send a strong message to the audience. The use of nonverbal clues is a tool used to connect with others. It is unavoidable with your audience when communicating. It applies when you are viewed on television or in person. It also applies when you are speaking over the radio and the tone of your written message.

A good communicator should be trusted and should have influence. As the PIO, Communication Specialist, and the designated spokesperson in a crisis, you operate in a gray area most of the time. A primary ingredient is your ability to influence others. As the spokesperson, you have been awarded the privilege of leadership. At some point in a crisis, you will be required to make direct, face-to-face contact whether in front of cameras with reporters or at the front door of your office before various stakeholders.

As the spokesperson, you are the face of the organization. The words you use are important. Any pictures used in association with these words are symbols that can be more important than what is said; for example, the background in the picture (indoors on a stage or outside in a field), how those in the picture are dressed (in a suit—formal, or in a polo shirt—informal), and where we are as the spokesperson (in the boardroom or on a soccer field).

It has been repeatedly mentioned in previous chapters that effective messaging is clear, concise, and easily understood. This helps your message stand out from the 35,000 messages a person is barraged with a day.[3] Your message is also competing with the 16,000 words people speak each day.[4] The communicator must also connect with his or her audience. According to John Maxwell, the founder of EQUIP, *connecting* is the ability to identify with people and relate to them in a way that increases your influence with them. The communicator's inability to connect will alienate people. Simply stated, connecting is essential. Having the capability to connect, combined with a clear concise message, makes a huge difference to the attention your message will command from your audience.

For the communicator, connecting goes hand in hand in conducting meetings. Connecting means being considerate of others' time. Avoid minor meetings; send written communiqués instead. Come prepared with a written agenda that you can share with others, and anticipate questions. Rehearse your responses, paying attention to the words used. The words you use have impact.

The crisis communications team is a valuable part of the emergency management family. It takes people working to make crisis communications and the mass notification process great. The team should be dedicated to continuously improving everything it does, better and faster than before. Learning how to make continuous improvement the underlying foundation is a key to the success of the crisis communications team and the organization.

The crisis communications team must emphasize health and safety first, the environment, diversity, and people. Most organizations and teams are good at looking at themselves vertically (the stovepipe) but not good at looking horizontally (across the organization and in the community). Crisis communications teams must share information and work to eliminate boundaries within the organization. This requires breaking down the old stovepipes and building an integrated team. Everyone on the team and in the organization has the responsibility of following the organization's communication plan. Every manager within the organization is responsible for helping his or her employees understand and follow the plan.

TIPS IN PREPARING YOUR EMERGENCY OPERATIONS CENTER

Having a plan and trained employees are critical components in effective execution of the plan. When preparing to be successful and having your operations prosper, know some of the tips, tricks, and traps that others have learned and are shared below:

1. *"ICE"*—"in case of an emergency"—is information all team members should program in their cell phones. There are a number of free mobile apps available. Also, you can enter a contact entry called ICE and input information that would be good for emergency staff to know, such as allergies, your emergency contacts, and the names of your physicians with their contact information. Emergency personnel will look for this information on your cell phone if you cannot respond.
2. *Two-way radios* are the primary means to communicate in an emergency among first responders. These devices are known for their reliability when other communications methods are no longer available.
3. *The telephone (landline)*—"is there a telephone in this room?" is a familiar question. Although most people carry a cell phone and many offices use VoIP, cable, or Internet phones, a landline is still a more reliable form of communication in an emergency

because it has its own power source. Before disconnecting this service, ask your telecommunications providers for the least expense package they offer and consider this as an alternative so that a landline remains available to you. In an office, the only landline may be the one you use for your fax machine. Consider at least one line on each floor and in your EOC.
4. *Crisis supplies* are recommended. If there is any possibility, your facility should become a shelter or a place for triage, such as in a mass casualty event or fire. Crisis supplies are also recommended for EOC operations.
5. *Visitor management.* Always have visitors and guests sign in and present a government-issued ID. Consider using the Walmart approach, "Hi, how can I help you?"
6. *Social media use department pages rather than individual employee pages.* Social media are an excellent resource for getting the message out. They are also subject to security breaches and employee abuses, whether malicious or intentional. Ensure that your Computer Use Policies include information on social media use in the workplace.
7. *Prepare to use your crisis communications plan for planned events* such as sporting events or fund-raisers where large crowds are anticipated. Planned events can quickly become unplanned events. Using the plan is an opportunity to test its effectiveness and to provide training.
8. *Safety is everyone's business, not just the police's.*
   a. For facilities with open parking, the first place to protect is the facility, followed by parking, on the grounds and for offsite events. This is during normal operations and emergencies.
   b. For each outside organization, have a security addendum that includes
      i. If you or someone else see a crime, call 9-1-1.
      ii. Provide a list of attendees at least 24 hours in advance.
      iii. Organizations that will use parking must park and enter in a specified location such as during voting or a conference.
      iv. Access to the rest of the facility should be prohibited.
9. *Keep your doors closed and consider security for your EOC.* In a crisis, you never know who may be in the area or interested in knowing more by looking through windows or walking in on the group unannounced.
   a. Consider adding some form of access control to this room to restrict access.
   b. When the room is not in use, it should be secured and monitored.
   c. Walk twice daily to check doors, and report, record, and lock doors.

10. *Evacuation maps.* Post evacuation maps by the door that include the assembly point and locations of any fire extinguishers in the area.
11. *Emergency preparedness*
    a. Post bomb threat and suspicious package instructions on the wall and by every phone.
    b. Use crisis flipcharts and color code them such as RED for fire and hazmat, BLUE for security, and SILVER for missing persons.
    c. Provide each member of the crisis communication team an emergency supplies backpack. You can assemble these backpacks yourself or purchase them for about $50. Items to include are
        i. Six bottles of water, cups, energy bars, orange safety vest, rain poncho, flashlight, batteries, mylar blanket, flashlight with hand-cranked energy source, clipboard, glow sticks, tarp, duct tape, toilet tissue, whistle, paper, and pen
    d. Conduct drills and exercises annually:
        i. Three fire drills (one monitored by the fire department)
        ii. One bus evacuation if you use buses
        iii. One shelter-in-place
        iv. One lockdown

MARKETING EMERGENCY SERVICES AND YOUR PLAN

> Man's mind, once stretched by a new idea,
> never returns to its original dimensions.
>
> Oliver Wendell Holmes
> *(1809–1894)*

Once an organization has experienced a crisis, it never returns to its original state. Your stakeholders may not know what a crisis communications team is, understand what you do, or anything about the crisis communications plan you have in place, but they do know if you are communicating with them in a crisis. Your stakeholders also know when you are not communicating. As long as they do not understand what you do, you will not get what you need in terms of support. It is time to embrace branding for the emergency management profession and the role you have as a member of a crisis communications team.

A big problem from your stakeholders' perspective is that it is hard to get financial and management support unless a big disaster occurs, disasters such as Hurricane Katrina or when the federal government opened spillways along the Mississippi River in a misplaced move to spare Louisiana's two largest cities from flooding in 2011. Flooding occurred along the Mississippi River affecting thousands of homes and acres of crops in less populated areas for the first time since the floods of 1927. National attention was cast on an area that has been underfunded for years—the nation's spillways. How you handle catastrophic events can restore or increase trust and integrity in your organization. It is time to prepare to prosper by changing your mindset, your philosophy.

The crisis communications team must know who their target audiences are. Addressing the communication requirements among the least of your audiences helps to ensure that you are attempting to reach all your audiences. With 24/7 media attention, it is critically important this is not overlooked. We must be careful in how we label these audiences yet meet the needs of the most vulnerable individuals. For example, the label "senior" is often stereotyped as being of low income and without financial resources. However, if the label "low income" or "poor," defined as someone without or with very little financial resources, is used instead, you may adjust the communication tools you use to reach the poor. Text messaging, Twitter, e-mail, or Facebook would be secondary communications media. By using only these tools, you are not reaching many of the poor. Instead, you would use television, word-of-mouth, flyers, or radio to reach the poor. Another example is the push for emergency kits. Many emergency kits are advertised for sale for $50, with pictures and statements on the items these kits contain. In trying to market the purchase of an emergency kit to the poor, you will not capture their attention, or, worst yet, it will become a reminder of what they do not have. They do not have the money to buy the basics; the emergency kit is out of the question. For the poor, the struggle is for the basics—making it to their next payday, choosing between paying the rent and utilities or buying food and medicine. Day-to-day existence is their focus; there is no discretionary income available to prepare for what might happen. Hurricane Katrina's aftermath was a reminder that the United States has large pockets of low-income people who do not have access to transportation during an ordered evacuation, nor do they have the resources to shelter-in-place without assistance.

Marketing is really targeting the middle class with money to leverage. The wealthy have money to invest and conserve; they pay less attention to marketing. Both audiences are better able to prepare and have access to the many methods you select for disseminating emergency messages.

The crisis communication plans must be culturally sensitive and take into consideration people who speak other languages. A message stated in English for a U.S. market is successful; the same message in another culture or where a different language is spoken is a fiasco. For example, "Got milk?" in the United States was cute and catchy. For the Spanish speaking, "Got milk?" takes on a radically different meaning. It is not funny not having milk for your children because it is a basic item that some cannot afford. The "Got milk" message was never translated into Spanish before it was launched. The same applies to your crisis communications plans and messages. Taking your message written in English and then using a free Internet translation tool to convert the message into Spanish may have your message taking on a different meaning, not be understood, or even be offensive. A professional translation service or a member on your team who is fluent in the second language are better resources to assist you with the translation of a message.

Crisis communications may include evacuation and shelter instructions, yet those with pets and animals are overlooked during planning. Some pet owners make life and death decisions based on their pets and livestock. They will stay behind if they cannot bring their pet or if they cannot move their livestock to safety. These people think, "... after today's disaster is over, what am I going to do?"

Mentally and cognitively disabled, physically disabled, and people fearful of receiving services from the government—how do you reach these audiences? Door-to-door communications is the method of choice; however, in going from door to door, you may miss some of these individuals. You miss them not because they are not on the other side but they may not be able to hear you knock or call to them, or will not respond to you at the door. If you send a police officer to the door, a person who is fearful of people they do not know, or a person who refuses government services, may not answer the door or even your telephone calls. This means the call made using Reverse 9-1-1 may receive no response from the listener. The answering machine may have picked up the call, the people may have even heard the message, but because of who you are, they will not take the desired action.

These same individuals, if viewing your message on television or receiving the same message from an intermediary or trusted source, such as a parent, guardian, family member or caretaker, will take action. You must first reach their intermediaries before you can reach them. These same factors must be considered for the culturally isolated, individuals having little interaction outside their chosen community, for example, religion, sobriety, lesbian, gay, bisexual and transgender (LGBT), geography-caused isolation, etc. Other subgroups to consider are those that are hospitalized, incarcerated, or transients. The transient subgroup include tourists, people needing replacement hearing aids or glasses, etc. The crisis communications team has to market differently to any group that is stigmatized.

Online advertising increased 192% from 2000 to 2008, and another 24% from 2008 to 2012.[5] Meanwhile, television, newspaper, magazines, radio, and outdoor media have declined over the years. In emergency management and crisis communications, the individual decides whom to trust.

The campaign for Viagra is considered the most successful marketing campaign ever. It used then-Senator Bob Dole. He was short-lived as the spokesperson due to age at the time; he was 75 years old. Viagra was selling to 35-year-olds, yet marketing to 75-year-olds. Viagra dominated the market in 2000; today they have 50% market share.[6] Once competition enters the marketplace, it is difficult to maintain the lead. The minute you have a choice, the numbers plummet.

Want, need, desire, fear—every message has at least two of these characteristics. When you think of a car that has had a previous owner, the car is marketed as either "preowned or "used." The dealer looks to appeal to a want, need, or desire. When selling exercise equipment, a picture of someone exercising on it will sell the product because it satisfies something the buyer wants, needs, or desires. A message stating a hurricane warning is in effect targets your fear and need to take action to protect people and property. What images would you use, what words would be said, what feelings would you try to evoke?

Products are marketed using images. A sample of an image is a bargain "beauty shampoo" in an attractive bottle with an attractive female shampooing her hair. This gives the illusion that you are smart, successful, beautiful, environmentally friendly, smell good, have an incredible sex life, save money, etc. Sales are up since the start of the campaign. You will spend money to feel good, so this means the marketing campaign is working. Conversely, would you sell a shampoo using the same woman washing her hair with a bottle of "ugly-b-gone" shampoo, for example? This image conjures up fears, concerns, and insecurities. What images would you use, what words would be said, what feelings would you be trying to evoke? Would you spend money to scare people? It doesn't work.

Context and framing are crucial. Your emergency communications should provide positive actions. Most of the preparedness message your see and hear say "... get prepared for disaster. ..." This is a negative reason because it is reminding you that something terrible is coming your way, for example, earthquakes, fires, floods, and flu. Positive messaging giving instructions presents you as loving and caring like toward your spouse by reminding yourself to do dishes or send flowers. Giving the same instructions from the standpoint that "you need to do this or else ..." are negative. The reminder "You need to send flowers and do the dishes because divorce is expensive, and you could lose your house, you will be financially ruined, and things will get ugly if you don't"—the same message is now negative.

Most disaster campaigns target their audiences as potential victims. This is negative messaging. Frame your communications and preparedness messages around people reflecting the messenger as caring and having empathy with the audience, not the crisis. This is positive messaging.

Using fear does not work long term. Consider the "Click It or Ticket" campaign. For this approach to work, there must be consequences. This campaign works because there are teeth involved, such as a citation for failing to buckle your seatbelt. Most emergency management plans do not work because there are no "teeth" for failing to adhere to the plan. The same is true of your crisis communications plan.

The use of threats to push products or increase brand recognition is another approach used. Political campaigns use this method regularly because they just scare you enough to vote for the person or cause they are representing. These campaigns are not about long-term change. The DARE program attempts to scare and threaten American students. An authority figure tells you what to expect, what to do, and explains the horrible things that will happen to you if you use drugs. Long-term, this program did not work. Abstinence-Only Education is another program with limited success. Teens forming sex circles are growing even in areas where Abstinence-Only Education is provided.

Using disaster imagery in consumer education stops people from taking preparedness action.[7] Your target audience may forget what you said, but they will never forget how you made them feel.

Components of Effective Messaging

The components needed for an effective crisis communiqué are

1. Message
2. Messenger
3. Medium/media of delivery
4. What to do
5. Why do it
6. How to do it
7. Timing

The outcome when each component satisfies the audience's requirements is an appealing and empowering message for all affected. Crisis communicators do a great job immediately after a disaster. The American Red Cross conducted a survey that found that, after the big California

2006 campaign "100 years after the big earthquake," only 6% of the population was prepared.

A part of many campaigns is to make your cell phone your best safety tool. Other useful opportunities to promote preparedness include all-hands staff meetings, potty poster (in each of the restroom stalls or on the wall), and the use of ICE. Be an example of preparedness. The cell phone is promoted as the way to quickly reach loved ones; it is your first line of defense, a requirement of the job. This is a continuous campaign. The underlying message this conveys is that we will be there, always ready; together we prepare, protect, and serve; three- to five-minute response; keeping America safe; help cannot wait. The true branding of response is "… if you can manage for three days, we will do all we can to get to you by then."

To get your preparedness efforts on track, spend one hour a week promoting preparedness from a positive outlook. Begin by

1. Asking the team hard questions
2. Learning as much as you can about the needs of the many subgroups that you need to service
3. Thinking in terms of reinventing your program by dissecting it and its messages
4. Starting with a single audience
5. Creating consumer profiles for every target market
6. Knowing your census and demographic data—become a specialist

Think of how you will deliver your promotional preparedness campaign and how you will package it. Brochures tell you what you should have done. A potty poster uses big image with few words. The same target audience may see both. The brochure will land in a stack of papers on the desk or, worst yet, in the trash. It may be read once. The potty poster is seen each time an individual goes to the restroom. People will see it more than once, and the information will be retained longer.

Embrace the trend of using social media. People today communicate this way. Educate senior management. Use tools such as LinkedIn to establish public/private partnership. Twitter is a favorite since it is faster than the people right there with you. Use Web 2.0 apps such as Facebook, WordPress, HootSuite (manages multiple platforms), and schedule postings on websites in advance. Use of social media tools require minimum levels of preparedness for staff, volunteers, etc., since many use these tools to communicate on a day-to-day basis. Use plain language, clear text, and common experiences in your messages. This gives your audience a way to relate. Most know how to cook a meal, few know how to be a pilot. This is the couch potato form of preparedness and messaging—sitting in front of a screen. Stay relentless by threading preparedness through everyday activities. Most importantly, minimize the use of fear in promoting preparedness and creating your crisis messages.[8]

OTHER CRITICAL PLANS TO SUPPORT EMERGENCY OPERATIONS

Crisis communication plans are one of several plans an organization should have in place. Most organizations have a disaster recovery (DR) plan. DR plans are primarily focused on IT operations—the graceful shutdown of systems, migrating systems to alternate locations, specifically mission-critical systems, and restoring services. Technology is important and needed in a crisis for gathering intelligence, sending messages, and tracking, in addition to its use in supporting day-to-day operations. Other plans include your Emergency Response Plan (ERP), Business Continuity Plans (BCP), Mitigation Plans, and any departmental plans.

The *ERP* is a tactical plan. It details your structure, who to contact, and what you are to do before, during, and after a crisis. An ERP is your statement to your stakeholders of your commitment to preparedness should a disaster occur. Core teams involved are security, emergency management, operations, maintenance, and health and safety. It includes the organization's actions as it relates to

1. Health and safety of its stakeholders
2. Loss of property and environmental affects
3. Preservation of data
4. How you will respond to an event
5. Links to the crisis communications plan
6. Satisfy regulatory obligations

The ERP is activated whenever a crisis occurs that threatens life, disrupts operations, or causes damage to the organization, its property, or the environment.

The BCP, also referred to as *Continuity of Operations,* is focused on quickly restoring your operations to normal. It includes

1. An assessment of your organization's functions that are deemed mission critical
2. Establishing procedures for succession management
3. Documenting your vendors and other resources and how to obtain them

4. What to do if your facilities are not available
5. Defining individual responsibilities
6. Links to your crisis communications plan and ERP

The core teams involved in the activation of the BCP are human resources, IT, operations, purchasing, and finance and administration. The BCP covers both tactical and long-term plans for the variables listed and the critical operational capabilities that give the organization the ability to

1. Receive classified and unclassified information from trusted sources
2. Assess the implications of threats and risks
3. Gather, generate, analyze, and share information

THE HUMAN FACTOR—MENTAL HEALTH AND FIRST AID

Mental health is a functional part of health. This becomes glaringly apparent in a crisis when "every head is another world." This saying holds true in our everyday lives, but the stress of a crisis can unveil the extremes in human behaviors. It is a place you have not been—someone else's head. The challenge first responders, emergency managers, and communicators have is learning the worlds that are in your universe before, during, and after a crisis. You have to look at everyone as individual worlds even though they have been clustered into audiences, some your target audiences. This book has reviewed people with disabilities and other functional and access needs as subgroups. The same holds true for other subgroups, but thus far the book has not reviewed who could be included within an audience, a subgroup. Everybody includes

1. Local residents
2. Residents from outside the county, parish, or state
3. Citizens from other countries
4. Physically fit and challenged
5. Employees with functional needs
6. Wounded warriors and veterans
7. Elderly
8. Children
9. Pregnant women

The world each person represents is complex and with many different focuses that constantly change in priority with or without a pattern or warning. The list of the many worlds that exist within each head is exhaustive, such as

1. Employment
2. Family back home
3. Financial matters
4. Friends
5. Mission propaganda
6. Partying
7. Politics
8. Religious convictions
9. Significant other issues
10. Special interest
11. Vacations
12. Health matters

How each person maintains his or her world is as varied as there are people, such as

1. Paychecks from work
2. Income from business
3. Loans, scholarships, inheritance, grants
4. Government benefits
5. Special interest groups
6. Status in society
7. Expectations
8. Support
9. Medication

At some point, these worlds will collide. Cool heads can maintain calm even during a crisis. However, when their safety or health is threatened, or that of someone they have something in common with, or there is possible loss of property, cool heads can panic and act irrationally. When these worlds collide, some experience unmanaged stress that introduces other issues, including trauma. The perception of the individual's world has changed and is suddenly different in a crisis whether this perception is real or not. Perception is calling it as we see it, not necessarily seeing it as it is.

In times of crisis, people are motivated by deficiencies in their basic needs. *Maslow's Hierarchy of Needs* states how our needs motivate us. Maslow states that our most basic needs are inborn, and we must satisfy life's most basic needs, biological and physiological needs, first for mere survival. *Biological and physiological needs* are air, food, drink, shelter, warmth, sleep, etc. As these needs are satisfied, we are then concerned

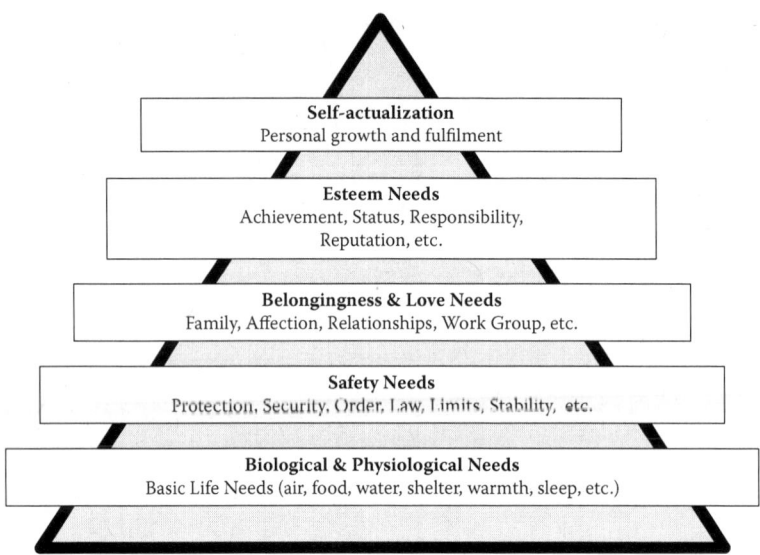

FIGURE 15.1 Maslow's Hierarchy of Needs.

with the higher-order needs of safety, followed by belongingness and love, esteem and, finally, self-actualization needs. Conversely, if what satisfies our lower-order needs are gone or suddenly stripped away, we are no longer concerned about the maintenance of our higher-order needs. People are motivated by their deficiencies in their basic needs. (See Figure 15.1 Maslow's Hierarchy of Needs.)

This is where organizations like the American Red Cross (ARC), which provides food, water, and shelter, are activated in a crisis. The role of the ARC is to first help us satisfy our lower-order needs. The ARC works with public safety groups and local officials to deliver the basic safety needs, the second order of needs. As these levels of needs are satisfied, we begin to return our focus on subsequent levels of needs. It is important to keep the human factor in the forefront when developing our communiqués in a crisis. With each message, the communicator should use a list of questions to validate that the message satisfies the requirements. The list must include the question "Will the message incite panic or calm?"

The response strategies of first responders are not to rush a person, crowd his or her personal space, or make any attempts to force an issue. The reason behind this approach is that anything could backfire causing unwanted behavior such as violence. As a communicator, you must remember that survivors are usually excited, alarmed, or confused. They need to feel in control. Your message must consider this. A crisis does not breed logic; rather, logic goes out the back door and emotions enter through the front door. Individuals think in terms of survival and what feels right to them now. This reaction in a crisis will lead people to do the unthinkable, such as risking harm to take in a stranger or rescue an animal.

The language and tone of your message makes a difference. Language has meaning. Your message should relay calm yet firmness by providing instructions. The message should be polite and first address basic needs, followed by security needs, then any remaining needs as defined by Maslow. In a crisis, it is also important to keep the message simple. When speaking in front of a camera or face to face, position yourself at eye level and maintain contact. Repeat your message for the sake of your audience to increase the likelihood that it has been heard and understood.

Above all, establish creditability with your audience. Begin by introducing yourself and showing that you care as you share your message. Once your message has been disseminated, then listen and be prepared to diffuse rumors and help people restore normalcy.

In summary, the first few minutes, more specifically, the first six hours, are the most critical in a crisis for an individual. Your message must take into consideration what motivates people. Each crisis is unique and will yield a different response; therefore, you must be flexible. Creditability is critical as the crisis evolves and eventually the situation returns to normal.

OPPORTUNITIES IN PUBLIC/PRIVATE PARTNERSHIPS

The recent economic crisis has had a ripple effect across the country largely because of budget shortfalls. These shortfalls have affected every level of government, with few organizations immune from its effects. The result has been a reduction in the public sector's ability to deliver services, as resources have been stretched to the limit. The effect on the private sector has decreased many organizations' ability to create jobs and boost competitiveness, as they look for clear signals that the economy has moved into recovery. Whether public or private, the pressures to reduce spending and find new opportunities to increase the bottom line require creative thinking.

The search for answers is reflected in trends indicating that the public sector is working with opportunities that may exist for emergency management. For example, the new FEMA PLAN program is an excellent example of a public sector–private sector parnership. Contractual agreements, called *Public Private Partnerships* (PPPs), have been implemented widely outside the United States, and are increasing with great success within.

PPPs are partnerships that involve the sharing of a project's rewards, costs, and risks. It is neither privatization nor procurement that the government is buying a good or service from the private sector. It is a collaborative venture that is mutually beneficial to both parties, and it supports the greater good of society. Efficiency gains and fiscal benefits of PPPs are assumed to be benefits of such a relationship. Additionally, PPPs are thought of as a better option than privatization, which, many believe, puts profits ahead of people and results in the marginalization of basic services.

The argument against PPPs is that transparency and accountability may be lacking, there is mistrust of the private sector participation, expressed by "what can they up-sell by providing this service," and the erosion of workers' rights. Sprint and Tracfone are two companies that have taken advantage of a type of PPP in the U.S. wireless communications industry. Both companies have a federally sponsored program providing wireless service to the poor. The user receives a free phone and 250 free minutes of talk time each month. Sprint's program is run under Assurance Wireless, and Tracfone's program is run under SafeLink Wireless. These programs are beneficial to consumers who qualify, and these businesses receive federal government subsidies to the tune of $10 per subscriber per month. These businesses also have new customers requiring little overhead for recruiting. The Universal Service Administrative Company (USAC) pays the carriers. USAC is the organization serving as the manager of a federal fund; U.S. telecommunication companies are required to contribute fees. The $10 is about the same spent by the average paying wireless user in the United States. This has been a win-win situation, and new entrants will soon enter the market as other carriers seek government approval to offer this free service.[9] PPPs create viable projects when combing the resources of the private and public sectors.

To learn if a PPP is a good fit for your organization, a feasibility study is recommended. Areas to visit are cost, technology, demand, value, and risks. Lessons learned from other countries that have used PPPs for an extended period are

1. All necessary stakeholders should be included in the process before approval.
2. Any government policy must have clarity.

The benefits of PPPs are

1. The use of competition and incentives can enlarge access, contain costs, and impel innovation.
2. Commercial risk can be distributed among partners.

3. When higher levels of transparency are present, the arrangement can provide a higher level of acceptance and trust by consumers.
4. If managed correctly, better quality by experienced private sector operators is provided at similar rates to the public sector.
5. PPPs enable the extension of services into remote or sparsely populated areas by requiring regional coverage. Subsidies may be offered to increase coverage and significant penalties levied if the private sector cannot fulfill this obligation.
6. PPPs offer the opportunity to transfer economic power to the local population through business ownership and better participation.

The concerns most cited were

1. Canceling or ending early a PPP deal that is poorly constructed could be costly for the public sector and could result in moving forward with a project that otherwise would be canceled.
2. Small operators could be excluded if one operator is awarded the agreement.

Politically, some view public/private partnerships as a way to address gaps in government services and an alternative to privatization. Privatization is thought of as a solution that eventually leads to job losses, higher prices, the sale of public assets, and long-term possible corruption. Some view privatization as a win–lose proposition—a win for business and a loss for consumers. This makes privatization a tough sell for the government. For the private sector, a PPP may not be advantageous if the potential financial gain is not attractive, but this is outweighed by the lack of strong political backing. Project viability and politics are affected by pricing.

To overcome these challenges, the parties need to gain buy-in from their stakeholders. Factors that can aid this process are transparency and consensus building. There are also risks, if demand is miscalculated, which impacts the bottom line, the financial outlook is negative. A thorough feasibility study is required to help mitigate the risk through intelligence. A third challenge is to ensure compliance with the deliverables outlined in an agreement. The public sector must identify ways to manage the relationship through monitoring and regulations to ensure compliance.

In summary, PPPs can present many of the same challenges inherent in any procurement process. PPPs must be cost effective, transparent, provide a venue to manage risks, and offer a balance between value and costs. The goal of PPPs is to provide better service at an affordable cost, using experienced private operators with a proven record of accomplishment of success.

THE FUTURE STARTS NOW

Crisis communications is an area of emergency management that can define success or failure in the management of a disaster. It uses a deeply interconnected ecosystem of diverse players, sometimes with competing interests, such as legal and public affairs or public and private organizations. The challenge is to have these stakeholders in the ecosystem to service the same goal from their various perspectives. Changes are needed to solve the mesh of connections and strike a balance among stakeholders' requirements. How this happens is through transformation.

Organizational leaders and the crisis communications team should support a strategic and comprehensive assessment of the hazards that could affect their organization from the most probable to the least likely. Radical options should be considered, which include significant use of different technologies as transformation enabler. Inspiration should come from other areas within an organization and the community that led to reinvention, innovations, and improved quality. This is recognizing that emergency management and crisis communications is about preparedness/prevention, response, recovery, and mitigation with the goal of maximizing the health and safety of the community and minimizing the loss of property. The diversity of solutions is remarkable as are the opportunities for public and private partnerships for integrating facilities, personnel, technology, processes, and distributing costs over a larger group of interests.

An informed strategic and comprehensive assessment led by the crisis communications team should address

1. Who you serve
2. Processes that should be used covering how and where
3. People needed to make it happen
4. Technology, equipment, and supplies
5. Funding

The assessment should take a fresh view of the scope and vision of the organization, how members of the crisis communications team should interact among themselves and with their external partners, and what is needed to achieve the desired outcome.

If an organization is to progress from where they are today to a higher-quality outcome, any incentives, processes, and priorities must be aligned. Consideration of a diverse set of stakeholders must be included in all that is decided, which is not a small task. Opportunities abound to thrive and provide cost containment coupled with higher quality. The result is better crisis communications that reaches more people quicker.

Organizations must be creative and identify opportunities to increase agility, prepare for the future, and be more decisive and responsive to change. Organizations must systematically accelerate their ability to adapt to change in the midst of a crisis. This can be achieved as organizations learn next-generation governance, change and risk management, and compliance that guide the decision-making process. What is learned in a crisis can be leverage throughout an organization to create advantage.

Integration has become a strategic imperative in crisis communications and successful emergency management. During a crisis, personnel learn to reexamine critically information and intelligence gathered, modifying messaging requirements, workflows, and processes on demand. The crisis communications team puts the affected audiences first. The benefits of planning, combined alerting systems, and other technologies, excellent intelligence and information, management, and dedicated personnel are realized in a crisis.

Another key to success is working in partnership—the organization's capacity to collaborate with its diverse set of stakeholders so that it can do its part to promote safety and good health and realize reduction in the loss of property. The ability of an organization to be customercentric and provide emergency messaging through pooled resources can speed the delivery of the crisis communiqué to a broader audience through PPPs. These partnerships have produced notable outcomes.

Sustainable change is possible through ongoing exchange of information among stakeholders, and change and risk management. The potential use of technology to facilitate crisis communications holds great promise. The crisis communications team needs to retain the ability to be inquisitive and should have an open mind as a prerequisite for success. The greatest challenge is to share risk and to devise a crisis communication program that supports and aligns with the collaborative efforts among stakeholders. Considering the world in which we live, the investment in crisis communications and mass notification and how effectively we communicate are the starting and end points for defining success for an organization in a crisis. We can do everything right, but if communication is lacking, the perception will be that we are at best short of success and at worst a failure.

SUMMARY

Communication is crucial in a crisis. It can be the key to defining success in how you handle a crisis. Organizations must be prepared to communicate regardless of whether technology is available. Communicators should ensure that the environment in which they will operate has all of the tools they will need to be effective in performing their responsibilities, such as an emergency kit with flashlights and batteries. Batteries for cell phones are also important, as well as a list of essential contacts programmed in your cell. It is important that you the communicator work constantly to build relationships since you may also be the spokesperson for your organization, the face of the organization.

In this chapter, a number of tips have been shared such as the ICE information on safety. This chapter has reviewed information on how to market your crisis communication plan and emergency services. Samples of good and bad marketing approaches have been provided. Most disaster campaigns target the potential audience as victims—which contributes to negative messaging. Your target audience may forget what you said but will remember how you made them feel.

Other plans that are linked to your crisis communications plan have been referenced. The Emergency Response Plan, Disaster Recovery Plan, and Business Continuity Plan are each significant in your overall emergency management program.

Mental health in a crisis is about every head is its own world, and there are many worlds in your universe. Understand that the many variations in the worlds that represent your stakeholders can define your message and will determine whether it will be successful. Shaping the message means verifying that you have addressed the needs of your audiences, starting with the most basic needs of biological and physiological and safety. As you move up the Maslow Hierarchy of Needs, you are moving away from the crisis stage to recovery and ultimately normalcy.

Today's world is about leveraging relationships and establishing partnerships between the public and private sectors. The FEMA PLAN program is an excellent example of such a partnership. These have many benefits when managed effectively. The future starts now through integration, partnership, and sustainable change.

ENDNOTES

1. Keane. (2010). Keane white paper. Getting it right the first time. Boston, MA.
2. Briese, Garry. (2010). *Do you know who you need to know?, Leadership in crisis situations: Five critical skills*. Fairfax, VA: Briese And Associates.
3. Elway Research, Inc. (2008). *Welcome to the Age of Communications*. http://www.elwayresearch.com (accessed February 1, 2011).

4. Mehl, Matthias R., Vazire, Simine, Ramirez-Esparza, Nairan, Slatcher, Richard B., Pennebaker, James R. (2007, July 6). Are women really more talkative than men? *Science*, 317 (5834): 82, DOI: 10.1126/science.1139940. http://www.sciencemag.org/content/317/5834/82.abstract (accessed May 5, 2011).
5. Jones, Ana Marie. (2010). *Marketing emergency services: A plan to prosper approach*. Oakland, CA: CARD—Collaborating Agencies Responding to Disasters.
6. Jones, Ana Marie. (2010). *Marketing emergency services: A plan to prosper approach*. Oakland, CA: CARD—Collaborating Agencies Responding to Disasters.
7. Lopes, Rocky. (1992). *Public Perception of Disaster Preparedness Presentations Using Disaster Damage Images*. http://www.colorado.edu/hazards/publications/wp/wp79.pdf (accessed May 12, 2011).
8. Jones, Ana Marie. (2010). *Marketing emergency services: A plan to prosper approach*. Oakland, CA: CARD—Collaborating Agencies Responding to Disasters.
9. Bloomberg. (2011). Free cell phone service for the poor. *Bloomberg Businessweek Magazine*. http://www.businessweek.com/magazine/content/11_04/b4212019667595.htm (accessed May 5, 2011).

CHAPTER 16

Conclusions

An increasing number of public tragedies have too often led organizations to learn in the midst of an emergency that there has to be a better way to warn people in a timely manner. Legislative requirements at every level—federal, state, territorial, tribal, local, industry—are asking organizations to do more to warn more people in time when their health and safety may be at risk, loss of property is possible, or the environment can be harmed. Organizations with inadequate emergency notification systems face a serious threat of failure in getting the right message out the first time to those affected in a timely manner.

Crisis communications and mass notification requires preparedness, and events around the world have triggered the need for crisis communications planning. From terrorism and cyber-sabotage of critical infrastructures to bad actors within your own organizations, the event you could face may be around the corner. There are natural disasters such as hurricanes, earthquakes, and floods that have brought large regions to a standstill. In the United States, these disasters led to a record 81 presidential declarations in 2010. The average is 34 a year. Crises are occurring with greater frequency and increased severity, and, more often than not, they trigger multiple events. The costs in dealing with such crises are staggering.

Many organizations only have disaster recovery plans that are focused on IT operations and restoration. If a crisis communications plan exists, it is tucked away on the shelf to demonstrate you have a plan. Only a few are privileged to know that it exists, and even fewer have received any training or have used the plan. In other situations, a crisis communications plan does not exist until a crisis occurs. If the organization can afford it, it scrambles and hires a consultant. For the rest, they usually wing it.

Key challenges organizations face in building their crisis communications plans are many—organizational structure, industry of operations, the community where they reside, the facilities and people you used to carry out your operations, and other stakeholders.

The solution chosen for crisis communications and mass notification must offer the maximum amount of flexibility. Organizations without flexible and multiple emergency communications methods will find themselves at a distinct disadvantage. In this book, we have supplied you with what you need to do, how to go after the most value for your dollar, and how to get started. This book reviews technologies you already have, no-technology solutions, alerting technologies, and high-end solutions. There is a solution waiting for you. This book also provides valuable information to help you recognize the importance and availability of existing tools, minimizing the extent of potential notification delays, and messaging under pressure.

This book has extended new knowledge on mass notification systems and crisis communications needs that face organizations today, and provided insights into trends and guidance on the actions organizations can take to identify and use a system designed specifically for you using what you already have, without incurring additional costs. Building policies and procedures around any solution is what helps to make it work.

The materials covered gives readers a broadminded view of crisis communications and mass notification. The information gives the reader an understanding of emergency response communications, associated technologies, and target populations historically serviced and a conceptual understanding for the future. A review of the current legal landscape, key decision-making consideration, and deliberation needed for effective planning and preparedness help ensure that the right message is delivered the first time and in a timely manner. This book provides an overview of what is needed to select, test, and install your new systems and integrate it with your existing system.

This book reviews issues related to crisis communications and mass notification that are often overlooked, such as mental health in a crisis and public/private partnership. Found throughout the book is an emphasis on training and incorporating a continuous improvement process for mass notification and crisis communications.

Discussed here is the importance of support from the top, such as senior management, and having adequate resources, training and readiness to participate as a member of the crisis communications team. Members on the team must have the authority to act and make decisions given the information and intelligence available to them. The spokespersons for your organization are the members of the crisis communications team and every employee.

In the midst of a crisis, it takes time for traditional media resources to arrive, to be set up, and to begin gathering information. They are outside the yellow tape. Anyone inside the yellow tape with a mobile device is a journalist. These people are talking, tweeting, taking and uploading pictures and videos, and giving live reports. They are the new subject matter experts.

Emergency management and preparedness work best when organizations operate in such a manner that everyone knows who is in charge. Having a structure with someone with the expertise and authority to manage the situation can lower panic in the midst of chaos.

The first chapter of this book begins with early humankind whose communication was limited to small groups over short distances. At the present time, communications can be tailored so that multiple population subgroups with varying needs and capacity can comprehend a single message using the tool that is most effective for them.

Disaster communications involves a *crisis requiring a rapid response*. Where a crisis intersects with inadequate warning, significant leaps in crisis communications and mass notification are made. Today, the Integrated Public Alert and Warning System (IPAWS) is the program of choice.

The richness of the communication options used improves the effectiveness of that communication. Time-limited decisions are to be made quickly, and the instructions needed by the target audience must follow. How the audience responds to the information will determine what improvements are needed to the plan and the associated training required.

The changing realities of communications must consider the essential services needed by the portion of the population with disabilities and others with functional and access needs. This segment of our population represents those who are physically or mentally disabled, medically or chemically dependent, the elderly, children, homeless, and non-English speaking. Understanding

- *who* the target audiences are
- *what* the message is you want to convey
- *how* you will deliver the message
- *why* it is important that the receiver pauses to understand the message
- *where* people will be to receive the message
- *when* they will receive the message

Each of these variables are the W's of crisis communication. These W's must be answered for each incident.

Working collaboratively and first responders talking to each other, regardless of the agency they work for, is important. Communications interoperability is critical and is mandated by the federal government.

Emergencies will happen. Any organization or community can expect to experience emergencies that could threaten the health and safety of persons and property, necessitating the implementation of protective actions. Effective communication is a critical step in the management of an emergency within each phase of emergency management. Solving problems and "winning" in crises is a function of speed, decision making, action, reaction, collaboration, and swiftly applied common sense. Any approach should take into consideration not only what is legal but also what is ethical and morally correct.

The number of disasters continues to increase, leading to more presidential declarations in 2010 than any other year. There are many laws that affect emergency communications, and some have been reviewed in this book. To facilitate all the changes in the world of emergency communications are the presidential directives and the National Strategy for Homeland Security, which include a comprehensive database of documents you can use to develop your communication plans.

Social media is the second generation of websites providing users their own mini-websites and web pages. There is a high degree of participation and interaction among users of such media. Using social media for mass notification is a challenge to organizations, but individuals have embraced them. Information regarding an incident can come quickly, as desired by individuals, yet an organization can quickly become deluged by thousands of messages on Facebook, Twitter, blogs, and other sites. Establishing a social media presence is easy; maintaining its currency and relevancy requires a team and is an ongoing effort. The rules of social media and crisis communications in general is that people want accurate information quickly from what is perceived to be is a trusted source.

Among the commonly used social media tools today for crisis communications and mass notification are Twitter, Facebook, blogs, widgets, mobile phone apps, and geotagging. The sharing of content has also made these tools both attractive and security risks.

There are a number of solutions available today for use in a crisis—"no-tech" to "high-tech." No-tech is the use of pen and paper; high-tech solutions require the aid of a specialist to use and administer, such as sonic buoys and highway alert digital signs. In a crisis, you begin with what you have and work your way toward what you need. The desired goals are combined alerting systems and using creative ways to integrate and use the systems you already have.

Your goal is to have an effective crisis communication system that is able to advise your target audience quickly. You also seek a system that strikes a balance between information "underload" and "overload," the need for efficient processes and quality data.

To succeed in selecting a solution that fits your budget, meets expectations, and delivered on time is a tall order yet achievable. Taking the time to gather your requirements, conducting adequate testing, and managing the project effectively from start to finish can improve your chances of success.

The purpose of writing a crisis communication plan is to use the resources available to you and understanding that communication is a tool to help solve problems. Organizations with a crisis communication plan are usually better prepared because they have begun thinking of what to say, why it should be said, and who is needed to say it, knowing that each factor could make the difference between keeping your employees and closing your doors. You can write your own crisis communication plan for your organization using the material covered in this book.

Getting the message right the first time can save time, energy, and money, and it may be your only chance to get your message out. The goal is to take a negative situation and crisis to your target audience and communicate to them what is going on around them or to them, and what action they should take. The information must be clear, concise, and easy to understand.

Communication is crucial in a crisis. It can be the key to defining success in how you handle a crisis. Organizations must be prepared to communicate regardless of whether technology is available. Other plans that are linked to your crisis communications plan are referenced such as the Emergency Response Plan and Business Continuity Plan.

Mental health in a crisis is about every head being in its own world, and there are many worlds in your universe. Understanding the many worlds among your target audiences is critical in creating your message. The better you know your audience, the likelihood increases that your message is received, understood, and appreciated. Shaping the message means verifying that you have addressed the needs of your audience, starting with the most basic biological and physiological needs and safety.

Today's world is about leveraging relationships and establishing partnerships between the public and private sectors. The FEMA PLAN program is an excellent example of such a partnership, which yields many benefits when managed effectively. Exploration and expansion of PPPs is envisioned in the future.

In reading this book, the future starts now when you learn that, through integration, partnership, and sustainable change, the desired outcome is possible. The materials covered give you what you need to select the right mass notification system for your organization to address your crisis communications needs. You may want to take action now when money may not be your biggest hurdle, considering the many free solutions available. Create your crisis communication plan using this book as your guide and then select the technology to support your needs beginning with what you already have.

I have witnessed a manufacturing operation experience significant power failures only to find that the backup generator, although tested, did not work. Hours of manufacturing operations were shut down, resulting in a significant loss in production and revenues. I have experienced an active shooter at work and witnessed the panic when no one knew what to do. As long as shots were not fired, people stood like deer caught in the headlights, waiting for instructions. Through the many audits I have conducted and participated in, I have seen what it costs an organization to not be prepared, and the positive remarks they receive when they have a plan that has been tested and is current.

Despite the horror stories, it is the success stories that I am most proud of in working with organizations that have a plan. It is rewarding to help an organization develop a plan and know that it had been tested before the organization faces its first crisis. Your customers remember when you sent them a message with an accurate and timely status of a situation. You build credibility with your stakeholders when you can get that message right the first time. Although they may forget the last crisis, they will not forget you.

Appendix A
List of Acronyms

Acronym/Abbreviation	Explanation
3D	Three dimensional
3G	Third Generation
4G	Fourth Generation
AAC	Augmentative and alternative communication
AAR	After Action Report
ADA	The Americans with Disabilities Act of 1990
AM	Amplitude modulation
AMBER	America's Missing: Broadcast Emergency Response
ANSS	Advanced National Seismic System
API	Application Programming Interface
APT	Advanced Persistent Threat
ARFF	Aircraft Rescue and Fire Fighting
ARPANET	Advanced Research Projects Agency Network
ASAT	Anti-satellite
ASRs	Automated Speech Recognition System
ATCT	Airport Traffic Control Tower
ATSC	Advanced Television Systems Committee
AWC	Aviation Weather Center
BCP	Business Continuity Plan
C4ISR	Command, control, communications, computers, intelligence, surveillance, and reconnaissance
CA	Communications Assistant
CAD	Computer-aided dispatch
CALVES	Community Activated Lifesaving Voice Emergency Systems
CAP	Common Alerting Protocol
CATV	Community Access Television
CBRNE	Chemical, biological, radiological, nuclear, and high-yield explosives

Acronym/Abbreviation	Explanation
CCITT	Comite Consultatif International Telephonique et Telegraphique
CEO	Chief Executive Officer
CERT	Community Emergency Response Training/Team
CFO	Chief Financial Officer
CIMS	Critical Incident Management Systems
CIO	Chief Information Officer
CLI	Calling line identification
CMAS	Commercial Mobile Alert System
CMS	Commercial mobile service
CNE	Computer network exploitation
CNO	Computer Network Operations
COFDM	Coded orthogonal frequency division multiplexing
CONELRAD	CONtrol of ELectronmagnetic RADiation
CONPLAN	U.S. Government Interagency Domestic Terrorism Concept of Operations Plan
COOP	Continuity of Operations Plan
COP	Common Operating Picture
COTS	Commercial-off-the-shelf
COWS	Community Outdoor Warning Sirens
CPR	Cardiopulmonary Resuscitation
CSAD	Communications Systems Analysis Division
CW	Continuous Wave
DARPA	Defense Advanced Research Projects Agency
DARS	Digital Audio Radio Service
DDoS	Distributed denial-of-service
DEF	Discrete Emergency Frequency
DHS	United States Department of Homeland Security
DIRS	Disaster Information Reporting System
DME	Distance Measuring Equipment
DOC	U.S. Department of Commerce
DOJ	United States Department of Justice
DR	Disaster Recovery
DRM	Digital Radio Mondiale
DTV	Digital Television
DVB-T	Digital Video Broadcasting-Terrestrial
E911	Enhanced 9-1-1
EAS	Emergency Alert System
EBS	Emergency Broadcast System

Appendix A: List of Acronyms

Acronym/Abbreviation	Explanation
ECPC	DHS Emergency Communications Preparedness Center
EISEC	Enhanced Information Service for Emergency Calls
EMC	Emergency Management Coordinator
EMS	Emergency Medical Services
ENSO	El Nino-Southern Oscillation
EOC	Emergency Operations Center
ERP	Emergency Response Plan
ESF	Emergency Support Function
ESMR	Extended specialized mobile radio
EW	Electronic Warfare
FAA	Federal Aviation Administration
FACTA	Fair and Accurate Credit Transactions Act
FAOC	FEMA Alternate Operations Center
FCC	Federal Communications Commission
FCO	Field Coordinating Officer
FEMA	Federal Emergency Management Agency
FEMA PLAN	FEMA Personal Localized Alerting Network
FERPA	Family Educational Rights and Privacy Act
FM	Frequency modulation
FNARS	FEMA National Radio System
FNSS	Functional Needs Support Services
FOC	FEMA Operations Center
FRERP	Federal Radiological Emergency Response Plan
FRP	Federal Response Plan
FSE	Full Scale Exercise
FSS	Flight Service Station
FTP	File Transfer Protocol
GDP	Gross Domestic Product
GETS	Government Emergency Telecommunication System
GIS	Geographic Information System
GLBA	Gramm–Leach–Bliley Act
GPS	Global Positioning System
GSA	U.S. General Services Administration
GSM	Groupe Speciale Mobile
HAARP	High Frequency Active Auroral Research Program
HCO	Hearing Carry-Over
HCOHSEM	Harris County Office of Homeland Security and Emergency Management
HD	High Definition

Acronym/Abbreviation	Explanation
HDTV	High Definition Television
HIPAA	Health Insurance Portability and Accountability Act
HMGP	FEMA's Hazard Mitigation Grant Program
HSPA+	Evolved High-Speed Packet Access
HSPD	Homeland Security Presidential Declaration
HTML	Hypertext Markup Language
Hz	Hertz
IAP	Incident Action Plan
IC	Incident Commander
ICE	In Case of Emergency
ICP	Incident Command Post
ICS	Incident Command System
IM	Instant Messaging
IMS	Integrated Multimedia System
IMT	Incident Management Team
IOM	International Organization for Migration
IP	Internet Protocol
IPAWS	Integrated Public Alert and Warning System
IPCC	Inter-Governmental Panel on Climate Change
IRAC	Interdepartment Radio Advisory Committee
ISO	International Organization for Standardization
ISP	Internet Service Provider
ISR	Intelligence, surveillance, and reconnaissance
IT	Information Technology
JFO	Joint Field Office
JIC	Joint Information Center
KHz	Kilohertz
LAN	Local Area Network
LAR	Legally authorized representative
LDAP	Lightweight Directory Access Protocol
LED	Light-emitting diode
LEP	Limited English Proficiency
LGBT	Lesbian, Gay, Bisexual, and Transgender
LORAN	Long Range Navigation
LSCS	Lone Star College System
LTE	Long-term Evolution
LWP	Local Warning Point
MHz	Megahertz
MMS	Multimedia Messaging Service

Acronym/Abbreviation	Explanation
MOU	Memorandum of Understanding
NASA	National Aeronautics and Space Administration
NATO	North Atlantic Treaty Organization
NAWAS	National Warning System
NBA	National Basketball Association
NCAA	National Collegiate Athletic Association
NCS	National Communications System
NECN	FEMA National Emergency Coordination Net
NEIC	National Earthquake Information Center
NFPA	National Fire Protection Association
NG911	Next Generation 9-1-1
NGO	Nongovernmental organization
NIFOG	DHS National Interoperability Field Operations Guide
NIMS	National Incident Management System
NOAA	National Oceanic and Atmospheric Administration
NOC	DHS National Operations Center
NORS	Network Outage Reporting System
NRCC	National Response Coordination Center
NRF	National Response Framework
NRP	National Response Plan
NS/EP	National Security Education Program
NSF	National Science Foundation
NTIA	National Telecommunication and Information Administration
NTSC	National Television System Committee
NWS	National Weather Service
OEC	Office of Emergency Communications
OEM	Office of Emergency Management
OS	Operating system
OSM	Office of Spectrum Management
PCS	Personal Communications Service
PCS	Picture Communication Symbol
PECS	Picture Exchange Communications System
PHI	Protected health information
PIN	Personal Identification Number
PIO	Public Information Officer
PMR	Professional Mobile Radio
POTS	Plain old telephone service

Acronym/Abbreviation	Explanation
PPD	Presidential Policy Directive
PPP	Public/Private Partnership
PSAP	Public Service Answering Point
PSHSB	Public Safety and Homeland Security Bureau
QAM	Quadrature amplitude modulation
QoS	Quality of Service
QR`	Quick Response
RACES	Radio Amateur Civil Emergency Services
RAM	Random Access Memory
RECC	Regional Emergency Communications Coordination
RFID	Radio Frequency Identification
RFP	Request for Proposal
RGB	Red Green Blue
RRCC	Regional Response Coordination Center
RSS	Really Simple Syndication
SCCT	Situational Crisis Communication Theory
SCEC	Southern California Earthquake Center
SCIGN	Southern California Integrated GPS Network
SCIP	Statewide Communication Interoperability Plan
SDLC	System or Software Development Life Cycle
SDO	NASA's Solar Dynamics Laboratory
SIM	Subscriber Identity Module
SMART	Specific, measureable, attainable, realistic, and timely
SME	Subject matter expert
SMR	Specialized Mobile Radio
SMS	Short Message Service (Text Message)
SOC	State Operations Center
SOP	Standard Operating Procedures
TACAN	Tactical Air Navigation System
TCL	Target Capabilities List
TCO	Total Cost of Ownership
TCP/IP	Transmission Control Protocol/Internet Protocol
TDD	Telecommunication Device for the Deaf
TETRA	Terrestrial Trunked Radio
TEWAS	Texas Warning System
TRS	Telecommunications Relay Service
TSP-R	Telecommunication Service Priorities for Radio
TTX	Table Top Exercise
TTY	Text Telephone

Acronym/Abbreviation	Explanation
TWC	Tsunami Warning Center
UASI	Urban Areas Security Initiative
UC	Unified Command
UHF	Ultra High Frequency
UMTS	Universal Mobile Telecommunications System
USAC	Universal Service Administrative Company
USGS	United States Geological Survey
USITA	United States Independent Telephone Association
USPS	United States Postal Service
UTC	Universal Time
UTL	Universal Task List
VCO	Voice Carry-Over
VHF	Very High Frequency
VOCA	Voice Output Communication Aids
VoIP	Voice over Internet Protocol
VOR	Very High Frequency Omnidirectional Range
VORTAC	VHF Omnidirectional Range/Tactical Aircraft Control
VPEAAC	Video Programming and Emergency Access Advisory Committee
VPN	Virtual Private Network
VRS	Video Relay Service
VSB	Vestigial Side Band
WAN	Wide Area Network
WHO	World Health Organization
WLAN	Wireless Local Area Network
WMD	Weapons of Mass Destruction
WPS	Wireless Priority Service
WWI	World War I
WWII	World War II

Appendix B

Sample Mass Notification System Activation and Criteria Guidance Sheet

Activation Level	Activation Decision Authority	Activation Criteria
Level 1 (Red Warning/Alert) Enterprise Emergency (All locations)	Vice President Security Director President Public Affairs	Catastrophic emergency immediately threatening the health and safety of employees and other stakeholders. MANDATORY: take action now.
Level 2 (Yellow Alert) Emergency Limited to one Location	Vice President Security Director Public Affairs Department Head Human Resource Manager	Emergency that directly affects a single location—health and safety matters. Evacuation, shelter-in-place, lockdown imminent
Level 3 (Green Advisory) Advisory	Vice President Security Director Public Affairs Department Head Human Resource Manager	Incident that may affect the location. Presents no immediate danger to the location or community. Could involve suspicious activities or materials.
Level 4 (White—Information Only) Advisory	Vice President Security Director Public Affairs Department Head Human Resource Manager	Situation requiring increased awareness such as traffic advisories that may directly or indirectly affect a location.

Appendix C

Sample Messages

Prescripted Emergency Communication Messages

Message	Text Message (<91 Characters)	Written Message (<301 Characters)
Test	THIS IS A TEST of [COMPANY] system. This system will keep you informed in an emergency. THIS IS ONLY A TEST.	This is a test of COMPANY Emergency Notification System. This system is used to keep you informed in the event of an emergency. If this had been an actual emergency, the message would have official information or instructions. This concludes this test of the COMPANY's Emergency Notification System.
Airplane Down	Aircraft down at [LOCATION]. Stay away from area and wait for further instructions.	Aircraft down at [LOCATION]. Stay away from area and wait for further instructions.
All Clear	ALL CLEAR. Normal operations have resumed.	ALL CLEAR. [LOCATION] has resumed normal operation.
Bomb Threat	Bomb Threat — evacuate [LOCATION] NOW! Stay away from the area and wait for further instructions.	A bomb threat has been received at [LOCATION]. As a precautionary measure, operations have been suspended. Evacuate the area now. If away from the area, stay away! Wait for further instructions. Go to our website at www.WEBSITE.xxx for details.

Prescripted Emergency Communication Messages

Message	Text Message (<91 Characters)	Written Message (<301 Characters)
Code Blue	CODE BLUE in progress at [LOCATION]. Emergency personnel are en route. Please stay away from the area.	CODE BLUE in progress at [LOCATION]. Emergency personnel are en route. Please stay away from the area.
Computer Disruption	Computer Alert: disruption in services at [LOCATION]. Please turn off your computer while we work to restore services.	Computer Alert! We are temporarily experiencing a disruption in services at [LOCATION]. Please turn off your computer while we work to restore services.
COMPANY Is Closed	The [COMPANY] is closed. Monitor your e-mail and www.WEBSITE.xxx for updates.	The [COMPANY] is closed. Monitor your e-mail and www.website.xxx for updates.
Evacuate Building	Emergency in progress at [LOCATION]. Evacuate NOW. Go to the assembly area. If away, stay away. Wait for further instructions.	An emergency is in progress at [LOCATION]. Evacuate NOW. Assist others as you leave. Take your valuables. Go to the assembly area outside. If away from the area, stay away and wait for further instructions.
Fire Alarm	Fire reported at [LOCATION]. Evacuate the area now! Wait for further instructions.	A FIRE has been reported at [LOCATION]. Evacuate the area now, and wait for further instructions before returning. Updates will be provided at www.WEBSITE.xxx and e-mail.
Flood Warning	FLOOD WARNING in effect for [LOCATION]. Do not walk or drive thru water. Monitor www.WEBSITE.xxx for updates.	A FLOOD WARNING is in effect for [LOCATION]. Be alert for high/rising water, and be ready to move to higher ground. Do NOT walk or drive into flooded areas. TURN AROUND...DON'T DROWN. Monitor current weather information. Updates will be provided at www.WEBSITE.xxx and e-mail.

Prescripted Emergency Communication Messages

Message	Text Message (<91 Characters)	Written Message (<301 Characters)
Hazardous Spill	Hazardous material release in progress at [LOCATION]. Seek shelter immediately, and wait for further instructions.	Hazardous material release in progress at [LOCATION]. If you are in the area, seek shelter now and stay there. Close all doors and windows. Turn off any heating or cooling system. Keep your telephone, radio, or computer on, and stay tuned for instructions.
Hostage Situation	Emergency in progress at [LOCATION] @ [TIME]. Police responding; stay away from the area. Details to follow.	Emergency situation in progress at [LOCATION] @ [TIME]. Police are responding. Stay away from the area. Details to follow.
Hurricane or Tropical Storm Warning	HURRICANE WARNING! All locations closed. Seek shelter now.	HURRICANE WARNING! All [LOCATIONS] are CLOSED. Please leave and seek shelter now. Monitor the weather, the www.WEBSITE.xxx website, and e-mail for updates. Additional weather information is available at www.weather.gov.
Hurricane or Tropical Storm Watch	A HURRICANE WATCH is now in effect. Stay alert, and be prepared to take action if a warning issued.	A HURRICANE WATCH is now in effect for the area. Stay alert, and be prepared to take action if a warning is issued. Monitor the weather, e-mail, and www.WEBSITE.xxx for updates. Additional weather information is available at http://www.weather.gov/.
Severe Winter Weather	Severe winter weather imminent. [LOCATION] closed. Stay alert for changing conditions. Monitor weather at www.weather.gov for updates.	Severe winter weather imminent. [LOCATION] is now closed. Regular operations are expected to resume (DATE/TIME). Please remain alert to changing weather conditions. Monitor the weather, e-mail, and www.WEBSITE.xxx for updates.

Continued

Prescripted Emergency Communication Messages

Message	Text Message (<91 Characters)	Written Message (<301 Characters)
Lockdown	LOCKDOWN! Armed person reported at [LOCATION]. Lock doors, and shelter in place NOW. If away, stay away. Follow police instructions.	LOCKDOWN! An armed person reported at [LOCATION]. Lock doors, and shelter in place NOW. If away from the area; stay away. Follow police instructions.
Medical Help Request	Immediate medical assistance needed at [LOCATION]. If a doctor, please report to the area immediately.	Immediate medical assistance needed at [LOCATION]. If a doctor, please report to the area immediately.
National Security Warning Attack	NATIONAL ATTACK WARNING issued by [AGENCY]. Go inside. Monitor TV and radio for information.	The [AGENCY] has issued an ATTACK WARNING. Threat of an attack on the United States is possible. Please go inside, take cover, and remain calm. Monitor your local television and radio stations for current information.
Power Outage	Power outage at [LOCATION]. Work in progress to restore services.	We have temporarily lost power at [LOCATION]. Please remain calm. Work is in progress to restore services.
Security Alert	An emergency is in progress at [LOCATION]. Exit building NOW. If away, stay away. Follow police instructions.	An emergency situation is in progress at [LOCATION]. Exit the building NOW. If away, stay away. Follow police instructions.
Shelter in Place	An emergency in progress at [LOCATION]. Seek shelter NOW, and wait for further instructions.	An emergency in progress at [LOCATION]. Seek shelter NOW, and wait for further instructions. Monitor the www.WEBSITE.edu website and e-mail for further instructions.
Suspicious Mail Alert	Suspicious mail alert. If you receive a suspicious letter/package, do NOT touch, open, smell, or taste. Call the police at 911.	Suspicious mail alert. If you receive a suspicious letter/package, do NOT touch, open, smell or taste. Call the police at 911 or immediately.

Prescripted Emergency Communication Messages

Message	Text Message (<91 Characters)	Written Message (<301 Characters)
Suspicious Person	Suspicious person reported near [LOCATION] @ [TIME]. [DESCRIPTION] Do NOT approach. Call the police at 911 if seen.	Suspicious person reported at or near [LOCATION] @ [TIME]. [DESCRIPTION] Do NOT approach. If seen, call the police at 911.
Tornado Warning	TORNADO WARNING for [LOCATION]. Seek shelter now. Stay away from windows.	A TORNADO WARNING has been issued for [LOCATION]. Seek shelter now — go to an interior hallway, and stay away from windows! If in a mobile building or automobile, leave and go to a strong building or lie flat in the nearest low spot and cover your head. Monitor the weather, www.WEBSITE.xxx, and e-mail for updates. Additional weather information is available at http://www.weather.gov.

Appendix D

List of Questions to Ask a Vendor before You Buy

1. Experience
    1.1 Credentials and registration to supply, install, and service the system
    1.2 Certificates of insurance
    1.3 Length of time providing mass notification systems
    1.4 Two to three references
    1.5 Will subcontractors be used?
    1.6 Hours of operations and service
2. Solution Requirements
    2.1 Alternate colored strobe light
    2.2 Alternate audible horn signals to differentiate between fire alarm and other emergencies
    2.3 Voice communications
    2.4 Sirens or horns
    2.5 Redundant amplifiers providing intelligible sound for broadcasts
    2.6 Enable VoIP announcements over loudspeakers
    2.7 Outdoor speakers and notification beacons
    2.8 Notification beacons to interface with external audiovisual devices such as sirens, strobes, televisions, and scrolling marquees
    2.9 Instant communications with mobile devices
    2.10 Sending of different messages to different areas simultaneously
    2.11 Audio commands able to override fire alarm signals
    2.12 Integrate with:
        2.12.1 Closed-circuit television/digital security camera system
        2.12.2 Voice messaging
        2.12.3 Message boards
        2.12.4 Visual signals

2.12.5 E-mail system
2.12.6 Text messages
2.12.7 Social media: Twitter, Facebook, web page, blogs, forums, etc.
2.13 Inputs to trigger the system to originate from:
2.13.1 Security system "panic" buttons
2.13.2 Firefighter phones
2.13.3 Fire alarm initiating devices
2.13.4 Company telephone
2.14 A point-to-point and point-to-multipoint system across the wide area or local area network
2.15 Support wireless networking
2.16 Narrowbanding integration
2.17 Adhere to NFPA 72, 2010
2.18 Type of licenses (by client installations, number of users, by location, for all locations and all users [enterprise])
2.19 Emergency Phone Tower System
3. System Requirements
3.1 Proprietary technology or open-architecture
3.2 Interoperability with other communications systems
3.3 Make a minimum of 1,000 contacts per minute
3.4 Capacity for handling a high volume of calls over a short period
3.5 Capacity limitations in handling a high volume of calls over a short period
3.6 Is the system able to adjust the volume of calls and alerts based on throughput?
3.7 Able to recognize human voice versus an answering machine
3.8 Able to wait until the answering machine or voicemail outgoing message has ended prior to leaving a message
3.9 Repeat redials for all unanswered numbers
3.10 Use a mapping tool to designate an area to be notified
3.11 Designate specific addresses within a radius around target areas
3.12 Prioritize notifications from closest to the event location and expand outward
3.13 System designed as an emergency mass notification system
3.14 Capable of sending notification events 7/24/365 with Internet access
3.15 Incoming phone number the same each time
3.16 Residents can call the number back and hear the last message sent with the date and time
3.17 Interface with National Weather Service (NWS):
3.17.1 Make automated Weather Warning calls without human interaction

Appendix D: List of Questions to Ask a Vendor before You Buy   469

        3.17.2 Use the Latitude/Longitude Polygon box issued by NWS with their warnings as a target for the systems automated Weather Warning calls
        3.17.3 Unlimited automated Weather Warnings with no price per call.
   3.18 Accept customer-supplied data (i.e., telephone numbers, cell numbers, e-mail addresses)
   3.19 Database of customer-supplied data easily maintained
   3.20 Easy-to-use web link to allow self-registration of additional contact numbers and devices, and change their information

4. System information
   4.1 Database management system (DBMS) used
   4.2 Supported systems and platforms
   4.3 Client operating systems
   4.4 Server operating systems
   4.5 Hardware platforms
   4.6 Hardware requirements and limitations
   4.7 Specifications on scalability
   4.8 Client/server architecture (describe supported tiers)
   4.9 Web browser/intranet architecture
   4.10 GUI interface
   4.11 Assign and revoke access
   4.12 User documentation, including documentation defining system errors and recovery procedures
   4.13 Online tutorial
   4.14 Any proprietary hardware or software technologies
   4.15 Licensing and ongoing support structure
   4.16 Additional license requirements for third-party applications

5. Security
   5.1 If a server or dialer fails, the system will self-heal and automatically roll over to another server or dialer.
   5.2 System redundancies, including redundant dialing facilities
   5.3 Security features preventing unauthorized access
   5.4 Security features of system software
   5.5 Interaction between the application software security, database security, and operating security

6. Database management and third-party application integration
   6.1 System can scrub and validate imported data to prevent duplication
   6.2 Population and updates to database
   6.3 Database limitation (e.g., number of records supported, storage requirements, etc.)
   6.4 Number of scenarios/prescribed messages that can be created

- 6.5 Describe online access (centralized and remote locations)
- 6.6 Real-time graphical monitoring of the system and assignment of rights
- 6.7 Availability of parts and backward compatibility
- 6.8 Ad hoc queries and searches accessing all available data
- 6.9 Third-party application integration
- 6.10 Messages initiation

7. Training, Maintenance, and Implementation
    - 7.1 Type and cost of training
    - 7.2 Duration and frequency of training
    - 7.3 Training manuals and documentation
    - 7.4 Technical support provided
    - 7.5 System updates
    - 7.6 Time to implement and "go live" from date of purchase

8. Reporting
    - 8.1 Real-time reporting
    - 8.2 Reports export to CSV or comma-delimited format
    - 8.3 System able to provide status on all connected and nonconnected calls
        - 8.3.1 Answered by a live person, answering machine, busy signal, or operator intercept
        - 8.3.2 Include each contact's name, phone number, address, city, state, and zip code
    - 8.4 Project plan, including regular status reports
    - 8.5 Key resources assigned to project
    - 8.6 Roles and responsibilities of vendor and purchaser
    - 8.7 Archiving capabilities

9. Pricing
    - 9.1 Does pricing include all costs?
        - 9.1.1 Acquisition through go-live
        - 9.1.2 Training
        - 9.1.3 Upgrades/updates (hardware and software) costs and fees
        - 9.1.4 Maintenance including new additions and other ongoing costs or fees for five years
        - 9.1.5 Potential out-year price increases the vendor may charge
        - 9.1.6 Payment terms

10. Warranties
    - 10.1 All warranty terms and conditions (i.e., period of coverage, on-site, carry-in, etc.)
    - 10.2 All manufacturers' warranty terms and conditions (minimum one year)
    - 10.3 Extended warranty and service availability
    - 10.4 Factory-approved service depots or alternate

Appendix D: List of Questions to Ask a Vendor before You Buy   471

    10.5 Estimated turnaround time for repairs
    10.6 Replacement parts and labor for repairs
    10.7 Telephone, online support during and after the warranty period, including costs
        10.7.1 Support hours
        10.7.2 Services offered
11. Delivery Requirements
    11.1 Delivery timeline
12. Respondent Requirements
    12.1 A currently authorized dealer with the authority to sell, service, and warranty the proposed goods and services in the installed area
    12.2 List of specifications for all standard features, including hardware and software, and purchase price
    12.3 Optional features with costs

Appendix E

Sample Social Media Strategy

At [COMPANY NAME], social media tools have the potential to transform our organization and the way we serve our stakeholders and the community we serve. Our company is committed to being open by promoting transparency, participation, and collaboration. Social media tools will help us accomplish our goals as we strive to make information about our company more accessible to you.

Social media is about community and conversations. Our social media strategy is based on six core values that will help transform [COMPANY NAME]. We focus on three primary communities that we seek to engage: our employees, our customers, and the communities in which we serve and operate.

OUR CORE VALUES FOR SOCIAL MEDIA

- **Collaboration:** Together as one company and as partners with the public to accomplish our mission
- **Leadership:** Out in front within our industry
- **Initiative:** An organization of leaders who are passionate, innovative, and responsible
- **Diversity:** Making [COMPANY NAME] a great place to work by valuing diversity and all voices
- **Community:** Caring about and focusing on our stakeholders and the community
- **Openness:** Creating an open [COMPANY NAME] with an authentic voice

REVOLUTIONIZE COMMUNICATION AND COLLABORATION BETWEEN OUR EMPLOYEES

We want to make [COMPANY NAME] a great place to work. We will use social media tools to increase information and knowledge sharing at and across all levels at [COMPANY NAME]. Peer-to-peer communications and networking can lead to better collaboration, more efficiency, and less friction.

As organic communities of practice emerge and communications tools are put directly in the hands of employees, we will have opportunities to be leaders and to influence the thinking of our workplace. Collaboration networks uncover experts in unexpected places and surprising solutions to problems. We will foster the spirit of innovation by making [COMPANY NAME] a safe place to discuss and try new things.

Strategies

- Empower employees to use social media tools to work effectively.
- Develop a cadre of social media leaders at [COMPANY NAME], who are subject matter experts with social media savvy.
- Implement and encourage the use of social media tools for collaboration.
- Implement and encourage the use of social media tools for professional networking.
- Implement and encourage the use of social media tools for information and status update sharing.

ENGAGE WITH OTHERS IN THE PUBLIC SECTOR

We will lead the way within the industry in focusing on the community, including all levels of government, and our stakeholders. We will strive to make [COMPANY NAME] more efficient and effective. Government community members' opinions, expert advice, and knowledge will inform and shape our plans of action. We will embrace, develop, and invest in new technology to further our mission. By promoting innovation, collaboration, and the smart use of technology, we will do more with less. We will develop our employees' knowledge of that technology. We will take the initiative and be leaders in the industry for social media. We will lead by example and develop best practices for the capture of social media records created by our own activities and work.

Strategies

- Create spaces and platforms for conversations with the community and stakeholder groups.
- Participate in online spaces and conversations that engage our employees, stakeholders, and the community.
- Develop and demonstrate best practices for social media.

BUILD AND STRENGTHEN OUR RELATIONSHIPS WITH THE PUBLIC SECTOR

Many of you are natural sharers, and we hope to foster that impulse and encourage you to use our social media sites to exchange information with [COMPANY NAME]. We hope to create online spaces and platforms where we can make available and collaborate on these kinds of pathfinder resources for information. By engaging in more conversations with you and getting to know you better, we seek to exchange our insights and thereby improve access to our company's information.

Strategies

- Participate in online spaces where stakeholders and employees spend time online.
- Make our resources and services more searchable and sharable.
- Create opportunities and platforms for stakeholders to help us and for them to help each other.
- Turn our information into a social catalog inviting the public to contribute to the online information available about our organization.

Source: U.S. National Archives and Records Administration, 12/8/2010. http://www.archives.gov/social-media/strategies/. Accessed August 27, 2011.

Appendix F
Emergency Call Numbers

The most commonly used emergency call numbers in the world are 112, 911, and 999. These are pre-programmed in many mobile phones. Countries continue to update their systems and covert to new universal emergency call numbers as they expand their emergency calling systems. Before using the table below, please confirm the number for the desired country before it is needed in an emergency.

INTERNATIONAL EMERGENCY
TELEPHONE NUMBERS – JULY 2011

Country	Ambulance	Fire	Police
A			
Afghanistan		local numbers only	
Albania	17	18	19
Algeria	21606666	14	17
American Samoa		911	
Andorra	118	118	110
Angola	118	118	110
Anguilla		911	
Antarctica (McMurdo Station)		911	
Antigua & Barbuda		999/911	
Argentina	107	100	101
Armenia	103		
Aruba		911	
Ascension Island	911/999/6000	911/999/6412	911/999/6666
Australia		000 (112 on mobile)	
Austria		112/122	
Azerbaijan (Baku)	03	01	02
Azores	112		

Country	Ambulance	Fire	Police
B			
Bahamas		911	
Bahrain		999	
Bali	112	118	
Bangladesh (Dhaka)	199	9 555 555	866 551-3
Barbados	511	311	211
Belgium		112	
Belarus	103	101	102
Belize		911	
Benin		local numbers only	
Bermuda		911	
Bhutan	110	112	113
Bolivia	118		110
Bonaire		911	
Bosnia-Herzegovina		911/112	
Botswana		997/911	
Brazil	192	193	190
Bosnia	112/104/94/124	112/105/93/123	112/107/92/122
British Virgin Islands		999	
Brunei	991	995	993
Bulgaria	150	160	166
Burkina Faso		local numbers only	
Burma/Myanmar		999	
Burundi		local numbers only	
C			
Cambodia, The Kingdom of (Phnom Penh)	119	118	117
Cameroon		local numbers only	
Canada (AB, MB, NB, NS, ON, PE, QU)		911	
Canada (BC, NF, SK)		911 local only1	
Canada (NT)		3 dig+2222	3 dig+1111
Canada (NU)		local only	
Canada (YK)	3 dig+3333	3 dig+2222	3 dig+5555
Canary Islands		112	
Cape Verde (Santiago Island)	130	131	132
Cayman Islands		911	

Country	Ambulance	Fire	Police
Central African Republic		local numbers only	
Chad		18	17
Chile	131	132	133
China, The People's Republic of	999/120 (Beijing, Shanghai)	119	110
Christmas Island		000	
Columbia		112/123 land & mobile	
Comoros Islands		local numbers only	
Congo		local numbers only	
Cook Islands	998	996	999
Costa Rica		911	
Côte d'Ivoire		110/111/170	180
Croatia	94/112	93/112	92
Cuba		26811	
Curaçao	112	114	444444
Cyprus		112/199	
Czech Republic	112/155	112/150	112/158
Congo, Democratic Republic of/Zaïre		local numbers only	

D

Country	Ambulance	Fire	Police
Denmark		112	
Djibouti	351351	18	17
Dominica, Commonwealth of		999	
Dominican Republic		911	

E

Country	Ambulance	Fire	Police
East Timor		112	
Easter Island	100-215	100-264	100-244
Ecuador	131		101
Egypt	123	180	112
El Salvador		911	
England		112/999	
Equatorial Guinea		local numbers only	
Eritrea		local numbers only	
Estonia		112	
Ethiopia	92	93	91

continued

Appendix F: Emergency Call Numbers

Country	Ambulance	Fire	Police
F			
Faeroe Islands (Denmark)		112	
Falkland Islands		999	
Fiji	911	9170	911
Finland		112	
France	112/15	112/18	112/17
French Guiana	112/15	112/18	112/17
French Polynesia	15	18	17
G			
Gabon	1300-1399	18	1730
Gaborone	997/911	998	999
Gambia, The	16	18	17
Georgia		022	
Germany	112	112	110/112
Ghana	193	192	191
Gibraltar	190	190	199/112
Guyana	911/913	911/912	911
H			
Haiti	118		114
Herzegovina	112/104/94/124	112/105/93/123	112/107/92/122
Honduras	1952/37 8654	198	119
Hong Kong		999/112 mobile	
Hungary	112/104	112/105	112/107
I			
Iceland		112	
India	102	101	100
Indonesia	118/119	113	110
Iran	115	125	110
Iraq		local numbers only	
Ireland, Republic of		112/999	
Isle of Man		999	
Israel	101	102	100
Italy	118/112	115/112	112/113
J			
Jamaica	110	110	119
Japan	119	119	110

Country	Ambulance	Fire	Police
Jersey		999	
Jordan	191	193	192
K			
Kazakhstan		03	
Kenya		999	
Kiribati	994		
Kosovo		911	
Korea, The Democratic People's Republic of (North Korea)		local numbers only	
Korea, The Republic of (South Korea)	119	119	112
Kuwait		112	
Kyrgyzstan		103	
L			
Laos		local numbers only	
Latvia	03/112	01/112	02/112
Lebanon	140	175/125	112/999
Lesotho	121	122	123/124
Liberia		911 mobile only	
Libya		193	
Liechtenstein		112	
Lithuania		112	
Luxembourg	112	112	112/113
M			
Macau		999/318	
Macedonia, Republic of	112/194	112/193	112/192
Madagascar		118	117
Madeira		112	
Malawi	998	999	997
Malaysia	999	994	999
Maldives Republic	102	999	119
Mali	15	17	18
Malta		112	
Marianas Island	911		

continued

Country	Ambulance	Fire	Police
Marshall Islands	625 4111		625 8666
Martinique	15	18	17
Mauritania		118	117
Mauritius		999	
Mayotte	15		
Menorca	112	112	112/091
México		066, 060 or 080	
Micronesia, Federated States of		local numbers only	
Moldavia	903	901	902
Monaco		112	
Mongolia	103	101	102
Montserrat	911		999
Morocco	15	15	19
Moyotte	15		
Mozambique	117	198	119
Myanmar	199	199/191	199

N

Country	Ambulance	Fire	Police
Namibia	2032276	2032270	1011
Nauru		local numbers only	
Nepal		101	100/103
Netherlands (Holland)		112	
Netherlands Antilles		112	
New Caledonia	18	18	17
New Zealand		112/911/111	
Nicaragua	128	115/911	118
Niger		local numbers only	
Nigeria		199	
Niue		999	
Norfolk Islands		000	
Northern Ireland		112/999	
Norway	113	110	112

O

Country	Ambulance	Fire	Police
Oman		999	

P

Country	Ambulance	Fire	Police
Pakistan	115	16	15/1122
Palau		911	

Country	Ambulance	Fire	Police
Palestine	101	101	100
Panama		911	
Papua New Guinea / Port Moresby*		1101	0001
Paraguay		911	
Peru		116	105
Philippines		117	
Pitcairn Islands		no telephone system	
Poland	112/999	112/998	112/997
Portugal		112 (117 for wildland fires)	
Puerto Rico		911	
Q			
Qatar		999	
R			
Réunion	15/112	18	17
Romania		112	
Russia		112	
Russian Federation	03/911/112	01/911/112	02/911/112
Rwanda			112
S			
Sabah (Borneo)		999	
Samoa		999	
San Marino	118	115	113
São Tomé and Principe		local numbers only	
Saudi Arabia	997	998	999
Scotland		112/999	
Scilly, Isles of		999	
Senegal		local numbers only	
Serbia		112	
Seychelles		999	
Sierra Leone	999	019	999
Singapore	995	995	999
Slovak Republic (Slovakia)	155	150	158
Slovenia	112	112	113
Solomon Islands		999	

continued

Country	Ambulance	Fire	Police
Somalia		local numbers only	
South Africa	10177	10177	10111
South Africa (Cape Town)		107	
S. Georgia Is./ S. Sandwich Is.		no telephone system	
Srpska, Republic of	124	123	122
Spain		112	
Sri Lanka	110	111	118/119
St Eustatius		140	
St Eustatius		911	
St Helena		911	
St Kitts & Nevis		911	
St Lucia		999/911	
St Maarten	911/542-2111	911/120	911/542-2111
St Pierre & Miquelon	15	18	17
St Vincent & the Grenadines		999/911	
Sudan			112
Suriname		115	
Swaziland			999
Sweden		112	
Switzerland	112/144	112/118	112/117
Syrian Arab Republic (Syria)	110	113	112
T			
Tahiti–French Polynesia	15		
Taiwan (Republic of China)	119	119	110
Tajikistan	03		
Tanzania		112/999	
Thailand	1669	199	191
Tibet	120		110
Togo			101
Tonga		911	
Trinidad & Tobago	990	990	999
Tunisia	190	198	197
Turkey	112	112/110	112/155

Country	Ambulance	Fire	Police
Turkmenistan		03	
Turks and Caicos Islands		999/911	
Tuvalu/Ellice Is.		911	
U			
Uganda		112 mobile/999 fixed	
Ukraine		112	
United Arab Emirates (Abu Dhabi)		998/999	
United Kingdom		112/999	
United States		911	
Uruguay		911	
US Virgin Islands		911	
Uzbekistan		03	
V			
Vanuatu		112	
Vatican City	113	115	112
Venezuela		171	
Vietnam	115	114	113
W			
Wake Island		no telephone system	
Western Sahara	150		
Western Samoa		999	
Y			
Yemen, Republic of	191	191	194
Yugoslavia (Serbia & Montenegro)	94		
Z			
Zambia	991/112 mobile	993/112 mobile	999/112 mobile
Zimbabwe	994/999	993/999	995/999

Source: Santa Clara County Fire Department. (2011). International Emergency Telephone Numbers. Reprinted courtesy of Santa Clara County Fire Department. http://www.sccfd.org/travel.html (accessed October 4, 2011).

Note: When roaming with a GSM telephone in the European Union, use 112 for emergency services.

* Papua New Guinea should be getting one number for all services in the near future.

Glossary

211: 2-1-1 is a telephone number in many communities across the United States providing free and confidential information and referral services for assistance with food, housing, employment, health care, counseling, and other nonemergency needs. This number is also used to register for emergency transportation services before an emergency occurs.

3G: The third generation of standards for mobile phones and mobile telecommunications that follows the IMT advanced specifications as set by the International Telecommunication Union.

4G: The fourth generation of standards for mobile phones and mobile telecommunications that follows the IMT advanced specifications as set by the International Telecommunication Union. 4G used a different frequency band over 3G, and the transfer rate is also faster. 4G networks include LTE, Wimax, and WiFi.

A0 solar flare: Is the weakest solar flare that can be categorized.

AAC: (Augmentative and alternative communication) It assists individuals in adopting and using the most effective communication methods for their unique circumstances by applying research, providing education, and offering clinical service delivery.

AAR: See After Action Report.

Advisory: An advisory is given when conditions could produce a situation such as a tornado in the area and the target audience should be prepared to take shelter immediately. The purpose of the advisory is to increase awareness of a potential threat.

After Action Report: (AAR) It is a retrospective analysis of a given sequence of actions previously taken.

All Clear: A term used by authorities to inform others that an imminent danger has passed or after a warning or alert has been previously issued.

AM: Amplitude modulation used for AM radio broadcasting. AM radio broadcasting was the primary method for broadcasting for the first 75 years of the 1900s and is used today.

Appraisal cost: Identification of defects in the message before it is disseminated to the targeted audience. Same as Inspection cost.

APT: (Advanced persistent threat). These are computer threats associated with cyber attack espionage assaults. It is a malware that has taken advantage of weaknesses in targeted network with the intent of stealing data, particularly intellectual property. APTs are persistent rather than a one-time event and can be stealthy.

Area Command: (Also Unified Area Command). An organization established to oversee the management (a) of multiple incidents that are each being handled by an ICS organization, or (b) of large or multiple incidents with several Incident Management Teams assigned. Area Command has the responsibility to set overall strategy and priorities, allocate critical resources according to priorities, ensure that incidents are properly managed, and ensure that objectives are met and strategies followed. Area Command becomes Unified Area Command when incidents are multijurisdictional. Area Command may be established at an emergency operations center facility or at some location other than an incident command post.

Attribution Theory: A view on how individuals interpret events and how their views related to their thinking and behavior.

Automated Speech Recognition system: (ASR). A technology that allows users of information systems to speak their entries rather than punching numbers on a keypad. ASR is used primarily to provide information and to forward telephone calls.

B0 solar flare: A B0 solar flare is ten times stronger than an A0 solar flare. (See A0 solar flare.)

Biometrics: Used as a form of identity access management and control to identify individuals to a system. Biometrics may include physiological (fingerprint, palm, ear lobe, iris recognition, DNA, face recognition, etc.) or behavioral (voice, typing rhythm, gait, etc.) inputs.

Black hat: Individuals who support illegal hacking activities.

Blackout: A loss of electrical power to an area for a short or long period.

Blogs: A type of website or part of a website used to provide commentary or descriptions of events sometimes using multimedia.

Bluetooth: Named after Harold Bluetooth, an ancient king in Denmark, it is the specification for using low-power radio communications to wirelessly link devices such as personal phones, headsets, computers, and other networked devices.

Botnet: A large collection of compromised computers used by hackers and spammers to generate spam, relay viruses, flood a network or web server with excessive requests such as that used in a DDoS attack. Botnets are also referred to as a zombie army.

Braille: Braille, created by Louise Braille in the 1800s, used the standard alphabet and reduces the number of dots by half so that a person can read by touch. Math and music symbols were added that were not included in the previous forms. Braille is the standard form of reading and writing used by those who are visually impaired.

Broadsides: Printed news pamphlets, the forerunners of newspapers.

Brownout: A drop in voltage in the electrical power supply.

Business continuity: Business continuity or continuity of operations is the processes, policies, and procedures related to keeping all aspects of a business functioning in the midst of a disruptive event.

CA: (Communications assistant). It serves as a link for the call, relaying the text of the calling party in voice to the called party, and converting to text what the called party voices back to the calling party.

Caldera: Occurs when the magma is removed from beneath a volcano. The ground subsides or collapses into the emptied space to form a depression. Also called a super volcano.

Captions: On-screen descriptions that are synchronized with the video to provide the dialogue, identity of speakers, and describe any relevant sound to the audio in text format.

C-class flares: These are small-size solar flares with few noticeable consequences on earth.

Cell Phones: See mobile phones.

Cellular phones: See mobile phones.

CERT: (Community Emergency response Team/Training). A program that educates individuals on disaster preparedness for hazards that may impact their area. Typically over an 8-week period, individuals learn the basics of disaster response skills (i.e., fire safety, light search and rescue), team organization, and disaster medical operations. Visit http://www.citizencorps.gov for more information.

Climate warfare: It is a weapon of mass destruction that is capable of destabilizing economies and agricultural and ecological systems globally. Manipulation of the climate using chemicals is a form of climate warfare.

Closed caption: Captions available if selected by the viewer.

CLR holding: Telephone equipment that could provide a method to hold the originating calling line long enough to check the line. This process worked even if the call was disconnected or the caller hung up, a function that could mimic what telephone operators had been able to do.

C0 solar flare: A C0 solar flare is 10 times stronger than a B0 solar flare and 100 times stronger than an A0 solar flare.

COFDM: (Coded orthogonal frequency-division multiplexing). A computer-aided system that makes and decodes signals, resists fading and ghosting, has error-correction coding, can resist interference, and is an adaptive system. It is used for Wi-Fi, LANs, DTV, and radio standards.

Cognitively impaired: Sometimes referred to as having an unsound mind, is the inability to perceive all relevant facts related a situation, condition, and treatment or actions to make an intelligent decision.

Command Staff: Members of the command staff report to the Incident Commander. It is composed of a Public Information Officer, Safety Officer, and Liaison Officer.

Commercial off-the-shelf: (COTS) is a nondevelopmental item (NDI) of supply that is commercial, sold in substantial quantities in the commercial marketplace, and can be procured or used under contract in the same precise form as available to the general public, such as computer software.

Common Operating Picture: (COP) is a single identical display of relevant information such as the position of resources and the status of important variables such as roads and bridges that will be used to transport the resources that is shared by more than on Command. It is the collaborative planning and used to assist with situational awareness.

Communicate: The act of conveying information.

CONELRAD: (CONtrol of ELectromagnetic RADiation) system. A method of emergency broadcasting for speaking to the public in the event of enemy attack during the Cold War error. CONELRAD was replaced by the Emergency Broadcast System in the 1960s.

Constraint: A factor or a subsystem that restricts an organization, project, or system from achieving its potential with reference to its goal—a bottleneck.

Continuous improvement process: The stage of the crisis communication program where an assessment of what went well and opportunities for improvement are reviewed and changes made to update the process.

Continuous wave (CW) telegraphy: An advancement in telegraph. CW uses a pure radio frequency that is produced by a vacuum tube electronic oscillator.

Cost of quality: The costs associated with preventing, detecting, and dealing with defects. It does not refer to the costs of using a higher-grade product such as Grade AAA eggs instead of Grade A eggs.

COTS: See commercial off-the-shelf.

Crowdsourcing: The act of outsourcing tasks, traditionally performed by one person or small group to an undefined, large group of people or community (a "crowd") through an open call.

Crawler: A line of text that scrolls at the bottom of the television screen such as those used for news headlines and severe weather warnings. Also called a ticker.

Crisis communications: The sending of messages and using channels to share information with internal and external sources in identifying a threat or crisis. It is the communications processes, procedures, and methods used early, during, and after a crisis.

Crisis recovery: The activities completed during the aftermath of a situation that are focused on restoring the organization to normal operations or situation to normalcy.

Crisis: A unique unpleasant occurrence characterized by the elements of surprise, threat, and short response time.

Culture: A full range of human behaviors related to values, beliefs, and practices shared by a group of people.

Dah: (Morse Code) Short marks and dashes (-).

DARS: (Digital Audio Radio Service). Any form of digital radio services.

DDoS: (Distributed denial-of-service). An attack in which many compromised systems attack a single target, causing a denial of service or denial of access to the target.

Decision paralysis: Extreme indecisiveness in the midst of a crisis.

Developmental disability: A broad range of conditions that interfere with the ability of an individual to function in every activity, such as spina bifida, autism, and cerebral palsy.

Disaster recovery: (DR) It is referred to as the process, policies, and procedures related to preparing for recovery or continuity of technology infrastructure critical to an organization after emergency. It is a subset of business continuity.

Disaster: A natural situation or event that overwhelms the local capacity to respond, recover, prevent, or mitigate damage and may require a request for external help.

Dit: (Morse Code) A group of dots (.).

Doppler effect: The change in the frequency of energy in the form of waves, such as light or sound, that is due to the motion from the source or the receiver of the waves. Some search radars use the Doppler effect to separate moving vehicles from clutter.

DTV: (Digital Television). A technology using digital signals to transmit television programs rather than analog signals.

EAS: See Emergency Alert System.

El Nino: A disruption of the ocean–atmosphere system in the Tropical Pacific. El Nino leads to increased rainfall in the southern United States and Peru, causing destructive flooding, and drought in the West Pacific sometimes associated with devastating brush fires in Australia. El Nino is referred to as the warm phase of ENSO. During El Nino years, temperatures in the winter are warmer than normal in the North Central States, and cooler than normal in the Southeast and the Southwest.

Elderly: Individuals who are 65 years of age or older.

E-mail: (electronic mail). A program to send electronic mail.

Emergency Alert System: (EAS). A national warning system that is coordinated and regulated by the FCC covering all states and several territories. It is designed to let the president speak to the nation within 10 minutes; mostly all telecommunication-related systems and providers are required to participate in the EAS by the end of 2007. EAS has been superseded by the FEMA program Integrated Public Alert and Warning System (IPAWS). EAS relies primarily on radio and television for communicating with the public.

Emergency Broadcast System: (EBS). This was a method of emergency broadcasting like CONELRAD, designed for use by the president to quickly communicate with the American public in the event of enemy attack, war, threat of war, or grave national crisis. Its use was expanded to state and local peacetime emergencies, civil emergency message, and severe weather hazard warning. EBS was replaced with the Emergency Alert System in 1997.

Emergency Operations Center: A central command and control facility for emergency preparedness, emergency management or disaster management.

Emergency Support Functions: (ESFs). These are annexes detailing the structure used for coordinating interagency support for a response to an incident. They are mechanisms for grouping functions most frequently used to provide support for emergencies under the Stafford Act and for non-Stafford Act incidents. Functions include, for example, transportation and communications.

Emergency: A situation, human caused or natural, that poses an immediate risk to the health, and safety of people, property, or the environment.

Enhanced 911: (E911). 9-1-1 services that could automatically retrieve the address assigned to the wireline telephone used to place the call in the computer. This service upgraded 9-1-1 services in the 1990s.

Environmental refugees: Individuals who are unable to have a secure livelihood in their homelands where drought, soil erosion, desertification, deforestation, and other environmental issues threaten their homeland with poverty issues and population pressures present.

EOC: See Emergency Operations Center.

ESFs: See Emergency Support Functions.

Event: A planned, nonemergency activity.

External Affairs: This department advances the goals of an organization by building and maintaining relationships with key stakeholders, such as legislative partners, the business community, and other key constituents.

External failures: Messages sent that result in issues, perceived or real by the receivers of the messages.

FCC: The Federal Communications Commission regulates interstate and international communications by radio, television, wire, satellite, and cable for the all 50 states, the District of Columbia, and U.S. territories.

FE: See Functional Exercise.

FEMA PLAN: FEMA's (Personal Localized Alerting Network) is a public safety system for individuals having an enabled mobile device to receive geographically targeted, text-like messages alerting them of imminent threats to safety in their area. This technology ensures that emergency alerts will not get stuck in highly congested user areas, which happens at times with standard mobile voice and texting services. Government officials can target emergency alerts to specific geographic areas through cell towers (e.g., lower Manhattan), that pushes the information to enabled mobile devices.

FEMA: The Federal Emergency Management Agency is responsible for ensuring that the United States works together to build, sustain, and improve the US capability to prepare for, protect against, respond to, recover from, and mitigate all hazards.

Flash message: Messages displayed in response to user interaction with a site or on a network providing priority messaging such as success or failure messages after performing an action which submits a form, or in response to an attempt to access a resource for which the user does not have permission.

Flash mob: A group of people who suddenly assemble in a public place to perform an unusual or pointless act for a brief time, then disperse. They are sometimes performed for entertainment, artistic expression, or violence.

FM: Frequency modulation used to provide high-fidelity sound over broadcast radio. Similar to AM broadcasting, FM radio uses electromagnetic radiation to broadcast sound information.

FSE: See Full Scale Exercise.

Full Scale Exercise: (FSE) is a multi-agency, multi-jurisdictional, multi-disciple exercise involving functional groups, such as the EOC or JIC, and first responders and emergency officials in response to a mock incident.

Functional Exercise: (FE) is an operations-based exercise used to validate plans, policies, agreements, and procedures, to clarify roles and responsibilities, and identify gaps in the operational environment. The functional exercise is focused on the coordination, command, and control among participating agencies. This type of exercise does not include the first responders or emergency officials responding to an incident real time.

Fusion center: A prevention and response center used to gather intelligence and information from public and private source and uses the data gathered to promote information sharing at all levels of government and within the private sector.

Gaps: (Morse Code). Pauses between letters, words, and sentences.

GDP: (Gross Domestic Product). The total market value of all goods and services produced in a country in a given year, equal to total consumer, investment, government spending, plus the value of exports minus the value of imports.

Geocoding: It is the process of finding associated geographic coordinates from other geographic data such as a zip code or street address that can be used for mapping and entering into a GIS system or embedding into digital photos. Reverse geocoding is the opposite. It will use the GIS and embedded information in a digital photo to find a street address, zip code, etc.

Geo-hash: It is a latitude/longitude geocoding system. It is similar to decoding but returns the interval for each coordinate.

Geo-intelligence: It is used to develop precise estimates of the threats posed by hazards and their ramifications.

Geo-location: It is a method used in the identification of the geographic location of an object, such as radars, mobile phones, or an Internet-connected computer terminal based on metadata and can tracked real time, providing a meaningful location such as a street address rather than just geographic coordinates.

Geotagging: This is process of labeling photos and video, "tagging" to depict where they were taken. Location information can be used to browse, search, and organize photos according to where they were taken.

Geo-targeting: It is a method used to determine the geo-location of a website's visitor and then use the information to deliver different content based on the visitor's location, such as city, zip code, ISP, etc.

Global Positioning System (GPS): A US space-based radio-navigation system that provides positioning, navigation, and timing services to civilian users on a continuous worldwide basis. This service is provided free to the public.

Governance: The process or part of management or leadership relating to decisions that define expectations, grant power, or verify performance.

GSM: (Groupe Speciale Mobile). It defines the internationally accepted digital cellular telephony standard. One GSM standard mandates that the user of a GSM phone should be able to dial the local emergency services number without unlocking the keypad.

Hams: Amateur radio operators licensed to transmit radio signals as a leisure-time interest or in cases of emergency and public service communications.

Hard Disk Drive (HDD): A random access digital data storage device.

HDTV: (High Definition television). A technology using DTV providing better-quality sound and higher-resolution pictures.

Hearing impaired: Individuals who are deaf or hard of hearing.

Hieroglyphic writing: A writing system used by the ancient Egyptians that uses pictures as characters.

High Frequency: Radio frequencies between 3 and 30 MHz.

Hiliography: Sending messages using mirrors or shiny metals and the rays from the sun by flashing reflected rays to another location up to 50 miles away.

Homeland Security Presidential Directive: (HSPD) are Directives issued by the President on matters pertaining to Homeland Security.

HSPD: See Homeland Security Presidential Directive.

IAP: See Incident Action Plan.

IC: See Incident Commander.

ICP: See Incident Command Post.

ICS: See Incident Command System.

IMT Advanced: (International Mobile Telecommunications-Advanced). These are the telecommunication requirements defined by the International Telecommunication Union Radiocommunication Sector.

IMT: See Incident Management Team.

Incidence of National Significance: An actual or potential high-impact event that requires a coordinated and effective response by a combination of federal, state, local, tribal, nongovernmental, or private sector entities for saving lives, minimizing damage, and providing the basis for long-term community recovery and mitigation activities. (Source HSPD-5)

Incident Action Plan: (IAP). An oral or written plan that includes the general objectives and overall strategy for managing an incident.

Incident Command Post: (ICP). The field location where the primary tactical-level, on-scene incident command functions are performed.

Incident Command System: A standardized, on-scene, all-hazards incident management approach that (a) allows for the integration of facilities, equipment, personnel, procedures, and communications operations within a common organizational structure,

(b) enables a coordinated response among various private and public jurisdictions and organizations, and (c) establishes common processes for planning and managing resources. ICS is a part of NIMS.

Incident Commander: (IC) The individual with the overall authority and responsibility for all incident activities including strategies, tactics, and the release of resources.

Incident Management Team: (IMT). A team composed of the Incident Commander and the appropriate Command and General Staff personnel assigned to an incident.

Information science: The study of the ways in which the human experience is structured, represented, managed, stored, retrieved, and transferred later.

Informationization: The extent to which a geographical area, economy, or society becomes information-based such as documented by the size of its information labor force, for example.

Ingratiation: A social psychological technique in which an individual attempts to become more attractive or likeable to their target.

Inspection cost: See Appraisal cost.

Internal failures: Failures related to effective crisis communications such as the failure of the message and its delivery conforming to the specifications outlined in the crisis communication plan. These are failures identified before a message is sent.

Internet Service Provider: (ISP) A company that provides access to the Internet.

Internet: A global system of interconnected computer networks that serves billions of users via TCP/IP.

Interoperability: The ability of a system or a product to work with other systems or products without significant effort on the part of the end user.

Interpretation: An oral service that uses the act of listening to a communication in one language and orally converting the communication into another language retaining the same meaning.

Interpretation: How well the receiver understands what has been sent.

IP: (Internet Protocol). The principal communications protocol used for relaying packets, basic transfer units, across an internetwork using the IP Suite.

IPAWS: (Integrated Public Alert and Warning System). A system coordinated jointly by the National Weather Service, FCC, and FEMA. It expands existing EAS by enabling emergency management officials to reach as many people as possible using as many communication devices that are compatible and available. IPAWS uses a common alerting protocol enabling the use of mobile phones, personal computers, and network attached speaker arrays, among other device types.

ISP: See Internet Service Provider.

Jargon: A special language, used within a particular trade, profession, or group.

JIC: See Joint Information Center.

Joint Information Center: (JIC). A central location that facilitates operation of the Joint Information System. A location where personnel with public information responsibilities perform critical emergency information functions, crisis communications, and public affairs functions.

Joint Information System: (JIS). A system that provides the mechanism to organize, integrate, and coordinate information to ensure timely, accurate, accessible, and consistent messaging across multiple jurisdictions or disciplines with nongovernmental organizations and the private sector. A JIS includes the plans, protocols, procedures, and structures used to provide public information.

Key fob: A USB flash drive used as an identification key or a security token or device that is given to an authorized user to log in to the network. The token may be like a credit card and the numbers entered into the computer, a device displaying a changing number that is type as a part of a password, or plugged directly into a computer via a USB port.

Kinematoscope: A machine that flashed a series of still photographs onto a screen.

Kinetograph: The early motion picture cameras and projectors that could photograph motion pictures.

La Niña: Unusually cold ocean temperatures in the Equatorial Pacific. It is sometimes referred to the cold phase of the El Niño-Southern Oscillation cycle. During a La Niña year, winter temperatures are warmer than normal in the Southeast and cooler than normal in the Northwest.

LAN: (Local area network). It is a computer network that connects computers and other devices that are in close proximity such as in a home, school, computer laboratory, or office building.

Legally authorized representative: (LAR). When an individual is unable to give informed consent, this person can have an individual, judicial, or other body authorized by law to consent on behalf of the person.

LEP: (Limited English Proficiency). Identifies individuals whose primary spoken language is not English.

Liaison Officer: A member of the Command Staff responsible for coordination with representatives from cooperating and assisting agencies.

Local Warning Point: An individual authorized to send warnings or other information on behalf of an organization or group to other warning points or the affected populations.

Long Term Evolution: (LTE). A standard in the mobile phone network technology that is a part of the 4G technology strategy.

Loudspeaker: An electro-acoustic driver, or transducer, producing sound in response to an electrical audio signal input and is a part of a speaker system.

LTE: See Long Term Evolution.

LWP: See Local Warning Point.

MACs: See Multiagency Coordinated Systems.

Magazine: A publication published periodically, less frequently than a newspaper.

Maslow's Hierarchy of Needs: This is a psychology theory named after Abraham Maslow that defines the levels of human needs.

Mass notification: The capability to provide information to all affected parties in a given area. Mass notification systems used for emergency situations provide the capability to provide real-time information to building occupants or others in the immediate area during event.

M-class flares: These are medium size solar flares that can cause brief radio blackouts that affect the earth's polar regions. Minor radiation storms sometimes follow an M-class flare.

Media: The resource used for sending a message, such as paper, television, or smoke signals.

Memorandum of understanding: (MOU). It is a document that defines a bilateral or multilateral agreement between parties indicating an intended common line of action and responsibilities. Mutual Aid Agreements, much like a MOU, are agreements between public and private entities.

Metadata: It is data about data, that is, information that characterizes data, like library cataloging of data.

Microbroadcasting: A campus or local radio stations running over carrier current, a secondary signal transmitted in a "piggy back" fashion along with the main program. The reach is generally less than their AM competitors.

Mimeograph: One of the early office copying machines.

Mitigation: The effort to reduce loss of life and property by lessening the impact of disasters. Mitigation includes prevention activities.

MMS: See Multimedia Messaging Service.

Mobile phones: An electronic device that transmits its signal to a local cell site (transmitter/receiver) wirelessly. It is also known as *cellular phones* or *cell phones*.

Mobility challenged: Individuals who use mechanical devices to move from one place to another within their resident, workplace, or places they visit. Mechanical devices include wheelchairs, walkers, walking canes, and crutches.

Morse Code: A method for transmitting textual information using a series of indentation marks (dots and dashes) on paper tape that can be directly understood by a skilled listener or observer without special equipment.

MOU: See Memorandum of understanding.

Multiagency Coordinated Systems: (MACs) provide the architecture to support incident prioritization and coordination, critical resource allocation, communications systems integration, and information coordination. The components of multiagency coordination systems include facilities, equipment, emergency operation centers (EOCs), specific multiagency coordination entities, personnel, procedures, and communications.

Multimedia Messaging Service: (MMS). It a standard way to send a message that include multimedia content between mobile devices. It is an extension of SMS, used for sending text messages.

Multiplexing: The process of combining multiple signals onto one composite signal so that the receiver reconstitutes the original signals.

Murder board: This is a committee of questioners established to assist an individual or individuals in preparing for a difficult oral examination such as a debate, hearing, presentation, or news conference.

Mutual aid agreement: See memorandum of understanding.

Narrowband: The bandwidth of a radio message not exceeding its channels' maximum bandwidth.

News conference: A media event where journalists are invited to hear and ask questions of the speaker about an event for the first time. A press briefing is provided to give updates.

Newspaper: A regularly scheduled publication containing news of current events, articles, advertising, etc., that is typically printed on low-grade paper.

NG911: (Next Generation 9-1-1). A system of hardware, software, data, policies, and procedures that provide standardize interfaces from call and message services, covering both voice and nonvoice messages, leverages additional data for targeted call routing and handling, incorporates various emergency organizations, supports data and communications needs for incident response, and management of the system and data in a secure environment needed for emergency communications.

NGO: See Nongovernmental organization.

NIMS: (National Incident Management System). A system used in the United States in the coordination of emergency preparedness and incident management among stakeholders from all levels within both the public and private sectors including nongovernment organizations and nonprofits.

Nongovernmental organization: (NGO) is a term commonly used to reflect a nongovernment organization such as a nonprofit organization or a private voluntary organization. It could also be used to include for-profit businesses.

NRF: (National Response Framework). The guiding principles enabling response partners to prepare for and provide a unified national response to disasters and emergencies. It is an all-hazards approach to domestic incident responses as managed under the US Department of Homeland Security.

Open caption: These are captions that are always on and available to all viewers.

Open source espionage: A form of social networking where confidential information is compromised through electronic spying and then sharing the information involving the security or an organization or nation.

PAN: (Personal Area Network). PANs are used by to temporarily connect an individual personal device over a short distance or join the network. Bluetooth-enabled devices use PAN.

Party lines: Telephone lines shared by more than one household.

PDD: See Presidential Directives.

Peak ecological water: The point where water is being diverted from the environment for human consumption at such a rate that the ecosystem can no longer function normally.

Pictograph: An ideogram conveying meaning through pictorial resemblance to a physical object such as that used by the Ancient Chinese and Egyptian civilizations.

PIO: See Public Information Officer.

Pocket dialing: This occurs when a cell phone, usually in a purse or pocket, accidentally dials a number without the knowledge of the phone owner.

Potable water: Water that is fit for consumption by humans and animals.

POTS: A global network of public circuit switched telephone network.

Preparedness: FEMA defines preparedness as a continuous cycle of planning, organizing, training, equipping, exercising, evaluating, and taking corrective action in an effort to ensure effective coordination during incident response. Preparedness includes prevention activities.

Presidential Directives: (PDD). Also called Presidential Decision Directives. they are a form of executive order issued by the U.S. President with the advice and consent of the National Security Council. It carries the full force and effect of law and is a statement of executive policy.

Press briefing: See news conference.

Prevention cost: Activities that aim to reduce the number of defects.

Prevention: The efforts used throughout the emergency management cycle. It is included in recovery and mitigation.

Propaganda: A form of communications used to influence the attitude or actions of a community toward a specific cause or position to benefit oneself.

PSAP: (Public Safety/Service Answering Points). A call center responsible for answering calls to an emergency telephone number and then dispatching services for police, fire, EMS by trained staff.

Public Affairs: This department communicates the priorities and goals of an organization to its stakeholders such as the media by disseminating formation, publications, and other services related to requests about the organization.

Public Information Officer: (PIO) is a member of the Command Staff responsible for interfacing with the public, media, and other agencies regarding incident-related information and information requirements.

Public payphones: Telephones that require a user to pay before placing a call using the device.

Quick Response Codes: (QR Codes) are two-dimensional matrix barcodes designed to be quickly read (decoded) by smartphones. The code contains black modules arranged in a square pattern on a white background. The barcode is encoded with text, a URL, or other data.

Radar: (Radio Detection and Ranging). It detects objects at a distance by bounding radio waves off the objects. The delay caused by the echo from the object measures the distance. The direction of the mean determines the direction of the reflection.

Radio Frequency Identification: (RFID). It is an electronic tag technology using radio waves to transfer data about the object it is attached to for purposes of identification and tracking.

Radio spectrum: Refers to the segment of the electromagnetic range of all possible frequencies of electromagnet radiation related to radio frequencies (frequencies 300 GHz or less).

Radio telephony: Voice communications using radio waves.

Radio: A term used in referring to the actual transceiver device or chip.

Radiograms: Written messages, telegram style messages, routed through a network of amateur radio operators who are on the air to relay messages over the radio.

Really Simple Syndication: (RSS) is a family of web feed formats used to publish work that is frequently updated in a standardized format such as blog entries or news headlines.

Receiver: The party that receives the message.

Reception: How well a message is received.

Recovery: Occurs in phases, beginning with efforts to address the immediate needs of individuals after a situation in the basic areas of housing, food, and water. Long-term recovery includes activities to restore individuals and communities back to normalcy with as cleanup and rebuilding efforts.

Regression testing: Also called iteration testing within the test plan. It is any type of software testing that is looking to detect new errors, or regressions, in existing functionality after changes have been made to a system, such as enhancements or patches. It helps a developer to test the interaction among the modules or lines of code within the software.

Reinsurance business: Insurance companies selling insurance to other insurance companies. It is an approach used by a primary insurer to protect against unforeseen or extraordinary losses, to increase individual insurers' capacity, to share liability when losses overwhelm the primary insurer's resources, and to help insurers stabilize their businesses in the face of the wide swings in profit and losses.

Response: The activities associated with saving lives and protecting property and the environment.

Reverse geocoding: See geocoding.

RFID: See Radio Frequency Identification.

Route alerting: Occurs when a public safety vehicle is dispatched to an area to provide warning information through their public address system.

RSS: See Really Simple Syndication.

Safety Officer: A member of the Command Staff responsible for monitoring and assessing safety hazards or unsafe situations, and for developing measures for ensuring personnel safety.

SCCT: See Situational Crisis Communication Theory.

Seminar: An informal discussion, designed to orient participants to new or updated information such as plans, policies, or procedures.

Sender: The party that triggers a message.

Sensory channel: Seeing, hearing, touching, tasting, and smelling are the five senses used in the basic channels of communications.

Service animal: Any guide dog, signal dog, or other animal trained to provide assistance to an individual with a disability.
Short Message Service: (SMS). A text messaging services of a mobile communications system that enables the exchange of short text messages between devices.
Situational Crisis Communication Theory: (SCCT) A framework for understanding how stakeholders will react to a crisis in terms of the reputational threat posed by the crisis.
SME: See subject matter expert.
SMS: See Short Message Service.
Snippet: An excerpt from a website that is shared by default.
Snowbirds: Seasonal residents who move from colder climates to warmer drier climates during the winter months and then return home during the summer months. This is commonly practiced by the elderly.
Social bookmarking: A method for Internet users to organize, store, manage, and search for bookmarks of resources online. This is not the actual sharing of information, only the sites that reference them.
Social media: The use of mobile and web-based technologies for interactive communications using user-generated content.
Social network: A node or social structure composed of individuals who are linked by one or more specific type of interdependency such as kinship, common interest, dislikes, friendships, events, etc.
Software/System Development Life Cycle: (SDLC) is a structured process for creating or altering information systems, and the models and methodologies that are used to develop these systems.
Solar flare: An explosion on the sun that occurs when energy stored in twisted magnet fields, usually above sunspots, is released suddenly. Solar flares generate a burst of radiation across the electromagnet spectrum from radio waves to x-rays and gamma rays.
Spear phishing: An e-mail spoofing fraud attempt that targets a particular organization with the intent of obtaining unauthorized access to confidential data.
Subject matter expert: (SME) is an individual or group of individuals who is an expert in a given area or topic.
Super storm: Any storm that is extremely and unusually destructive, often referred to in doomsdays types of events.
Super volcano: See caldera.
Surprise: The discrepancy between the desired or expected, the actual state, and the probability of loss, or the thought of suddenly confronting an unanticipated and unfamiliar circumstance.
Switchboard: A device that connected several telephones together manually or to an external telephone exchange.

Tabletop Exercise: (TTX). It is an exercise involving key personnel discussing simulated scenarios in an informal setting. TTXs are used to assess current plans, policies, and procedures.
Target audience: Generally, everyone who can benefit from the information.
TCO: See Total cost of ownership.
TDD: (Telecommunication Device for the Deaf). See TTY.
Telecopier: Equipment used to transmit and receive facsimile copies of documents.
Telephone: A device that can transmit speech electrically.
Television: (TV). A telecommunication medium for receiving and transmitting moving images with sound.
Terrestrial: To send over the air.
Test plan: This is a document defining a systematic approach to the testing of a system such as equipment or software.
TETRA: (Terrestrial Trunked Radio). A digital-trunked mobile radio standard developed by the European Telecommunications Standards Institute as an advancement in mobile telephony solutions.
Text orthography: A method defining the correct way of using a writing system to write a language.
Text Relay: A national telephone relay service that connects a caller who is using a textphone with a person who is using a telephone or another textphone.
Textphone: A telecommunication tool used by the hearing impaired to access emergency services. It is a telephone with a keyboard.
The Americans with Disabilities Act of 1990 (ADA): Legislation outlining the requirements for ensuring equal opportunity for persons with disabilities, with regard to access to commercial facilities and transportation, employment, government services, and public accommodations. ADA also required the establishment of TDD/telephone relay services.
Threat: A threat is driven by circumstances beyond the capabilities to respond by an individual or organization.
Ticker: See Crawler.
Total cost of ownership: (TCO) is a financial estimate to assist people or organizations in determining the direct and indirect costs of a product or system throughout its lifecycle.
Transmission: The method used for sending a message, such as the radio.
Triode: An electronic amplifying tube used to enhance telephone and radio communications.
TRS: (Telecommunications Relay Service). A service that enables a person having hearing or speech disability to place and receive telephone calls anywhere in the United States by dialing 7-1-1. It uses trained operators who type what is said by the caller so

that the receiver can read the message on a TTY display and then type a response.

Tsunami: A series of ocean waves that is generated by a sudden displacement in the sea floor, landslides, or volcanic activity. The waves in the deep waters may be only a few inches high. As these waves come ashore, they may increase in height, becoming a fast-moving wall of turbulent water several feet high.

TTX: See Tabletop Exercise.

TTY: (Text telephone). A device enabling the hearing impaired or speech impaired to type messages back and forth with the caller rather than talking and listening. The device is sometimes referred to as TDD (Telecommunication Device for the Deaf).

UHF: (Ultra high frequency). The radio frequency range of electromagnetic waves between 300 MHz and 3 GHz.

Unified Area Command: See Area Command.

Unified Command: Within ICS, it is defined as having more than one agency with incident jurisdiction or when incidents cross political jurisdictions. These agencies work together through the designated members to establish a common set of objectives and strategies and a single Incident Action Plan.

Unit test: A method within a test plan where individual units of source code are tested to determine if they are fit for use.

Universal Emergency Calling Number: A single emergency telephone number used by a country or region to request emergency services.

Use case: It defines the steps or actions between a user (or "actor") and a software system that leads a user toward something useful. A use case is used by software developers to select features to implement and to resolve errors.

User acceptance testing: Also called the final release testing of the test plan, this method of testing is conducted to determine if the requirements of a specification or contract are met before accepting transfer of ownership.

UTC: Coordinated Universal Time. The primary time standard the world uses to regulate clocks and time. Although different from Universal Time (UT) and Greenwich Mean Time (GMT), it is used interchangeably when subsecond precision is not required.

VHF: Very High Frequency, in the 30 to 300 MHz radio spectrum.

Visually impaired: Individuals who are blind or unable to see at night or distinguish different colors or shapes.

VOR: (Very High Frequency Omni-directional Range). These are systems used by aircraft that have an antenna array transmitting two signals simultaneously: a directional signal rating like a lighthouse at a fixed rate and an omnidirectional signal that pulses when the directional signal is facing north. An aircraft can determine its bearing by measuring the difference in phase of these two signals to establish a line of position.

Warning: A warning is imposed when an event is evident, such as when a tornado has been spotted in the area and the target audience must seek shelter immediately. A warning requires a response.

Water stress: Occurs when the demand for water exceeds the available amount for a given period or when poor quality restricts its use.

Widget: On the web, it is a small application that can be installed and executed within a web page by an end user.

WiFi: (Wireless Fidelity). A wireless means of connecting electronic devices to the Internet using a wireless network access point or hotspot. Coverage may be limited to 65 feet, campus-wide, or city-wide.

Wiki: A website enabling the creation and editing of interlinked web pages via a web browser using a text editor or markup language and used collaboratively by multiple users such as Wikipedia.org.

Wimax: (Worldwide Interoperability for Microwave Access). A wireless broadband access technology that is IP based and used across cellular networks.

Wireless: A term used to refer to the system or method used for radio communications.

Workshop: This consists of sessions used to build a specific product such as a draft plan or policy in support of an initiative.

Worm: A type of computer virus that is able to spread from computer to computer without human action. A worm uses file or information transport features on a computer to travel unaided.

X-class flares: These are huge solar flares, a major event that can trigger planet-wide radio blackouts and long-lasting radiation storms.

Zero day exploit: (0-day). An attack or threat that attempts to take advantage of the vulnerabilities a computer application may have that are unknown to its users or developer or have not been corrected by the developer.

ZIP Codes: Zoning Improvement Plan (7 + 4 digits). It is used to leverage technology in the sorting and distribution of mail.

Additional Resources

Community Emergency Response Teams (CERT)
http://www.citizencorps.gov/cert/
Email: cert@dhs.gov

Ready America
Resources for getting a kit, making plans, and to stay informed.
http://www.ready.gov
Email: ready@dhs.gov

Federal Emergency Management Agency
500 C St., SW
Washington, DC
1.800.480.2520

Federal Communications Commission
http://www.fcc.gov
Email: fccinfo@fcc.gov
445 12th street, SW
Washington, DC 20554
1.888.225.5322 (Voice)
1.888.835.5322 (TTY)

Institute for Public Relations
http://www.instituteforpr.org
Email: info@instituteforpr.org
2096 Weimer Hall
Gainesville, FL 32611-8400
1.352.392.0280

University of Maryland
Department of Communications
http://www.comm.umd.edu
Email: eltoth@umd.edu
2130 Skinner Building
College Park, MD 20742-7635
1.301.314.9471

Public Relations Society of America
http://www.prsa.org
Email: pr@prsa.org
33 Maiden Lane, 11th Floor
New York, NY 10038-5150
1.212.460.1452

Federal Emergency Management Agency (FEMA)
http://www.fema.gov
Email:
U.S. Department of Homeland Security
500 C Street SW
Washington, DC 20472
1.800.621.3362 or 1.202.646.2500 (Voice)
1.800.462.7585 (TDD/TTY)

U.S. Department of Homeland Security
http://www.dhs.gov
1.202.282.8000 (Voice and TTY via Federal Relay Service)
Washington, DC 20528

U.S. Department of Commerce
http://www.commerce.gov
Email: thesec@doc.gov
1401 Constitution Ave., NW
Washington, DC 20230
1.202.482.2000

U.S. Geological Survey
http://www.usgs.gov
12201 Sunrise Valley Drive
Reston, VA 20192, USA
1.888.275.8747 or 1.703.648.5953

National Oceanic and Atmospheric Administration
http://www.noaa.gov
Email: outreach@noaa.gov
1401 Constitution Avenue, NW, Room 5128
Washington, DC 20230

Government Emergency Telecommunications Services (GETS)
http://gets.ncs.gov
Email: gets@dhs.gov
1.866.627.2255 or 1.703.760.2255 (Voice)
1.800.231.6702 or 1.703.488.4048 (TTY)

Wireless Priority Service (WPS)
http://wps.ncs.gov
Email: wps@dhs.gov
1.866.627.2255 or 1.703.760.2255 (Voice)
1.800.231.6702 or 1.703.488.4048 (TTY)

National Institute on Deafness and Other Communication Disorders (NIDCD) Information Clearinghouse
http://www.nidcd.nih.gov/
E-mail: nidcdinfo@nidcd.nih.gov
1 Communication Avenue
Bethesda, MD 20892-3456
1.800.241.1044 (Voice)
1.800.241.1055 (TTY)

Index

A
accelerometers, 320
acceptance process in product selection, 364–365
acronyms, 451–457
activation of emergency communications, 166–171
Adams, Franklin, 55
administrator training, 375–376
advanced persistent threat (APT), 139
Advanced Research Projects Agency Network (ARPANET), 31–32
Advanced Television Systems Committee (ATSC), 28
advertising, 430
advisories, tsunami, 124
after-action reports (AAR), 401–402, 409–410
aging infrastructure, 135–138, *139*
alarm systems, 294–295
alerting, route, 170
alert systems
 emergency automated telephone notification system, 302–303
 public address systems, 303–306
all clear messages, 164
alternate reality games, 279–280
amateur radio, 14–17
AM broadcasting, 17–18
American Radio Relay League (ARRL), 15–16
American Sign Language (ASL), *99*
animals, service, 94
AOL, 298

appendices, communication plan, 395–398
Apple, 31, 256, 267, 291, 423
Application Programming Interfaces (APIs), 284
appraisal costs, 83
ArcGIS, 281
Arch Re, 117
area command, 216
Armageddon, 74
Armstrong, Edwin Howard, 18–19
Armstrong, Neil, 29
Assange, Julian, 270
assumptions, project, 337
Asteroids, 129–130
atmospheric conditions, 125–126
AT&T, 46–47, 49
attainment, 337
attribution theory, 192, *193*
audiences, target, 69, 332–335, 392–393
augmentative and alternative communication (AAC), 89–90
Australia, 57
Automated Speech Recognition systems (ASRs), 97–98

B
backhaul, 325
Bain, Alexander, 7
bandwidth and speed, 292–294
Barbier de la Serre, Charles, 34
Beard, John Stanley, 42
Bell, Alexander Graham, 8–9, 33
Bellanger, Amanda, 54

Ben Ali, Zine El Abidine, 133
benefit analysis, 354–356
Bevill, Tom, 47
bilingual employees, 103
billboards, 307–310
Bing Maps, 273
bin Laden, Osama, 264, 266–267, 276
biometrics, 149
BlackBerry, 267, 268, 298
BlackBoard, 320
black hat programmers, 140
blocking inbound calls to 9-1-1, 54–56
blogs, 266–267, 276
Blogs.com, 266
Bluetooth, 25
Bluetooth, Harold, 25
bookmarking, social, 276
Braille, Louis, 34
Braille system, 34
broadsides, 4
Brousell, Lauren, 148
budget, 339–342
bulletin boards, 279
Bush, George W., 210
Business Continuity Plans (BCP), 433–434
Buzz up!, 262

C

calderas, 122
capacity building, 184–185
captions, 96
Carbonite.com, 275
Castro, Carmen, 146
c-class flares, 129
cellular phones. *See* mobile phones
Center for Personal Assistance Services, 87
centralized monitoring, 323
challenges to effective communications, 62–66, 81–83
  with children, 107
  cost of quality and, 83–87
  with culturally and geographically isolated people, 104–105
  with the elderly, 107–108
  with hearing impaired people, 95–98
  with homeless people, 105–106
  with incarcerated people, 105
  with Limited English Proficiency (LEP) people, 102–104
  with medically or chemically dependent people, 106
  with mobility challenged people, 101
  with non-English speaking people, 101–102
  for people with disability and others with functional and access needs, 87–108
  with visually impaired people, 98–101
Chang Heng, 318
channels, sensory, 65–66
Chappe, Claude, 4
children, 107
Citizens' Band Radio, 25
clarity in messages, 417
Clark, Greg, 134
classrooms, 301, *302*
Clery Act, 251–252
climate warfare, 125
closed captions, 96
CLR holding, 45
codecs, 28
coded orthogonal frequency-division multiplexing (COFDM), 25, 28
cognitively impaired persons, 90–95
combined alerting system software, 320–321
comets, 129–130
Comité Consultatif International Téléphonique, 8
Commercial Mobile Telephone Alerts (CMAS), 244–245
common air interface, 243
common alerting protocol (CAP), 324
common allerting protocol (CAP), 40
common operating picture (COP), 324–326

communications
  continuum, 221–222
  needs, 335–336
Communications Act of 1934, 241
communications dynamics, 61–62
Communications Systems Analysis Division (CSAD), 243–244
community access television (CATV), 307–310
community activated lifesaving voice emergency systems (CALVES), 305–306
Community Emergency Response Training (CERT), 146
community outdoor warning sirens (COWS), 305–306
computer labs as EOCs, 323
CONELRAD, 39
conference rooms, 301, *302*
confusing cease-fires and negotiations, 75
confusion, 75
Connell, Joseph Bernard, 47
CONPLAN, 211–212
constraints, 337–338
content, shared, 274–275
context, 338
continuity of operations, 433–434
continuous improvement process, 71, 204–206
continuous wave (CW) telegraphy, 24
conversations, forum, 279
Cooke, William, 4
Cooper, Martin, 21
copy machines, 9–10
costs of quality, 82–87
crawlers, television, 98
crisis communications, 66–68.
  See also disaster communications; messages
  after-action reports (AAR), 401–402, 409–410
  challenges to effective, 81–82
  characteristics, 71–74
  costs of quality in, 82–87
  defined, 37–38, 68
  examples of poor, 67–68
  future of, 440–441
  in the midst of violent events, 74–78
  planning, 68–71, 85–87, 402–403, 445–446, 449
  process, 70
  propaganda and, 77–78
  recognizing when a message is not being communicated in, 73–74
  sample messages, 461–465
  team expectations, 388–392
  for war preparation, 76–77
cross patch switch, 238
crowdmapping, 262
crowdsourcing, 320
culturally or geographically isolated people, 104–105
cultural norms, 404–405
cyber attacks and cyber warfare, 139–141
cyber-espionage, 139
cyber-sabotage, 139

D

data
  elements, 230–231
  requirements, 344
debriefing, 222–228
decisional paralysis, 70
decontextualizing violence, 74
dedicated telephone landlines, 323
Defense Advanced Research Projects Agency (DARPA), 30
de Forest, Lee, *13*, 14
department heads, 177
design brief
  budget, 339–342
  goal setting, 336–337
  problem statement, 337–342
  solution analysis in, 342–345
  system requirements specification, 345–353
desired outcomes, 365
Desperate Journey, The, 8
developmental disabled persons, 90–95
dialing, pocket, 54
Dickson, William, 8

Digg, 262
digital audio radio service (DARS), 152
digital radio, 23–25
Digital Radio Mondiale (DRM), 25
digital rights management (DRM), 271
digital signage, 307–310
digital television, 28, 29
direct impact threats, 135
direct installation, 370
direction and control of emergency communications operations, 179
disabilities/functional and access needs, people with
   challenges to effective communication with, 90–101
   communications methods for, 108–109, *110–111*
   emergency communications activations for, 169–170
disaster communications
   9-1-1 and, 44–57
   around the world, 57–58
   emergency broadcasting, 39–40
   in the future, 148–151
   mass notification and, 37, 38–39
   new mobility and, 147–148
   universal emergency calling number in, 40–58
Disaster Reporting System (DIRS), 244
disasters, 6, 447–448
   aging infrastructure, 135–138, *139*
   atmospheric condition, 125–126
   catastrophe statistics, 116–117
   changing demographics and, 115–116
   comet, asteroid, and meteor strike, 129–130
   cyber attacks and cyber warfare, 139–141
   earthquake, 119–121
   food shortages due to, 133

   health, 135–138, *139*
   mental health and first aid in, 434–437
   new mobility and, 147–148
   planning, 143–152
   power shortages due to, 133–134
   reinsurance companies and, 117–118
   shortages of key resources due to, 130–134
   solar flare, 127–129
   superstorm, 126–127
   tsunami, 123–125
   types predicted over the next 25 years, 119–130
   volcanic disruption, 121–123
   war/conflict, 141–143
   water shortages due to, 130–132
Discrete Emergency Frequency (DEF), 240
disorders, communication, 32–35
   effective, 87–108
   hearing impairment, 95–98
   primary groups, 89–101
   speech disabilities and, 89–90
dissemination of warnings to the public, 166
Distance Measuring Equipment (DME), 22
distributed denial-of-service (DDoS) attacks, 139
documentation, user, 374–375
Dolbear, Amos, *12*
Dolby, 10
Dole, Bob, 430
domain names, 54
Donna Prince L. v. Waters, 145
Doodle, 281
doppler effect, 23
Dropbox, 275
Drucker, Peter, 200
drug abuse, 106
dualism, 74
Dudley, Robert, 381
Duffys, Thomas, 42
dynamics of communications, 61–62
   effectiveness and, 62–66

E

earthquakes, 119–121, 318–320
echo immigrants, 118
echolink, 39
Edison, Thomas, 8, 9–10, 33
effective communications, 62–66
   after-action review for, 409–410
   getting the message right in, 407–419
   how of, 196–198
   as a human issue, 407–409
   lessons learned and continuous improvement in, 71, 204–206
   message and messenger in, 198–201
   mistakes and, 403–406
   preparation for, 203–204
   situational crisis communication theory and attribution theory, 192, *193*
   what of, 191–194
   what to avoid saying in, 194
   when of, 194–195
   where of, 195–196
   who of, 190–191
   why of, 189–190
   Ws of, 188–198
Ehret, Cornelius, 18
El-Baz, Farouk, 131
elderly, the, 34–35
   challenges to effective communication with, 107–108
electorn, 89
electronic bulletin boards, 260
electronic larynx, 34
El Niño, 125–126
e-mail, 32, 262, 296–298
Emergency Alert System (EAS), 40
emergency authorities, 43
emergency broadcasting, 39–40
Emergency Broadcast System (EBS), 39–40
emergency communications framework, 157–158. *See also* crisis communications
   direction and control of, 179
   dissemination of warnings to the public, 166
   emergency management phases, 171–176
   interrelationships, 158–160
   needs, 335–336
   organization and responsibilities, 176–179
   put into operations, 160–162
   readiness levels, 179–181
   receiving warnings from trusted sources, 164–166
   record keeping and, 182
   security, system maintenance, and capacity building, 184–185
   sending messages using, 162–164
   training and education, 182–184
   warnings to outside agencies, 170–171
   when to activate, 166–171
Emergency Management Coordinator (EMC) or Director, 177–178, 384
emergency management phases, 171
   mitigation, 174–175
   preparedness, 172–173
   prevention phase, 175–176
   recovery, 174
   response, 173
Emergency Operations Center (EOC), 162, 166, 178, 179, 227
   computer lab as, 323
   critical plans to support, 433–434
   fusion center and, 331–332
   readiness levels, 180
   tips for preparing, 425–427
Emergency Response Plan (ERP), 433–434
emergency signaling, 3
Emergency Support Functions (ESFs), 222–223, *224–225*
Emmanuel, Rahm, 37, 58
Enhanced 9-1-1, 49, 51–52
   in tribal lands, 243
ENIAC, 30
enthusiasm, 417, 418

environmental refugees, 133
Ericsson, Lars Magnus, 9
ethical and moral reflection, 201–203
Eureka 147, 25
Europe, emergency calling number in, 57–58
evacuation
  maps, 427
  warning systems, 315–316
evaluation process for product selection, 363–364
excluding and omitting the bereaved, 75
external affairs teams, 383
external entities, 178
external failures, 84
eye contact, 418–419

F
Facebook, 147, 255–257, 260–262, 265, 280, 300, 320, 385, 432
  Chat, 298
  maps, 276
  mobile, 267
  Places, 270
facsimile, 7–8
fact sheets, 388
failures
  common communication, 403–406
  to explore peace, 75
  to explore the causes of escalation, 75
  external, 84
  implementation, 372–374
  internal, 84
  software, 359
Fair and Accurate Credit Transactions Act (FACTA), 250
Family Educational Rights and Privacy Act (FERPA), 250–251
Family Radio Service, 25
fax, 7–8
Federal Aviation Administration (FAA), 239–241

Federal Communication Commission (FCC), 241
  2-1-1 and, 148
  Communications Systems Analysis Division (CSAD), 243–244
  Emergency Alert System and, 40
  enhanced 9-1-1 and, 52
  new mobility and, 152–153
  television and, 29
  universal emergency calling number and, 45
Federal Emergency Management Agency (FEMA), 16, 69, 87–88, 164, 227
  National Emergency Coordination Net (NECN), 238
  National Radio System (FNARS), 238
  Operations Center (FOC), 164
federal laws. *See also* federal Communication Commission (FCC)
  Clery Act, 251–252
  Fair and Accurate Credit Transaction Act (FACTA), 250
  Family Educational Rights and Privacy Act (FERPA), 250–251
  Gramm-Leach-Bliley Act (GLBA), 249–250
  Health Insurance Portability and Accountability Act (HIPAA), 190, 248–249
  Incident Command System (ICS), 215–228
  National Emergency Communications Plan (NECP), 228–233
  National Incident Management System (NIMS), 210, 212–215
  National Response Framework (NRF), 164, 210–212, 227
  national telecommunications and information administration, 233–247
  Stafford Act, 227, 247–248

systems establishment after September 11, 2001 terrorist attack, 209–210
Federal Radiological Emergency Response Plan (FRERP), 212
federal spectrum use, 234–237
Fessenden, Reginald, 11, *13*
Field Coordinating Officer (FCO), 227
final release testing, 368
Finance and Administrative Section Chief, 220
financial considerations, 360–362
Financial Services Modernization Act, 249
fire and gas detector and alarm systems, 294–295
Firefox, 280
first aid, 434–437
First Amendment rights and 9-1-1, 52–53
fixed satellite services (FSS), 152
flash messages, 298–299
flash mobs, 285
Flickr, 258
FM broadcasting, 18–20
food shortages, 133
Forever and a Day, 8
FORTRAN, 31
Forty-Ninth Parallel, The, 8
forums, 279
Foursquare, 280
Fox, Michael, 255
Freeh, Louis, 55
free radio, 26
Friendster, 260
full-scale exercises (FSE), 394
functional and access needs, communications for people with. *See* disabilities/functional and access needs, people with
functional exercises, 394
functional requirements, 343–345
fusion centers, 331–332
future communications, 148–151, 440–441

G

gadgets, 267
Gadhafi, Mu'ammar, 141–142
Gallagher, Bob, 47
Galtung, Johan, 74
games, video, 31
gas detectors, 294–295
Genachowski, Julius, 158
generalizations, 66
geocoding, 273
geo-location systems, 270–273
geotagging, 272–273, 285
geotargeting, 271–272
Gerke, Friedrich, 6
Gleick, Peter, 132
Global Positioning System (GPS), 23, 24, 30, 151–152
 geocoding and, 273
 geotagging and, 273
 measuring earthquake potential with, 319–320
 mobile, 268
goal setting, 336–337
Google, 203, 266–267, 270
 Buzz, 280
 Calendar, 281
 Docs, 274–275
 Gmail, 281, 297
 Maps, 273, 281, 286
 Talk, 298
 Wave, 275
governance, 230
Government Emergency Telecommunications System (GETS), 235, 237
Gramm–Leach–Bliley Act (GLBA), 249–250
Gray, Elisha, 8–9
green level, 180–181

H

hackers, 53
hams, 16
hard disk drives, 11
harmonic telegraphy, 9
Harvard-IBM Mark computers, 31
Hayward, Tony, 380–381
hazards of social media, 284–287

health disasters, 135–138, *139*
Health Insurance Portability and
    Accountability Act of 1996,
    190, 248–249
Health Map, 281
hearing, 63
    aids, 33, 96
    effective communications for
        people with impaired,
        95–98, 170
hearing carry-over (HCO), 92
hearing impaired persons, 33–34
heliographs, 3
Herrold, Charles, 17
Hertz, Heinrich, 11, *12*
hieroglyphic writing, 2
high-definition television (HDTV),
    28, 29
High-frequency Active Auroral
    Research Program
    (HAARP), 127
high tech emergency
    communications, 161
highway alerting, 316–317
Higinbotham, William, 31
history of communications
    copy machine in, 9–10
    digital radio, 23–25
    facsimile in, 7–8
    heliography in, 3
    information science and computers
        in, 30–32
    loudspeakers in, 10
    magnetic recording in, 10–11
    mail and parcels in, 6–7
    mobile phone in, 9
    navigation in, 22
    paper, newspapers, and magazines
        in, 3–4
    for persons with disaiblities and
        functional/access needs,
        32–35
    process, 1–2
    radar in, 22–23
    radio in, 11–20
    telegraph service and Morse code
        in, 4–6
    telephone in, 8–9, 20–22

television in, 27–29
3500 BC to 1 BC, 2–3
Hitachi, 10
Holmes, Oliver Wendell, 427
homeless people, 105–106
HootSuite, 432
Hotmail, 297, 298
How of effective communications,
    196–198
Hulu.com, 301
Hurricane Ike, 329–330
Hussain, Saddam, 76–77

I

IBM computers, 31
illegal drug use, 106
immigrants, echo, 118
implementation, system, 369–376
in-building voice announcements,
    303–304
incarcerated, the, 105
Incident Commander (IC), 196,
    216–217, 284
Incident Command Post (ICP), 182
Incident Command System (ICS),
    215–228, 284
    communications, 221–222
    debriefing, 222–228
indirect impact threats, 135
inference, 66
InformaCast, 320
informationization, 140
information science and computers,
    30–32
information security, 214
Information Technology Department
    (IT), 384
infrastructure, aging, 135–138, *139*
Inmarsat, 21–22
inspection costs, 83
installation, system, 369–376
instant messaging, 298
Institute of Electrical and Electronics
    Engineers (IEEE), 25
institutionalized persons, 90
Integrated Automated Fingerprint
    Identification Systm (IAFIS),
    149

Integrated Public Alert and Warning
    System (iPAWS), 40, 447
Interdepartment Radio Advisory
    Committee (IRAC), 233
internal failures, 84
international Morse code, 6
International Organization for
    Standardization (ISO), 7
International Telecommunication
    Union (ITU), 21
Internet, the
    9-1-1 address purchased for profit,
        54
    crisis communication and, 39
    future communications and,
        149–151
    hearing impaired people and, 97
    origins of, 32
    radio, blogs, and talk radio, 276
    war/conflict and, 142–143
Internet Protocol (IP) relay service, 92
interoperability continuum, 229
interpretation, 1, 103
intranets, 147–148
invisible Internet, 149–151
IP-captioned telephone service, 93
iridium, 21–22

J
Jackson, Michael, 269
Jeanne Clery Disclosure of Campus
    Security Policy and Campus
    Crime Statistics Act,
    251–252
Jeansonne, David, 56
Jobs, Steve, 423
Johnson, Lyndon, 45
Joint Information Center (JIC),
    178–179, 219, 284
journalists, emergencies related to
    violence by, 74–75
judgment, 66

K
Kaywa, 274
Kellogg, Edward, 10
Kennedy, John F., 29
key fobs, 149

kinematoscope, 8
kinetograph, 8
Kinsey, Dan, 293–294
Knightley, Phillip, 76, 77
Krums, Janis, 260–261
Kuwait Re, 117

L
landlines, telephone, 323, 425–426
Land mobile radio (LMR), 312
La Niña, 125–126
larynx, electronic, 34, 89
Lawhorn, Bob, 359
Lawrence, Nick, 54
legal departments, 385
Legally authorized representative
    (LAR), 90–91
Lenovo, 31
lessons learned and continuous
    improvement, 71, 204–206
Liaison Officer, 218
licensing, 364
Limited English Proficiency (LEP)
    people, 102–104
LinkedIn, 257, 280, 432
listening skills, 63–64
local area network (LAN), 25, 31,
    272
local officials, notification of, 165
Local Warning Point (LWP), 163, 166
Lodge, Oliver, 11, *12*
Logistics Section Chief, 219–220
Long Term Evolution (LTE) mobile
    communications standard,
    158, 243
Loomis, Mahlon, *12*
LORAN, 29–30
loudspeakers, 10
low-/medium-tech emergency
    communications, 160–161

M
MacDonald, Norman, 41
magazines, early, 3–4
magnetic recording, 10–11
mail and parcels, early, 6–7
MakeoutClub, 260
manicheanism, 74

MapBuilder, 281
MapQuest, 273
maps, 276–277, 286
    evacuation, 427
Marconi, Guglielmo, 11, *13*, 14, 23
marketing emergency services and plans, 427–433
mashups, 262, 281–282
Maslow's Hierarchy of Needs, 435–436
mass media, 77
mass notification, 37, 445
    requirements specifications, 346–353
    system activation and criteria guidance sheet sample, 459
    system integration with fire alarm systems, 322–323
    trusted sources, 164–166
    when to activate, 166–171
Maxim, Hiram Percy, 15
Maxwell, James Clerk, *12*
m-class flares, 129
media, 1
    news conference guidelines and, 396–397
    preparing to manage the, 387–388
    relations reminders, 397–398
medically or chemically dependent people, 106
megabits, 38
Memorandum of Understanding (MOU), 238
mental health, 434–437
mentally disabled persons, 90–95
messages. *See also* crisis communications
    after-action review of, 401–402, 409–410
    components of effective, 431–433
    effective, 198–201, 407–419
    flash, 298–299
    guidelines for, 413–419
    instant, 298
    preparing to manage, 387–388
    sample, 461–465
    sending emergency communications, 162–164

text, 277–278
timing, 405–406
warning, 168
what to avoid in drafting, 401–402, 411–413
messengers, effective, 198–201
meteor strikes, 129–130
microbroadcasting, 20
Microsoft, 256, 267
    Windows Live, 297, 298
microwave power transmission, 27
mimeograph, 9–10
Minow, Newton, 28
mistakes, communication, 403–406
mitigation phase, 174–175
MMS messaging, 277–278
mobile media centers, 323
mobile phones, 9, 20–21
    hearing impaired people and, 97
    pocket dialing on, 54
    quick response (QR) codes and, 274
    social media applications and web widgets, 267–268
    solutions for crisis communications, 290–292
mobile satellite services (MSS), 152
mobility, new, 147–148, 152–153
mobillity challenged people, 101
mobs, flash, 285
monitoring
    centralized, 323
    earthquake and tsunami, 318–320
    social media, 280
Monitter, 280
moral and ethical reflection, 201–203
Morello, Henry, 146
Morse, Samuel, 5–6
Morse code, 5–6
Motorola, 21
Mount St. Helens, 122, *123*
mudflows, 121
multiagency coordination systems, 213
multiplexing, 19
    orthogonal frequency-division, 25
Multi-Use Radio Service, 25–26
Munich Re, 117–118

murder boards, 406
Musser, Lori, 205
MyOtherDrive.com, 275
MySpace, 260
MythBusters, 285

N
Napolitano, Janet, 165
narrowband, 16
narrowbanding, 245
National Communications Service (NCS), 235
National Emergency Communications Plan (NECP), 228–233
National Hazard Total Risk Index, 117–118
National Incident Management System (NIMS), 210, 212–215, 331
National Information Exchange Model (NIEM), 325
National Interoperability Field Operations Guide (NIFOG), 234
National Planning Scenarios, 225–226
National Preparedness Guidelines, 223, 226
National Preparedness Vision, 223, 225
National Response Framework (NRF), 164, 210–212, 227
National Response Plan (NRP), 210
National Security/Emergency Preparedness (NS/EP), 234–235, *236–237*
National Telecommunication and Information Administration (NTIA), 233
National telecommunications and information administration, 233-
National Traffic System, 15
National Warning System (NAWAS), 164
National Weather Service (NWS), 40, 164–165

navigation systems, 22
negotiations, 361–362
new conference guidelines, 396–397
New Energy and Industrial Technology Development Organization (NEDO), 10
newspapers, early, 3–4
Nextel, 310–311
Next Generation 911 (NG911), 153
9-1-1, 152–153
  blocking inbound calls to, 54–56
  emergency operator procedures, 51
  enhanced, 49, 51–52
  First Amendment rights and, 52–53
  history and development of, 44–49, *50*
  Internet address purchased for profit, 54
  next generation, 57
  reverse, 317–318
  silent calls, 53–54
  in tribal lands, 243
non-English speaking people, 101–102
nongovernmental organizations, 227
noninstitutionalized persons, 90–91
no tech emergency communications, 160

O
Obama, Barack, 142, 267, 379, 423
Office of Emergency Management (OEM), 177
  readiness levels, 180
Office of Spectrum Management (OSM), 233–234
o-Files, 281
omitting reconciliation, 75
online advertising, 430
open captions, 96
open source espionage, 140
Operations Section Chief, 219
operators, emergency, 51
organization and responsibilities, emergency communication system, 176–179

orientation seminars, 393
Orman, Suze, 423
orthography, text, 65
outdoor sirens and speaker arrays, 304–305
OutWit, 280
overload, 330

P
Pacific Institute, 132
packets switching, 31–32
paper, early use of, 3–4
parallel installation, 370
paraphrasing, 417
parcels and mail, early, 6–7
partnerships, public/private, 437–440
passwords, 149
patriotism, 8
Pauken, Tom, 359
payphones, public, 42, 45
peak ecological water, 132
Pegasus, 43
Perskyi, Constantin, 28
Personal Area Network (PAN), 25
Personal Localized Alerting Network (PLAN), 245
Peter, Laurence Johnston, 289
phased installation, 371–372
Phillips, 10
pictographs, 2
pipelines, 135–136
pirate radio, 26
Pirnie, Robert M., 47
plain old telephone service (POTS), 21
planning
  crisis communications, 68–71, 85–87, 402–403, 449
  disaster, 143–152
  Section Chief, 219
plans, communication
  appendices, 395–398
  elements of, 379–381
  followed when drafting messages, 415
  marketing emergency services and, 427–433
  preparing to manage the message and the media using, 387–388
  purpose and objectives of, 381–382
  putting team in place, 383–387
  reviews and updates, 395
  target audience, 392–393
  team expectations and, 388–392
pocket dialing, 54
Pony Express, 6, 38
Popov, Alexander, 11, *12*
postal service, 2, 6, 273
potable water, 132
Poulsen, Valdemar, 10–11, *13*
power shortages, 133–134
preparation for effective communication, 203–204
preparedness phase, 172–173, 213
prevention phase, 175–176
Prince, Donna, 145
problem statements, 337–342
propaganda
  crisis communications and, 77–78
  movies, 8
prototypes, 342–343
ProxTalker, 94
proxy servers, 271
public address systems, 303–304, 303–306
public affairs teams, 383
Public Information Officer (PIO), 178–179, 196–197, 218–219, 284, 406
  successful, 424
public information systems, 213
public media, 300–301
public payphones, 42, 45
public/private partnerships, 437–440
Public Safety and Homeland Security Bureau (PSHSB), 242–243
public safety answering points (PSAPs), 49, *50,* 243
public switched telephone network (PSTN), 20–21
pyroclastic flows, 121

Q

QRStuff, 274
quadrature amplitude modulation (QAM), 24
quality, costs of, 82–87
quick response (QR) codes, 273–274

R

radar, 22–23
radio, 11–14
   amateur, 14–17
   AM broadcasting, 17–18
   control, 26
   coordination and use of emergency networks, 238
   digital, 23–25
   emergency use of nonfederal frequencies, 238
   energy autarkic, 27
   Federal Aviation Administration (FAA) and, 239–241
   federal spectrum use, 234–237
   FM broadcasting, 18–20
   Internet, 276
   interoperability between federal entities and nonfederal public safety licensees, 238–239
   navigation, 22
   shortwave, 15
   spectrum, 11
   talk, 276
   telephony, 14, 20–22
   two-way, 310–312, 425
   unlicensed services, 25–26
   weather, 313
   wireless, 26–27
Radio Amateur Civil Emergency Services (RACES), 16
Radio Corporation of America, 14
Rainy Day cases, 343
rapid response, 447
REACT! System, 320
readiness levels, 179–181
Reagan, Ronald, 423
reality games, alternate, 279–280
Really Simple Syndication (RSS feeds), 262, 268–269, 320
receivers, communication, 1, 61
   effective communication and, 64–65
reception, 1
   of warnings from trusted sources, 164–166
reconciliation, omitting, 75
recording, magnetic, 10–11
record keeping, 182
recovery, crisis, 70
   phase, 174
Red Flag Rule, 250
red level, 181
refarming, 245
reflection, moral and ethical, 201–203
refugees, environmental, 133
Regional Emergency Communications Coordination (RECC), 232–233
regional resources, 316–320
reinsurance companies, 117–118
relay services. *See* telecommunications Relay Service (TRS)
resource management, 214
response, crisis, 187–188
   lessons learned and continuous improvement in, 71, 204–206
   message and messenger in, 198–201
   moral and ethical reflection in, 201–203
   phase, 173
   preparation for, 203–204
   rapid, 447
   Ws of effective communications in, 188–198
responsibilities and organization, emergency communications system, 176–179
Reverse 9-1-1, 317–318

reviews and updates, communication plan, 395
Rice, Chester, 10
Riggs, Robert, 53
risk analysis, 354–356
RNID Typetalk relay service, 43
Roach, Stephen, 187
roadblocks to listening, 64
roads, bridges, and transit systems, 138, 139
Robert T. Stafford Disaster Relief and Emergency Assistance Act, 227, 247–248
route alerting, 170
Ruben, Robert J., 89

S
SAFECOM Interoperability Continuum, 229–230
safety Officer, 217–218
Sarget Capabilities List (TCL), 226–227
satellite(s)
    mobile services, 151–152
    phones, 21–22
    television, 29–30
Savage, Adam, 285
SchoolMessenger, 320
Scott, Jill, 264
Second Life, 279–280
security
    information, 214
    system maintenance, and capacity building, 184–185
Selecommunication Service Priorities for Radio (TSP-R), 234
Selecommunications Relay Service (TRS), 33–34, 91–94
senders, communication, 1, 61
    effective communication and, 64–65
sending of emergency communications messages, 162–164
sensory channels, 65–66
September 11, 2001 terrorist attack, 209–210, 273
service animals, 94

sewer lines, 137–139
ShakeMap, 318–319
shared content, 274–275
shared non-English language relay service, 93
SHAred RESources (SHARES), 237
shared video/streaming video, 278–279
ShareThis, 262
shortages of key resources, 130–134
short answers, 418
shortwave radio, 15
Siemens, Ernst, 10
signage, digital, 307–310
silent 9-1-1 calls, 53–54
single location installation, 370–371
Sirius (radio), 27
situational crisis communication theory (SCCT), 192, 193
skepticism, 417
Skype, 142, 279
SMART goals, 336–337
smartpens, 97
smartphones, 97, 267–268
    solutions for crisis communications, 290–292
SMS text messaging, 277–278
snippets, 267
snowbird populations, 107
social bookmarking, 276
social media, 426
    alternate reality games, 279–280
    blogs, 266–267
    commonly used applications, 263–287
    Facebook, 147, 255–257, 260–262, 265, 267, 270, 298, 300, 320, 385
    forum, 279
    fundamental change with, 282–284
    geo-location systems, 270–273
    growth, 259–262
    gurus, 384–385
    Internet radio, blogs, and talk radio, 276
    introduction to, 255–259
    maps, 276–277

mashups, 262, 281–282
mobile phone applications, 267–268
monitoring, 280
quick response (QR) codes, 273–274
Really Simple Syndication (RSS feeds), 262, 268–269, 320
shared content, 274–275
shared video/streaming video, 278–279
social bookmarking, 276
social storage, 275
strategy sample, 473–475
successful use of, 432–433
as a technological hazard, 284–287
Twitter, 255–258, 261–262, 263–265, 267, 298, 300, 320, 385
used before it is needed in an emergency, 280
widgets, 267
wiki sourcing, 269–270
social storage, 275
software development life cycle (SDLC), 332
software integration, 280–281
solar flares, 127–129
solutions, 322–324
   alert systems, 302–316
   analysis, 342–345, 354–356
   billboards, video displays, digital signage, and community access television, 307–310
   cellular phones and smartphones, 290–292
   classroom or conference room, 301, *302*
   combined alerting system, 320–321
   common operating procedure (COP), 324–326
   3-D meeting software, 321–322
   e-mail and instant messaging, 296–298
   evaluation process for, 363–364
   financial considerations with, 360–362
   fire and gas detector and alarms systems, 294–295
   flash messages, 298–299
   public address systems, 303–306
   public media, 300–301
   regional resource, 316–320
   selecting a product for, 362–369
   selection for organizations, 332–336
   speakers and video surveillance emergency phone towers, 306–307
   system installation and implementation, 369–376
   with systems probably already in place, 290–301
   test plans, 367–369
   trends in crisis communications and, 289–290
   two-way radio, 310–312
   use cases and, 365–367
   voicemail and voice systems, 295–296
   web page, 299–300
Sony, 10
Southern California Integrated GPS Network (SCIGN), 319
SPARQCode, 274
speakers and video surveillance emergency phone towers, 306–307
spear-phishing, 140
special needs populations, 32–35
speech impaired persons, 34, 89–90, 95
speech-to-speech (STS) relay service, 93
speed and bandwidth, 292–294
Spitzer, Eliot, 381
spotlighting individual acts of violence while avoiding structural causes, 74
Sprint, 292, 310–311
Stafford Act, 227, 247–248
stakeholders, 332–335, 383, 426

standard operating procedures (SOPs), 230
State Operations Center (SOC), 165, 233
Statewide Communication Interoperability Plans (SCIPs), 228–229
streaming video, 278–279
stress, water, 132
stress testing, 343
Stubblefield, Nathan, 13
success
    future crisis communications and, 440–441
    implementation, 372–374
    keys to, 421–425
    marketing and, 427–433
    preparing emergency operations centers for, 425–427
    public/private partnerships and, 437–440
Sunny Day use cases, 343
superstorms, 126–127
super volcanoes, 122
Swiss Re, 117
switchboard, 41
Switzerland, 133–134
Sysomos, 264
system design projects
    challenge in learning for, 329–331
    design brief, 336–353
    effective governance of, 353–354
    emergency communication, 335–336
    fusion center and EOC, 331–332
    solution selection for organizations, 332–336
system installation and implementation, 369–376
system maintenance, 184–185

T

tabletop exercises (TTX), 231–232, 393–394
talk radio, 276
target audiences, 69, 332–335, 392–393
teams, crisis communications
    expectations, 388–392
    roles, 383–387
    successful, 425
technology, 230
telecopiers, 7–8
telegraphy, 3, 4–6
    continuous wave, 24
telephony, 8–9
    cellular, 9, 20–21
    co-ops, 44
    emergency automated telephone notification system, 302–303
    hearing impaired people and, 96–97
    landline, 323, 425–426
    for mentally disabled, developmentally disables, and cognitively impaired persons, 91–94
    radio, 14
    satellite, 21–22
    text, 33–34
    universal emergency calling number, 41–58
television, 27–29
    crawlers or tickers, 98
    hearing impaired people and, 96
    satellite, 29–30
TELSTAR, 29
Tencent, 203
Tephra falls, 121
Terrestrial Trunked Radio (TETRA), 21, 27
terrorism, 209–210, 212, 273
Tesla, Nikola, 11, 12, 26
test plans, 367–369
text messaging, 277–278
text orthography, 65
textphone, 43
text telephones (TTY), 33–34
text-to-voice TTY-based TRS, 93
Thoughts.com, 266
3-D meeting software, 321–322
Thurber, James, 329
tickers, television, 98
tiered response, 212

timelines, project, 342
timing, message, 405–406
Tirado, Christina, 133
Titanic (ship), 11
T-Mobile, 292
tone alerts, 314
training
  communication plan exercises and, 393–395
  disaster education and, 182
  and exercises to practice communications interoperability, 231–232
  for users and administrator, 375–376
translation, 103
transmission, 1
tribal lands, 243
triode, 14
Truman, Harry, 39
tsunamis, 123–125, 318–320
Turtle Talk, 258
TweeDeck, 280
21st Century Communications and Video Accessibility Act, 246
Twitter, 255–258, 261–262, 263–265, 280, 286, 287, 298, 300, 320, 385, 432
  maps, 276
  mobile, 267
2-1-1 service, 91, 148, 246–247
two-way communication devices, 101, 322, 425
two-way radios, 310–312

U
U.S. Geological Survey, 119
U.S. Postal Service, 7, 273
U.S. Radio Act of 1912, 11
underload, 330
Unified Area Command, 216
Unified Command, 216
United Kingdom, the
  emergency calling number in, 41–44
  power shortages in, 134
United States Independent Telephone Association (USITA), 46

United States v. Riggs, 53
UNIVAC, 31
universal emergency calling number
  blocking inbound calls to, 54–56
  9-1-1 emergency operator procedures, 51
  enhanced 9-1-1, 49, 51–52
  First Amendment rights and, 52–53
  history, 40–41
  list of commonly used, 477–486
  North America, 44–49, 50
  silent 9-1-1 calls and, 53–54
  United Kingdom, 41–44
  vulnerabilities, 52–56
Universal Task List (UTL), 226
unlicensed radio services, 25–26
urbanization, 118
usage, 232–233
USC Geocoder, 273
use cases, 343, 365–367
user acceptance testing, 368
user documentation, 374–375
user training, 375–376
Ustream.tv, 279

V
Vail, Alfred, 6
vendors, questions to ask, 467–471
Verizon, 292
Very High Frequency Omni-directional Range (VOR), 22
video
  displays, 307–310
  games, 31
  relay service (VRS), 93
  shared/streaming, 278–279
  surveillance, 306–307
violent events, 74–75
virtual private network (VPN), 272
visitor management, 426
visually impaired persons, 34
  effective communication with, 98–101, 170
voice carry-over (VCO), 93
voice elements, 231
voicemail and voice systems, 295–296
VOIP technology, 39

voIP technology, 57, 322
volcanic disruptions, 121–123

W

war/conflict, 141–143
warfare, climate, 125
warnings, tsunami, 124
waste treatment facilities, 137
wastewater sewer lines, 137–139
watches, tsunami, 124
water shortages, 130–132
water stress, 132
Water: Sustainable Management of a Scarce Resource, 131
Waters, Mike, 145
WaveTube, 275
weather radios, 313
webmasters, 384
web pages, 299–300, 309–310, 416
webTV, 55–56
Weibrecht, Robert, 33
Weimer, Carl, 135
Went the Day Well?, 8
what of effective communications, 191–194
Wheatstone, Charles, 4
when of effective communications, 194–195
where of effective communications, 195–196
who of effective communications, 190–191
why of effective communications, 189–190
widgets, 267
wi-fi, 25
WikiLeaks, 270
Wikipedia, 269
wiki sourcing, 269–270
Wilson, Woodrow, 14
WinLink, 39
wireless, 26–27
wireless Communications and Public Safety Act of 1999, 51
Wireless Local Area Network (WLAN), 26
Wireless Priority Service (WPS), 235
WordPress, 255, 266
Wordr, 432
World Health Organization, 133
WorldSpace, 27
World Wide Web (WWW), 32
Wright, Marcia, 403
Wyman, Oliver, 204

X

x-class flares, 128–129
Xerox, 7
XM (radio), 27

Y

Yahoo!, 281, 297, 298
    PlaceFinder, 273
YAMMER, 258
yellow level, 181
Yellowstone National Park, 122
YouTube, 258, 279

Z

Zarouni, Alfred, 47
zero day exploit, 140
Zhang, David, 132
zip codes, 7
Zoning Improvement Plan, 7
Zuse, Konrad, 30
ZXing, 274